TRADITIONAL
HOUSEHOLD
HINTS

TRADITIONAL HOUSEHOLD HINTS

Tried and trusted ways that work today

Published by The Readers Digest Association, Inc.
London • New York • Sydney • Montreal

Contents

Note to readers

The creators of this work have made every effort to be as accurate and up to date as possible. Many traditional home remedies use plant substances. Plant-based products (infusions, tinctures, capsules, essential oils) may cause immune reactions and should therefore be used with care. Pregnant women and anyone taking prescription drugs or undergoing surgery should consult a doctor before using herbal medicines. Essential oils should not be used by children under the age of two. For a specific health problem, consult your physician. Health advice in this book should not be substituted for, or used to alter treatment without your doctor's advice. The writers, researchers, editors and publishers of this work cannot be held liable for any errors, omissions or actions that may be taken as a consequence of information contained within the book. The mention of any products in this book does not imply or constitute an endorsement by The Reader's Digest Association, Inc.

continued...

Contents

Yesterday's top tips for today's busy lifestyles

Almost every day, new – and usually expensive – consumer products that promise to make our lives easier and simpler are introduced. But are they all really necessary? Despite their initial hype, few of them live up to expectations.

Our grandparents' generation had abundant knowledge about how to solve everyday problems easily, inexpensively and effectively, without running to the shops – whether the issue was storing fresh produce, making skin look younger, relieving toothache pain or getting laundry whiter. Wouldn't it be a shame if all that expertise was lost to future generations?

To prevent that from happening and to preserve yesterday's top tips for today's busy families, the editors at Reader's Digest gathered a wealth of traditional wisdom to create this book. Traditional Household Hints is a comprehensive collection of the best time-honoured solutions from generations past. From your medicine cabinet to your freezer, your vegetable patch to your morning shower, here are clever suggestions and solutions that show you how to apply the good old ways from the good old days to improve the way we do things now.

It's not about harking back to a nostalgic past, but about presenting old-fashioned ideas in such a way that they can be easily used in today's world. It makes the wealth of our collective experience – built up over generations – accessible to people who need a helping hand today. Advice that is proven, timely, economical and environmentally friendly.

And all these clever solutions, useful hints, practical tips and helpful remedies from days gone by have been researched and tested for their practical applications in the 21st century. They offer alternatives to expensive products or the use of chemical additives and you'll often find the ingredients close to hand in your kitchen cupboards.

Organised into six main chapters that focus on health, beauty, home management, cooking, home decor and the garden, this book is packed with more than 1,900 practical hints and tips that our parents and grandparents trusted and relied upon. Every tip is guaranteed to help make life easier as well as less expensive. And each entry is presented alphabetically within each chapter, making solutions easy to find.

Traditional Household Hints steps back into the past to help you achieve a more rewarding way of living today – because the old ways still work best!

FROM THE EDITORS OF READER'S DIGEST

Good health ... naturally

Most of us regularly use over-the-counter pills and potions to treat our ills. In the past, people turned to plants for relief. Today clinical science supports many of these gentle remedies, confirming that a variety of ailments will respond to nature's help alone.

Acid reflux

The searing pain of heartburn can be brought on by eating spicy, fatty or acidic foods, or just by eating too much too quickly. Luckily, it is easy to combat – or to avoid altogether.

When stomach acid backs up into the oesophagus, you feel the burning pain of acid reflux. At the first sign of heartburn, try one of these natural solutions.

GOOD TO KNOW ✓

Causes of acid reflux

A trapdoor of muscular tissue called the lower oesophageal sphincter usually keeps stomach acid where it belongs – in the stomach. However, heartburn occurs when it allows the contents of your stomach to flow back into the oesophagus, causing a burning pain behind the breastbone. This is known as acid reflux.

HOME remedies

● A little bicarbonate of soda can be highly effective at neutralising excess stomach acid (but do not use it if you have high blood pressure). Stir 1 teaspoon into a glass of room-temperature water and drink.
● A mixture of 1 teaspoon of crushed juniper berries in 250ml water should improve things noticeably.
● Drink ginger tea: boil 1½ teaspoons of fresh ginger or ½ teaspoon of powdered ginger in 250ml water for about 10 minutes before drinking.
● Chew a liquorice tablet before meals. This encourages mucin production in the oesophagus, providing a protective barrier against stomach acids.
● Season foods with juniper berries or lovage, which makes them more digestible and soothes heartburn.
● Eat a piece of dry white bread or toast to neutralise stomach acids, or a tablespoon of dry oatmeal.
● Sleep with your upper body slightly elevated at night to prevent acid from entering the oesophagus.

PREVENTION

● Eat sensibly, avoiding foods, drinks and combinations you know give you heartburn – perhaps fatty or acidic foods, chocolate or wine.

● Avoid alcohol, nicotine and caffeine; they increase gastric acid secretion, which can cause heartburn.
● Eat slowly and always opt for smaller, more frequent portions.
● Watch your weight. If you are carrying excess pounds this increases pressure inside your stomach, which can lead to acid reflux and heartburn.
● Eat early in the evening to give your stomach time to digest the meal – this takes about 3 hours. If you go to bed shortly after eating, stomach acid can flow back into your oesophagus.

WHEN TO CONSULT A DOCTOR If you are a frequent victim of heartburn, seek professional advice.

DRINK A glass of FRESH CARROT JUICE to soothe HEARTBURN

Acne

Spots are the bane of many an adolescent's existence, but they don't just occur during puberty. About four out of every five adults between the ages of 20 and 30 suffer mild to moderate acne, and more than half of all women, regardless of age, get pimples or blackheads.

The causes of acne vary, from hormonal changes that lead to increased sebum production and plugged sebaceous glands to sun exposure, medications and oil-based cosmetics. Long before people sought a solution to unsightly blemishes at the chemist, spots were treated with a range of masks and tinctures made from active natural ingredients. But remember: don't squeeze spots or you could be left with scars.

HOME remedies

● Apply a healing mask. Mix together 3 tablespoons of fuller's earth (available at health-food stores) with an equal amount of water to produce a thick paste. Apply it to the face, leaving plenty of space around the eyes. After 30 minutes, rinse off thoroughly with warm water. Use three times a week.
● Or try this recipe: mix together 2 tablespoons of yoghurt and 1 tablespoon of honey and spread it on your face and neck. The yoghurt has a cooling effect and honey disinfects. Rinse with warm water after 30 minutes. Apply three times a week.

● Dilute tea tree oil with water in a 1:4 proportion and apply with a cottonwool ball to heal pimples and blackheads quickly.
● Use a juice extractor to squeeze the liquid from fresh leaves of the herb plantain, or crush them with a pestle, then dab the juice directly onto blackheads and pimples with a cottonwool ball.
● Drink one or two cups a day of stinging nettle tea (available at health-food stores). It detoxifies blood and promotes skin healing.
● Have a chamomile facial twice a week to open pores. Pour boiling water over chamomile flowers and position your face over the steam for 5 minutes.

TO cleanse the skin

● Soap is too harsh for acne-prone skin but soap-free cleansers with a pH of 5.5 (which match the skin's acid-protection coating) can soothe and protect.
● Follow up by treating oily skin and acne-prone areas of your face with small amounts of a facial cleanser, such as a facial toner containing alcohol.

Papaya mask

one Peel a papaya, remove the seeds and purée the fruit.

two Stir 2-3 tablespoons of plain yoghurt into the puréed fruit.

three Apply the purée to a compress, place it on your face and leave for 30 minutes.

four Rinse off the blemish-fighting papaya mask with warm water.

Homemade face masks cleanse your skin and leave it toned.

Back pain

Most people are affected by back pain at some point in their lives. Spending long periods slumped in front of a television or bent over a computer puts our musculoskeletal systems at greater risk. While good posture is no longer associated with moral superiority as it once was, it is still one of the best ways to prevent back pain.

In addition to good posture, lifting and carrying things correctly and using a chair with a backrest can all help keep your back healthy. However, regular exercise is the top tip for a strong back. Not only will it help you to control your weight (being overweight is linked with back problems), but physical activities such as walking, swimming or cycling will strengthen back muscles, taking pressure off your joints and tendons.

GOOD TO KNOW ☑

Cherry-stone relief

The benefits of cherry-stone bags were discovered during the manufacture of Kirsch liqueur. Workers found that if they put the leftover cherry stones in bags and warmed them, the soothing heat relieved their aches and pains.

HOME remedies

If you suffer acute back pain, a short period of bed rest may help but more than a couple of days will do more harm than good. Gentle exercise, such as cycling or swimming, improves circulation and helps prevent further problems.

● Apply heat to a sore back. Good choices include compresses with rosemary or thyme tea; warmed cherry-stone or spelt bags; heat packs (available from a chemist); or infrared light treatment.
● Relax tense muscles and promote circulation with moor mud and sulphur baths.

● Take hot oil baths with rosemary or thyme extracts to soothe pain, followed by massages with hand-warmed massage oil. Oils that contain lavender, rosemary and ginger are particularly effective.
● When you can, bathe in a natural hot spring. It can be just as good for your psyche as for your back.
● Rub your back with spirits of lemon balm. To prepare: steep 200g fresh lemon balm leaves for ten days in a tightly sealed container with 1 litre of surgical spirit, and leave in a warm place. Strain and dilute with water in a 4:1 proportion.
● Try a hot wheat pack. Boil about 1kg wheat grains until soft. Put the hot mix into a linen bag and leave it to rest on your sore back for 15 minutes.
● Take a therapeutic seawater bath. The main components of thalassotherapy are algae, littoral deposits and sea salt – available from health-food stores and chemists.

A GENTLE BACK MASSAGE relaxes cramped MUSCLES.

The 'child's pose' in yoga, which stretches the hips, thighs and ankles, can help relieve back pain.

● Place a thick cushion under your legs at night so that your thighs point straight up and your knees are bent at a right angle in order to take strain off your spinal column.

PREVENTION

● Don't carry lopsided loads. Lift heavy weights with your knees bent and always keep your back straight.
● Keep your back warm. Avoid exposing it to cold and draughts.
● Replace saggy mattresses and don't penny pinch when you buy a new one. It should be made from high-quality materials and neither too hard or soft.
● Avoid sitting in chairs without proper back support. Replace old, worn-out chairs.
● Use a wedge pillow to encourage erect posture while sitting.
● If you have a sedentary occupation, change your sitting position frequently and stand up and stretch every 30 minutes.
● Avoid high heels – wear comfortable shoes as often as possible.

WHAT is lumbago?

Lumbago is actually a blanket term for mild or severe pain in the lower back (or lumbar region). It differs from 'normal' back pain in that it can happen suddenly. A cold draught, jerky movement or combination of bending and twisting can trigger an abrupt pain deep in your back muscles. However, a slipped disc has similar symptoms, so if there's no improvement after a few days consult a doctor. First, try these remedies.
● Large adhesive bandages used with compounds like capsaicin, which stimulate circulation, continue

to warm muscles for a long time. But never use them on irritated or broken skin.

MAKING a wrap

Moist heat is particularly helpful for lumbago. Here are two wraps that might ease back pain.
1 Mix together 10 drops of lavender oil, 8 drops each of chamomile oil and cedar oil, plus 4 drops each of juniper oil and clary sage oil with 200g body lotion.
2 Pour 2 tablespoons onto a cloth soaked in hot water and wrung out, then apply it to the painful area.
3 Spread a dry cloth over it and cover with a wool blanket. Repeat several times daily.

For a verbena wrap:

1 Stir together a handful of verbena, one egg yolk, 1 tablespoon of flour and 2 tablespoons warm water.
2 Fold a cotton towel to fit the size of the painful area of your back. Sprinkle it with the verbena mixture, place it over a pot lid and warm it over steam.
3 Place a hot spelt cushion onto an exercise mat, spread the cotton cloth over it (with the verbena facing upwards), and lie down carefully with your lower back on it.
4 Cover up with a blanket and lie for as long as possible on the hot underlay.

Rub your back with spirits of lemon balm.

Bladder and kidney disorders

The kidneys and bladder are part of the urinary system, which transports harmful waste products out of the body. A good fluid intake is essential to ensure they function correctly, so drink plenty to avoid a bladder infection or kidney stones.

BLADDER weakness & incontinence

Do you check where the toilets are at the shopping centre before you get there? Or try to avoid sneezing or laughing because you are not sure you'll stay dry? You're probably suffering from bladder weakness or incontinence. Both men and women suffer from this embarrassing condition, but the physical stresses of childbirth and a decrease in oestrogen at menopause make women three times more susceptible.

Frequently, there's a psychological component to a weak bladder, so stress-reduction programmes such as yoga or autogenic training may help.

The problem isn't new – even the ancient Egyptians developed remedies to deal with it. As a result there are many time-tested solutions, including the following teas, that can help to strengthen bladder muscles.

● Blueberry tea is a well tested remedy for a weak bladder. Mix 20g of blueberry leaves with about 250ml water and drink three times a day.

● This tea is said to be effective within three weeks, if taken three times a day: mix 50g lady's mantle, 30g rosebay willowherb and 20g fennel seeds. Pour 250ml boiling water over 1 teaspoon of this mixture and steep for 10 minutes before drinking.

Drink two large glasses of cranberry juice a day to prevent infection.

PREVENT kidney stones

It has been said that the pain of passing a kidney stone is comparable to that of giving birth. Stones might pass in a few hours, but sometimes it takes days. It's best to try to avoid developing them.

Dehydration is a key factor. Lack of fluids prevents mineral salts in urine from being dissolved, causing them to clump together as grit and slowly form kidney stones. To ensure your kidneys are well-irrigated and healthy, drink plenty of herbal and fruit teas, non-carbonated water and diluted fruit juice.

● If you have a tendency towards kidney problems, avoid apple and grapefruit juices that can increase the risk of kidney stone formation.

● Drink enough fluids to ensure you produce about 2 litres of urine a day. Most fluids are all right, however avoid cola, beer and black tea.

● Salt and meat can aggravate kidney stone formation. Consume them in moderation.

HOME remedies

If grit or small stones have already formed in your kidney, the following home remedies may help to flush them out.

● Flush out small stones with diuretic (flushing) teas made from birch leaves, goldenrod or marshmallow root. Kombucha (available at health-food stores) has the same effect. This cold drink is fermented using the kombucha or tea mushroom and contains live bacteria and yeast, similar to yoghurt.

● Drink plenty of water (2.5-3 litres per day).

● If your kidney stones are causing pain, try a potato wrap made from boiled, mashed and still-hot potatoes wrapped in cloth. Place it over the kidney area. The warmth will soothe your afflicted organs, as will exposure to infrared light.

● Our grandmothers were familiar with the soothing properties of heat and regularly took soothing baths or placed a hot-water bottle on their lower abdomen at bedtime to ease kidney or bladder pains. Both are simple ways that can bring relief.

WHEN TO CONSULT A DOCTOR If you are experiencing serious pain in the kidney region and/or a build-up of urine, contact a doctor immediately.

To keep your bladder
and kidneys healthy,
drink about 2 litres
of liquid a day.

TREATING a bladder infection

About 50 per cent of women will experience a
bladder infection at some point in their lives, and
many will have multiple infections. Women suffer
from the problem more often than men because
their urethra is shorter, so bacteria can enter the
bladder more easily.

Dehydration is often a factor, but there is a
host of traditional remedies to combat pain and
the continual urge to urinate. But bear in mind
that these remedies are most effective when
applied early.

● Hot herbal tea can help flush out the bacteria
causing inflammation and soothe the pain.

● Take a high-strength cranberry supplement or
drink two large glasses of cranberry juice a day to
relieve the infection. Cranberry may also help to
prevent infections occurring.

● Heat reduces the pain caused by a bladder
infection. Place a warm spelt or cherry-stone bag
between your legs and/or over your bladder, or
soak in a hot bath.

WHEN TO CONSULT A DOCTOR If your bladder infection is
extremely painful or is not noticeably better after
three days, consult a doctor.

Bladder tea

This tea soothes pain, disinfects and
flushes out bacteria when you have a
bladder infection.

Mix together:

50g bearberry leaves
15g green beans
15g horsetail
5g each of fennel, pot marigold flowers
and peppermint

Boil 1 teaspoon of this mixture
with 250ml cold water for
5 minutes. Steep for
10 minutes, strain and
drink one cup three
times a day.

Blood pressure

Age, weight gain, lack of exercise, smoking and alcohol abuse are just a few of the lifestyle factors that can subject your heart and circulatory system to tremendous strain. That, in turn, causes damage to blood vessels, sending your blood pressure soaring. The good news: it is never too late to do something about it.

Regular exercise and gentle endurance sports such as cycling, Nordic walking and swimming can help both low and high blood pressure.

HIGH blood pressure

High blood pressure is one of the main risk factors for having a stroke. Get your blood pressure checked regularly at a pharmacy or health centre.

Many plants contain ingredients that can lower blood pressure effectively, relax the muscles of your blood vessels and have a calming effect on your nervous system. With a little help from nature, it is easy to bring slightly elevated blood pressure under control or complement conventional medical therapy if your blood pressure is very high.

HOME remedies

● Mistletoe and hawthorn have long been used to regulate blood pressure. Brew up these herbal teas individually, or use a blend that includes both ingredients.
● Try this recipe for olive tea from the Mediterranean: pour about 250ml boiling water over 2 teaspoons of dried, minced olive leaves; steep for 10 minutes then strain; press 3 garlic cloves and mix into the tea with 1 teaspoon of honey.
● Eat plenty of bear's garlic, one of the oldest medicinal herbs known to man. The tasty but odourless plant contains substances such as adenosine that lower blood pressure. Use its leaves to give a boost to salads, in pesto or on pasta.
● Chew a clove of garlic every day or add one to salads and other dishes. Raw garlic and onions keep blood vessels elastic and lower blood pressure.

Regular exercise is beneficial whether you have high or low blood pressure.

NUTRITION for high blood pressure

- In April, May and June enjoy fresh asparagus. This vegetable acts as a natural diuretic and can lower blood pressure by removing excess salt and water from the body.
- Reduce salt consumption as much as possible, as too much salt raises blood pressure. Season food with fresh herbs instead.
- Avoid alcohol, nicotine and coffee as they may increase blood pressure.
- Eat fresh, oily fish such as mackerel or salmon once a week. They contain valuable fish oils that may lower blood pressure.
- Use plant oils for cooking and frying.
- Use butter or margarine sparingly, especially if you are predisposed to high blood pressure or lipid metabolic disorder.
- Cut out fat, but eat plenty of fruit, vegetables and wholegrain products.

FATTY ACIDS in nuts CAN HELP REDUCE blood pressure.

LOW blood pressure

Fatigue and exhaustion, feeling faint or even actually fainting, especially just after getting up from a lying or sitting position, are typical symptoms of excessively low blood pressure. Get your blood pressure checked regularly at your local doctor's surgery or health centre.

HOME remedies

- Pump up the fluids. Dehydration reduces blood volume, which can lead to low blood pressure.
- Drink black tea; it is a stimulant. But don't let it steep for longer than 3-5 minutes.
- Sleep with your upper body slightly elevated to stimulate circulation and make it easier to bounce out of bed in the morning.

- Have a shot glass of rosemary wine with a meal at midday and in the evening to help boost blood circulation. To make it, pour about 750ml of white wine over 20g of rosemary leaves. Strain and bottle after five days.
- Indulge a sweet tooth with liquorice. Eat just one small piece a day (no more than 15g) as the active ingredient in it, glycyrrhizin, can have undesirable side effects if consumed in large quantities.
- Soak 30 raisins in water overnight. Raisins may help to regulate blood pressure. Eat the raisins in the morning, then drink the raisin water.
- Alternate hot and cold water during your morning shower. This practice has long been a tradition in some parts of Europe. It forces blood vessels to contract then expand, and helps blood pressure return to normal. Begin with warm water. After 2 minutes, turn the temperature to cold for 15 seconds. Repeat the procedure three times, always ending with cold water.
- Get moving. All physical activity increases blood pressure, so you can benefit from a regime of light exercise, as well as activities such as walking, swimming or cycling. However, you should first clear an exercise programme with a doctor.

Brush massage

1 Massage may help those with high blood pressure to cope with stress. Start with a natural-bristle brush on the back of your right foot and brush your right leg up to the buttock using a circular motion, first on the outside, then the inside. Repeat on the left leg.

2 Brush your buttocks, upper body and arms, again using a circular motion.

3 Finally, ask your partner to massage your back using the same technique.

Breath problems

Halitosis is unpleasant. Luckily, much can be done to prevent it. If you've failed the breath test and people back away when you stop to chat, rely on these remedies from the herb garden to freshen your breath quickly.

You can't usually tell if you have bad breath and people are often reluctant to point it out. An easy test is to hold cupped hands in front of your mouth, exhale into them and then move your hands quickly to your nose to check the exhaled breath.

HOME remedies

- Use mouthwash regularly after brushing your teeth. Add a couple of drops of chamomile, peppermint, clary sage or lemon balm oil to a glass of water and rinse your mouth with it.
- For morning breath, rinse your mouth with cider vinegar (1 teaspoon in a glass of water) as soon as you get out of bed.
- For continuing bad breath, chew parsley or mint leaves – both will freshen breath in an instant.
- Mix together dill, anise and fennel seeds and chew a few of them occasionally.

Anise mouthwash

one Bring 2 tablespoons of anise seeds to the boil in about 100ml water and leave to cool.

two Strain the mixture through a coffee filter and squeeze out the seeds.

three Mix the remaining liquid with about 50ml vodka and about 50ml rose water and pour the solution into a dark bottle.

four After brushing your teeth, put a dash of the mouthwash into a glass of water and rinse out your mouth thoroughly.

- If stomach problems are the cause of your bad breath, try this old and trusted home remedy: chew a coffee bean to neutralise the acid smell. But spit it out afterwards, don't swallow it.
- Both apples and yoghurt taste good, freshen your breath and provide a healthy snack.
- When you are on the go, suck on peppermint or eucalyptus sweets.

Gargle regularly with lemon water to avoid developing bad breath or a dry mouth.

PREVENTION

If digestive disorders or gum disease are the cause of your halitosis, a visit to the doctor or dentist should provide a solution. Otherwise, brush regularly, and clean the spaces between your teeth with dental floss.

- Avoid smoking as this is a prime cause of halitosis.
- If you drink a lot of coffee (which can also cause bad breath), rinse your mouth frequently.
- Eat and drink on a regular basis – halitosis often occurs when your stomach is empty.
- People who eat yoghurt regularly are less prone to halitosis than people who do not.

Burns and scalds

Burns are the result of direct contact with a hot object while scalds are produced by hot fluids or steam. Both damage the skin's tissues, causing blisters or charred skin, and should be treated the same way.

There are major distinctions between first, second and third-degree burns. Only first-degree burns and scalds should be treated at home. For more serious burns, see a doctor immediately. First-degree burns are characterised by reddened, painful skin but can be treated effectively with natural remedies (assuming you don't have an open wound).

WHEN TO CONSULT A DOCTOR More serious burns (blisters start forming with second-degree burns), large or deep burns and all chemical and electrical burns require medical attention. People who may be at greater risk from the effects of burns – such as children, pregnant women or anyone aged over 60 – should also see a healthcare professional immediately. Second and third-degree burns that cause blistering and tissue damage are serious, with a high risk of infection resulting from germs entering the body through the damaged skin.

HOME remedies

● To cool skin, reduce the pain and clean the wound, hold the affected part under cold running water (not ice water) for at least 30 minutes.
● Apply marigold salve to the burn if the skin is unbroken. It soothes and can help damaged skin to heal more quickly.
● Take vitamin C to build and maintain new skin.

● Squeeze a few drops of the juice from a cut piece of an aloe vera plant onto the burn. It will help to soothe it and prevent infection.
● Apply fresh sauerkraut, a traditional German burn remedy rich in vitamin C, directly to the burn.
● For healing without scars, pour 1.5 litres of water over 1 tablespoon of flaxseeds. Boil until scum forms on the surface, then strain and leave to cool. Soak a linen cloth in the liquid, wring it out and apply.

CABBAGE poultice

Check with a healthcare professional before trying this traditional remedy.
1 Rinse white cabbage leaves and remove the central vein.
2 Roll the leaves with a rolling pin until soft.
3 Place them onto burned skin and secure with a gauze bandage. Change the bandage after several hours (replace twice a day).

GOOD TO KNOW ✓

An outdated recommendation
You can't heed every home remedy from your grandmother's time – some of them can actually be harmful. For example, never treat burns or scalds with butter, which can be a breeding ground for bacteria. Also, under no circumstances should you pierce or burst blisters, as there's a danger of infection.

HOLD the affected PART UNDER cold running water

Colds

Blocked noses, sore throats, aching limbs and fever are symptoms we have to contend with each winter, with antibiotics powerless to help us fight off the season's viral invaders. However, most colds can be treated effectively with traditional home remedies.

In the past, people often used the terms cold and flu interchangeably. Today, doctors distinguish between the two. If symptoms come on gradually and include a sore throat, headache, achy limbs, coughing, a runny nose, elevated temperature or slight fever, you probably have a cold. By contrast, flu comes on fast and hits hard, accompanied by a high fever and chills – you will feel too ill to get up. But drink plenty to flush out your system and prevent dehydration.

TAKING ECHINACEA cuts the risk of GETTING a cold

Chicken soup

1 chicken
1 large onion, quartered
Salt and pepper
3 carrots
3 celery stalks
1 kohlrabi (if available) or
1 medium-sized cabbage
1 bunch of parsley

Simmer chicken, onion, salt and pepper in 2 litres of water for 1 hour. Add the washed, chopped vegetables and boil for 1 hour more. Remove chicken from the pot, debone and cut the meat into pieces. Pour soup through a strainer before returning chicken to the pot. Garnish soup with parsley and serve.

TO reduce fever

Fever is the body's response to illness and actually serves to fight infection. But a temperature higher than 39.5°C will make you miserable. While the local pharmacy offers a wide range of often expensive relief, you may find the following traditional remedies equally effective.

● Apply leg compresses to reduce fever over time. Dip two linen cloths in cold water, wring them out and wrap them tightly around the calves with a warm towel on top. Repeat as needed.

● Drink plenty of fluids. Good choices include: fruit juices rich in vitamin C and antioxidants, such as orange juice, blackcurrant and cranberry juice; non-carbonated mineral water with a dash of fruit juice; or herbal or fruit teas, especially vitamin C-rich rose hip tea. Another classic: mix the juice of a lemon with 1 teaspoon of honey in 250ml hot water.

● Lime or elderflower tea, which are often referred to in traditional medicine as 'fever teas', can help bring on sweating – the body's natural way of cooling itself – to help reduce a fever. To get the maximum benefit, sip a few cups then have a hot bath before snuggling up under a pile of blankets. When you begin to sweat, wait 2 hours then dry off. Change your clothes and, if necessary, the bed linen. Drink some fluids and return to bed.

TO relieve aches and pains

Traditional wisdom has it that a hot-water bottle can relieve pain, promote circulation and help you to feel relaxed. Scientists have discovered why: heat can physically shut down the normal pain response that triggers aches and pains. 'It deactivates the pain at a molecular level in much the same way as pharmaceutical painkillers,' says one senior researcher in physiology at University College London. But heat brings only temporary relief, so frequent applications may be necessary.

● Evidence dating back to 4500BC reveals that the ancients favoured warm compresses of peat, mud and fuller's earth. Now we have the luxury of a hot bath or a heating pad to ease pain.

● Apply a mustard plaster, a traditional congestion remedy. Crush a few tablespoons of mustard seeds (or use mustard powder), add the powder to 100g of flour and mix with a little water to form a paste. Apply to the chest and leave on for 15 minutes.

TREATING a head cold

Few things will make you as miserable as a head cold, and nothing soothes the misery better than a helping of chicken soup. US researchers from the Nebraska Medical Center recently found that chicken soup contains 'a number of substances with beneficial medicinal activity', including an anti-inflammatory mechanism that may ease upper respiratory tract infections. See the box on the left for a healing chicken soup recipe that has stood the test of time. Here are some other home remedies to try.
● In the past, people sometimes placed hot or warm moist compresses with fuller's earth, mashed potato or flaxseeds on the sinuses – an effective and economical remedy. If you have an infrared lamp, direct it onto your sinuses. Or simply warm a wet cloth in the microwave (don't make it too hot) and drape it across your face for 10 minutes at a time.
● To relieve nasal congestion, pour boiling water into a bowl, cover your head and the bowl with a towel, and inhale deeply. To make it even more effective, add six drops of eucalyptus oil or chamomile to the boiling water.
● Spice it up. Foods that contain chilli peppers, hot mustard or horseradish can remedy congestion. If it makes your eyes water, it'll make your nose run.
● Try a nasal rinse. Irrigate your nose with a saltwater solution (from any pharmacy) to soothe stressed nasal mucous membranes.
● To make a nasal rinse, dissolve 1 teaspoon of salt in 500ml water. Use a nasal dropper or pipette to drop it into your nostrils and then blow your nose gently.

PREVENTION

As yet there is no permanent cure for the common cold, but you can reduce susceptibility significantly by boosting your immune system.
● Get plenty of exercise in the fresh air and increase your intake of vitamin C (fruits and vegetables).
● Avoid stress, nicotine and alcohol.
● Eat lots of pungent onion, garlic, radish or horseradish, which have an antibacterial effect and cleanse the blood.
● Take echinacea (purple coneflower, pictured left), a traditional remedy, available from pharmacies and health-food stores. Recent studies support its ability to cut cold risk by as much as half.

To reduce a fever, offer plenty of liquids, including fruit juices.

Constipation

Lack of exercise, not eating enough fibre or being over or underweight are just some of the causes of irregular bowel movements and hard, painful stools. However, harsh laxatives shouldn't be necessary to get things moving again.

When irregularity occurs, that traditional solution – a tablespoon of castor oil – is still a useful weapon to dispel stubborn constipation. But it is not the only home remedy that works.

EAT FRUIT, vegetables and whole grains – HIGH-FIBRE FOODS.

HOME remedies

● Drink a glass of prune or elderberry juice, diluted apple cider vinegar or warm water with honey in the morning on an empty stomach.
● Dissolve 1 teaspoon of sea salt in 500ml of warm water and drink the solution. This should soften stools.

Prunes and other dried fruit relieve constipation naturally.

● Try a tried-and-trusted German cure. For centuries, Germans have touted the curative effects of the pickled cabbage dish sauerkraut. In fact, sauerkraut is rich in lactobacilli bacteria, which helps soften stools and keeps intestinal flora healthy.
● Give yourself a daily stomach massage with a cold facecloth (working in a clockwise direction) to stimulate a sluggish intestine.
● If constipation fails to respond to any of these remedies, try a warm water enema (enema devices and instructions are available from pharmacies).

PREVENTION

● Consume only moderate amounts of fats and sugars, which slow the operation of the intestines.
● Make room in your daily diet for additional digestion-regulating fibres, such as wheat or oat bran and flaxseeds. Sprinkle them onto cereal.
● Combine a high-fibre diet with ample fluid intake.
● Avoid the tannins contained in dark chocolate, cocoa, black tea and red wine as they disable digestive muscles. Stay away from these foods if you are prone to constipation.
● Exercise, such as a bike ride or a regular walk, is often all it takes to get a sluggish intestine going.
● Take a probiotic. Good bacteria keeps things moving along nicely. Yoghurt containing acidophilus may help, or opt for a supplement.
● Relax and take your time. Being pressed for time on the toilet doesn't help.

Laxative tea

These herbs regulate digestion.

50g senna leaves
15g fennel seeds, ground
15g elderflowers
10g chamomile flowers

Mix the ingredients together and pour 250ml boiling water over 1 teaspoon of the mixture. Leave to steep for 5 minutes, then strain. Drink a freshly prepared cup three times a day.

Depression

Our grandparents faced the blues with a stiff upper lip and little discussion, but they did have some effective home remedies to fall back on. Mild depression can often be treated successfully with these simple, natural techniques.

If your mother suggested a regimen of fresh air and sunshine to combat the blues, she was right. Exposure to natural sunlight has an elevating effect on mood. Portable light boxes have the same effect and can be used to combat seasonal depression during darker winter days. The therapy seems to work best when used every morning for about 30 minutes. If you're still feeling a bit down, try some of these tips.

HOME remedies

● Have a cup of tea. Teas made from St John's wort, hops, valerian or powdered liquorice (from a pharmacy) help stabilise mood.
● Use essential oils to ease the blues. Mix 2 drops each of rose and lemon balm oil and 3 drops of lavender oil and use in a fragrant oil burner.
● A scented sachet placed under the pillow may help you to sleep better. Preferred scents include valerian, lavender, primrose, elder and hops.
● When is the last time someone told you to eat a biscuit? Adding ½ teaspoon each of nutmeg and cinnamon to the recipe might lift your spirits.

● Sip a glass of milk with fennel and honey half an hour before bedtime. Bring 2 teaspoons of crushed fennel seeds and about 250ml of milk to the boil, let it steep briefly, strain and sweeten with honey.

BEHAVIOUR tips

● Light and colour can have an impact on mood: a bright, friendly environment in warm colours, such as yellow, red or orange, can lift your spirits.
● There is increasing evidence that fish oil, which contains a type of fatty acid called EPA, can help to chase the blues away, especially when it is combined with pharmaceutical antidepressants.
● Cut back on fizzy drinks and soft drinks that contain caffeine. Some research links caffeine, which suppresses serotonin production, to depression.
● Go away for a few days. A break from routine and discovering somewhere new will give you a boost.

WHEN TO CONSULT A DOCTOR Depression can become serious and start to affect your way of life. See a doctor if depression persists or becomes more extreme. A doctor can check for a medical cause and advise about treatment.

Use lavender oil to ease the blues.

Diarrhoea

Whether the cause of your discomfort is yesterday's lunch special or a viral infection, 'Montezuma's revenge' can be most unpleasant. But it is also extremely effective at expelling whatever it is that ails you. Rather than trying to halt the condition immediately, it is now considered better to allow it to run its course. Start eating – plain food, little and often – when your appetite returns.

When diarrhoea strikes drink plenty of clear liquid (including rehydration fluids) to replace the lost water and the salts essential for retaining water in the body.

Thyme tea

This traditional remedy can ease the unpleasant stomach cramps that often accompany diarrhoea.

2g dried common thyme
1 cup boiling water

Infuse the dried plant for 5 minutes, then strain. Drink three cups a day.

FIRST aid

● Sip non-carbonated mineral water or black tea flavoured with sugar and a pinch of salt.
● Buy a pack of special glucose-electrolyte mixtures (from a pharmacy) to offset salt loss. These solutions are especially important for children, pregnant women and older people for whom major fluid loss is particularly dangerous.
● To make up a quick and inexpensive electrolyte solution, mix 500ml non-carbonated mineral water with 7 teaspoons of sugar, 1 teaspoon of table salt and 500ml orange juice or fruit tea (to provide potassium and flavouring). Drink throughout the day.

● Drink at least 250ml of tea three times a day. Teas made from oak bark or mulberry leaves contain tannins that have a soothing effect on the intestines.

WHEN TO CONSULT A DOCTOR Infants and small children with diarrhoea should see a doctor straight away. Adults should seek medical treatment if the diarrhoea continues for more than three days.

HOME remedies

If diarrhoea persists for a day or two, it is important to restore a healthy balance in the intestinal flora. There are plenty of long-standing home remedies available for this purpose.
● Dissolve 1 teaspoon of powdered charcoal or fuller's earth in a glass of water and drink it.
● Eat yoghurt that contains 'good bacteria' to combat any harmful bacteria that may have caused diarrhoea in the first place. Replacing it can help you feel better faster. Look for yoghurt that contains live bacterial cultures or probiotics. It must be a product with billions of bacteria in it, as you will need this many to colonise your intestine effectively.
● Eating 1-2 teaspoons of dried blueberries is a time-honoured Swedish cure for diarrhoea. The berries act as an astringent, contracting tissue, reducing inflammation in the intestine and ultimately slowing diarrhoea.
● Dark chocolate contains a high percentage of cocoa and flavonoids that ease diarrhoea.
● Apply a little heat. Heat calms the intestine and makes you feel better. A hot water bottle or a spelt or cherry-stone bag should do the trick.
● Put two handfuls of crushed marigold flowers boiled in water into a cloth bag and place it on your stomach while warm. Test it with a fingertip first to make sure that it is not too hot.
● Calm your nerves. Nervous diarrhoea can be treated with an aromatherapy massage. Mix 3-4 drops chamomile, sandalwood, juniper or lavender oil with 2 teaspoons of cooking oil and massage your lower abdomen in a circle.

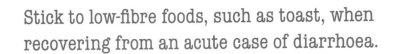

Stick to low-fibre foods, such as toast, when recovering from an acute case of diarrhoea.

● Next, treat yourself to a clear vegetable broth or a potato-carrot soup (see recipe, below) and dry toast.
● Try some cooked carrots, which are also high in pectin – just cook and purée.
● Gradually broaden your menu with a little fat and easily digestible protein.
● Avoid milk and dairy products until the symptoms disappear. Some of the organisms that cause diarrhoea can temporarily impair the ability to digest milk.
● During this time, also avoid coffee and alcohol. Instead, treat yourself to a cup of peppermint tea, which can help soothe intestines.

SLOWLY introduce a bland diet

When diarrhoea subsides, gradually return to a normal diet. Start with low-fibre foods such as crackers, toast, rice, boiled potatoes and chicken. Doctors will often recommend a diet of bananas, rice, apple sauce and toast, also known as the BRAT diet. These binding foods are suggested as the first to try after an episode of diarrhoea. Apple sauce contains pectin and other nutrients your body needs. Because the apples are cooked, they are easier to digest. Bananas are easily digested and contain high levels of potassium, which helps replace the electrolytes you lose when you have diarrhoea.

Potato-carrot soup

This soup delivers fluid and minerals without further irritating the intestinal mucous membrane.

250ml water
2 medium potatoes
1 carrot
1 pinch salt

Bring the water to the boil. Add the peeled and chopped vegetables. Cook over a low heat until tender, then purée and season with salt.

EAT PLAIN yoghurt to help RESTORE 'GOOD BACTERIA'.

Earache

Heat and medicinal plants are the traditional favourites for treating earwax and earache. Home medicine also offers useful advice to combat annoying tinnitus and to alleviate discomfort.

While often the bane of childhood, earache can strike anyone at anytime. It often originates in the middle ear, which is the tiny space located behind the eardrum. A thin tunnel called the Eustachian tube, which runs from the middle ear to the back of the throat, allows fluid to drain. It is also a passage where the pressure inside your ear adjusts to meet external air pressure. A build-up of fluid caused by a cold can accumulate in the Eustachian tube, causing significant pain.

Heat, which alleviates pain, is the earache treatment of choice.

EARACHE

Don't leave home in cold weather without a hat or scarf over your head – or at least a little cottonwool in your ears – if you are susceptible to earache. Wear a bathing cap or use wax or silicon earplugs when swimming to avoid getting water in your ears.

● Heat alleviates pain. Press a warmed flaxseed pillow gently against a painful ear, treat the ear with infrared light or lay your head on a hot-water bottle overnight.

● Pour clove oil onto a cottonwool ball and insert it into the entrance to your ear canal to relieve pain.

● Stir 3 tablespoons of mustard powder into warm water to form a thin paste, spread onto a handkerchief and place behind a sore ear for 15 minutes.

● Try eardrops. For homemade eardrops, mix 3 drops of tea tree oil and 1 tablespoon of olive or almond oil and warm the mixture in a bain-marie to body temperature. Tip your head to the side and use an eyedropper or pipette to drip the drops into the affected ear.

● Place slices of raw onion behind your ear and hold them in place with a headband. It has long been known that onions have formidable anti-inflammatory properties. It is a quick (if smelly) traditional home remedy.

WEAR a hat in the wind and cold TO PREVENT EARACHE.

Onion wrap

1 Cut two onions into thin rings and put them into a cloth bag.

2 Place the bag over boiling water and warm it in the steam.

3 Crush the onions with a rolling pin until the bag is soaked with juice.

4 Place the bag over an aching ear and secure.

● Even more effective, but a bit more work, is a chopped onion wrap (see panel, above). The heat has an additional role in soothing pain. Hold the wrap in place with a wool scarf or cap and leave it on the ear for 30 minutes. Apply three times a day.

WHEN TO CONSULT A DOCTOR See a doctor immediately if your child has earache, or if an adult's earache does not respond to home treatment within a short time.

EARWAX drops

The purpose of earwax is to protect the eardrums, keep the skin moist and guard against dust and other particles. Normally the ear cleans itself, but sometimes wax builds up in your auditory canal and hardens into a plug. If water gets in, it will swell up and may partially or completely block your auditory canal.

Note: removal of a wax plug should be left to healthcare professionals – under no circumstances should you try to remove an earplug by inserting any kind of object into your ear.

● Mix a few drops of olive oil with a little lemon juice and water. Use a dropper to drip the mixture into your ear, softening the plug. Eardrops from a pharmacy are also suitable for this purpose.

TINNITUS

If you have a constant noise in your ears, you may have tinnitus. The term refers to any ringing, rustling, humming or roaring sound that originates in the head, rather than from an external source of noise. Sometimes tinnitus goes away on its own, but usually these annoying auditory hallucinations can't be cured – they can only be reduced. Causes for the condition range from infection to noise, stress,

hearing loss and circulatory disorders. But if tinnitus sets in, there are a number of things you can do about it.

● Make time for relaxation exercises and regular breaks in the daily routine.

● Improve blood circulation by alternating hot and cold water in your morning shower and consider starting a light exercise programme.

● Drink lots of fluids to thin your blood and improve circulation to the inner ear.

● Enjoy a cup of lemon balm tea after a meal. It can have a positive effect on tinnitus.

● Ginkgo extracts promote circulation and can help some sufferers, but it takes several weeks for them to take effect.

● Avoid exposure to noise to prevent worsening the condition. Wear earplugs or earmuffs if your workplace is excessively noisy. However, total silence is also inadvisable as it makes the tinnitus more noticeable.

● Have a subtle source of noise to hand, such as a radio or sleep sounds CD, particularly at night. When you are outdoors, the rustling of the treetops or the splashing of a stream can mask the sounds in your head.

WHEN TO CONSULT A DOCTOR If tinnitus is accompanied by sudden difficulty hearing, it is advisable to see a doctor as soon as possible to eliminate rare but serious conditions.

Gingko extracts can help some tinnitus sufferers.

ESSENTIAL HOME MEDICINES

Next time you have an ache or pain, try some traditional cures before rushing to the pharmacy. A home medicine cabinet can be equipped relatively cheaply, and nature's simple remedies can treat a wide range of ailments.

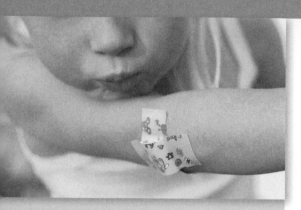

HERBS ARE VERSATILE AND can complement OTHER MEDICINES.

Recommended commercial preparations

For ailments that require quick intervention, have the following medications to hand:
- Antihistamines in case of allergic reactions
- Antiseptics for minor injuries
- Pills to control pain and fever
- Medications for constipation and diarrhoea
- Gel for burns and cuts
- Hydrocortisone cream to soothe itchy insect bites

Medicine cabinet must-haves

The following belong in every well-prepared home pharmacy:
- Hot-water bottle
- Thermometer
- Disposable gloves
- Tweezers, scissors
- Adhesive bandages in several sizes
- Sterile compresses
- Gauze bandages, elastic bandages
- Bandage clips, safety pins
- Triangular bandage
- Eye patch

Emergency numbers

Keep a list of the most important emergency numbers handy or on speed dial: emergency services (police, fire and ambulance), family doctor, nearest hospital with an accident & emergency department and pharmacy.

Scissors plus bandages in several sizes are essential.

At the pharmacy, always ask for less expensive, generic medicines.

A soothing cup of herbal tea can help with a wide range of ailments, from fevers to nausea.

Natural remedies

When it comes to stocking a home medicine chest, supplement your basic equipment with these natural solutions.

- The following essential oils have a place in every medicine cabinet: tea tree oil (healing of wounds); eucalyptus oil (respiratory passages). Use 3-5 drops of each for wraps, as additives to bathwater or for inhaling. Pick up essential oils from pharmacies or health-food stores.
- The most important tinctures for a medicine cabinet come from marigold flowers and chamomile. Use them externally or internally. Chamomile soothes ailments such as stomach aches and colds. A rule of thumb: for internal applications, use 10 drops three times a day in water or juice. For external treatment of things such as skin injuries, dilute the tincture in a 1:4 ratio and use with compresses or add to bathwater (for example, chamomile for insomnia or stress). Tinctures can be pricey but it is easy to make your own supply. You need about 15g of herbs per 100ml surgical spirit and dark, sealable glass bottles for storage. Kept cool, they will last for about a year.
- Prepare curative teas from 1 teaspoon of dried flowers and leaves of various herbs and 250ml hot water (see panel, below).

Herbal teas

Chamomile • Stomach and digestive ailments

Lime • Feverish cold and illnesses, stomach and intestinal cramps, nervousness, headaches (particularly migraines), infections

Lemon balm • Sleep disturbances, queasy stomach and intestinal problems, nervousness

Peppermint • Nausea, vomiting, inflammation of the stomach lining, intestinal wind

Plantain • Cough, hoarseness, whooping cough

Proper storage

Store your essential home medicines in a dry, dark, cool place in a lockable cabinet – preferably out of the reach of small children. Check the contents regularly to ensure that the expiry dates of medications have not passed and that there is enough on hand should you need them.

Eye problems

Of our five senses, sight is the one people fear losing the most. So it is important that we take care of our eyes, especially as they are exposed to numerous environmental factors that can affect vision. Fortunately, traditional rinses and compresses will help with a range of problems, from simple eyestrain to a stye.

Eyes are buffeted by countless irritants, including wind, smoke, dust, sun and even bacteria and viruses, which can result in eyestrain or eye ailments. Here are some ways that you can help protect them.
● Wear glasses to protect your eyes against direct sunlight, wind or dust.
● Reduce or avoid draughts and spending long periods in places with high humidity, such as a greenhouse.
● Use an adjustable reading lamp with a wide emission angle for reading.

Refreshing eye compresses

10g cornflour
10g yellow sweet clover (yellow melilot)
20g plantain (the herb)

Pour about 300ml boiling water over the herbs and strain after 10 minutes. Apply gauze pads soaked in the cooled solution onto tired, closed eyes.

● Get plenty of sleep at night, and strengthen your body's defences with relaxation, exercise and nutritious food.

WHEN TO CONSULT A DOCTOR When the layer between the eyelid and eyeball, the conjunctiva, becomes inflamed, the result is itchy, red and watery eyes. If the cause is a viral or bacterial infection, the ailment is highly contagious, so it is important to consult a doctor or pharmacist immediately.

RELIEVING conjunctivitis
● You can help by keeping your eyes clean. Carefully remove the discharge caused by the inflammation several times a day with a cottonwool ball soaked in distilled water.

GIVE yourself an eyebath
After cleansing, soothe inflamed eyes with eyebright (euphrasia) compresses. But first, check with a healthcare professional.
1 Finely chop 1-2 teaspoons of the herb eyebright and pour 250ml of boiling water over it (or use 1 teaspoon of dried flowers).
2 Let the mixture steep for 2 minutes before straining it.
3 When it is just lukewarm, soak two sterile gauze pads in the liquid and apply them to your eyes for several minutes.

LONG HOURS AT the computer CAN CAUSE EYE STRAIN

Squeezed-out black or green tea bags cooled in the refrigerator work wonders for swollen eyelids.

RED eyes

Windburn, making a bonfire or barbecuing can leave you with red, burning eyes. But don't rub irritated eyes – you might make them worse.

● Apply cucumber slices. They not only help with swollen eyelids, but also with reddened eyes.

SWOLLEN eyelids

Since swelling results from a build-up of tissue fluid in the eyelid, anything cold can help soothe the inflammation by contracting the blood vessels and stimulating circulation.

● Scoop a little cold, plain yoghurt onto a cloth to make a poultice and place it over closed eyes for 15 minutes. Take care not to get any in your eyes.

● Apply slices of cold cucumber to your eyelids for 10 minutes.

● Apply a cold pack to swollen eyelids. Crushed ice in a cloth works just as well, as does a metal spoon cooled in the refrigerator (not in the freezer) and laid carefully onto your eyelids.

TIRED, strained eyes

Long hours at the computer, poor lighting, lack of sleep – all of these things can result in eye strain, the symptoms of which include burning, itching and watery eyes.

● For relief, rub your hands together until they are warm and place them gently over closed eyes.

● Make sure that you blink frequently – this should happen unconsciously about 13 times a minute but people tend to blink less when using a computer. Blinking spreads a tear film over the eyes that clears away dust and dirt particles. It also keeps eyes moist.

● Then, try this natural remedy. Dampen a clean cloth with freshly boiled warm water. Leave the cloth over your eye for 15 minutes. Throw away the cloth after use.

STYE

The most common cause of that angry-looking pustule on the edge of your eyelid is bacteria. You should never squeeze a stye as you risk causing a severe infection. However, with a little help from heat and a compress you may be able to bring it to a head so that the pustule opens on its own.

● Use infrared light to hasten the ripening of a stye.

● Dampen a clean cloth with freshly boiled warm water. Leave the cloth over your eye for 15 minutes. Throw away the cloth after use.

DRY eyes

As people get older, their eyes tend to produce less moisturising tear fluid and even the fluid itself is less rich in oils so they can't lubricate the eye as well. Eyes may feel itchy and gritty at times.

● One over-the-counter solution is artificial tears but your eyes will also benefit from a healthy diet that contains walnuts, oily fish and other sources of omega-3 fats.

● If you suffer from dry eyes, it is also best to avoid smoky atmospheres, air-conditioned rooms and too much sunlight or wind. Blink frequently and take regular breaks from close-up work such as reading or sewing, or when using a computer.

Fatigue

It is normal to feel tired after a long, strenuous day at work. But if exhaustion is a constant companion, sapping your energy and your zest for life, it is time to do something about it.

Many people don't feel fully awake in the morning until they have had their first coffee. Unfortunately, coffee has only a short-term effect. Instead, rediscover some traditional techniques for getting moving. Open the window wide and breathe in the fresh air. Get your circulation going with a couple of deep knee bends and then move your arms in a circle. Follow this up with a healthy breakfast. If you are still not feeling energised, here are a few other things to try.

HOME remedies

● Alternate hot and cold water in your morning shower. A cold arm shower (see box, right) will give you a kick-start, especially if low blood pressure is causing your fatigue. If you are in a hurry, take a cold arm bath for a few seconds: just dip your arms up to the elbows in a sink filled with cold water.
● Need a siesta after a big meal? Eat smaller, more frequent meals. High blood sugar levels can switch off the brain cells that keep you alert, making you feel sluggish after a big meal.

Cold arm shower

one Direct a cold stream of water along the outside of your arm, moving slowly from the fingers of your right hand up to the shoulder.

two Go back down with the water, this time on the inside of your arm.

three Do the same with the left arm.

● If fatigue is a result of stress, get moving and take a walk. Exercise releases 'feel-good' endorphins, leaving you revitalised and more positive.
● Drink 500-750ml of stimulating stinging nettle or ginger root tea a day.
● Try a little ginseng. As the Chinese will tell you, ginseng tea can reduce feelings of stress and anxiety and combat fatigue. Pour a cup of boiling water over 1½ teaspoons of finely chopped ginseng. Leave it to steep for 10 minutes then strain. Drink 500ml a day.

Open the window wide and breathe in some fresh air.

● Go herbal. For thousands of years, rosemary has been treasured for both its aroma and its medicinal effects. It is thought to help blood circulation and improve memory and concentration. When added to a cool, brief bath, rosemary can provide an effective remedy for fatigue and exhaustion. Alternatively, try spruce needles in the bathwater.

DRINK an energy potion

A trusted home remedy is treacle with apple cider vinegar.
1 In a cup, stir together 2 teaspoons of treacle and 4 teaspoons of apple cider vinegar.
2 Fill the cup with honey and mix.
3 Take 2 teaspoons when you get up and before you go to bed, and 1 teaspoon before lunch and dinner.

NUTRITION to combat fatigue

● Increase your iron levels. If your diet is low in iron, blood cells aren't able to carry their usual load of oxygen around the body and your energy level plummets. Eat high-iron foods such as red meat, liver, whole grains and green, leafy vegetables.
● Iodine deficiency can also cause ongoing fatigue. You can counter it by eating oily fish and sprinkling food with iodised table salt.
● Buy whole-grain products. Whole grains break down slowly in the body, releasing sugars into the bloodstream evenly. This means your body gets a constant energy supply and blood sugar levels won't fluctuate dramatically, causing fatigue.

● Fresh vegetables, plus milk and milk products, contain a wide spectrum of vitamins and minerals crucial to well-being. They can boost the body's performance capacity, so include them in your menu.
● Enjoy a tasty, healthy vitamin and mineral boost. Spread a slice of whole-grain bread with cream cheese and add avocado, alfalfa sprouts and chives.
● Avoid coffee, cola, champagne and any other caffeine-containing drinks. They are often promoted as pick-me-ups, but the boost doesn't last long and often has a boomerang effect.
● Avoid sweets such as chocolate and toffee. They contain 'simple sugars' that quickly elevate blood sugar levels and performance capability – but this high is followed by a crash just as quickly, sending you into an energy slump.
● Snack better. Instead of sweets or fast food, eat a pot of yoghurt or a piece of fresh fruit.

WHEN TO CONSULT A DOCTOR If fatigue persists, see a doctor. A blood test can determine if you are suffering from a condition such as anaemia (low iron levels) or hypothyroidism (low thyroid levels) that can cause extreme tiredness.

EAT FRESH FRUIT, rich in VITAMINS and MINERALS, DAILY.

Flatulence

Everyone has it, but wind can cause bloating and discomfort, and passing it in public is embarrassing. Diet, lack of exercise and stress can all contribute to these unpleasant feelings of abdominal pressure.

If it is any consolation, Hippocrates (the father of medicine) proclaimed that passing wind 'is necessary to well-being'. But even the ancients sometimes turned to nature to restore the health of their intestines and eliminate wind. A distressed intestine needs soothing. When the stomach is distended, it is wise to consume less and try one of the following natural ways to reduce flatulence, using medicinal plants or heat applied to the abdomen.

GOOD TO KNOW ✓

Wind in babies and small children

Tiny clenched fists, a scrunched up face and an earth-shattering wail are the telltale signs that a baby has colic. Fortunately, there are some gentle, time-honoured remedies that may help. Even babies can benefit from the digestive properties of fennel tea. Mix 1 teaspoon of the tea into formula milk for bottle-fed babies. For breast-fed infants, use an eyedropper or pipette to administer the tea three times a day. Alternatively, place a warm hot-water bottle on the child's stomach or gently massage his or her tummy in a circular motion to provide a little relief.

HOME remedies

- Mix together 15g each of caraway seed, fennel and anise. Pour about 250ml boiling water over 2 teaspoons of the mixture. Let the tea steep for 10 minutes, strain and drink unsweetened.
- Finely grind caraway seed and coriander and take ½ teaspoon with a little water before every meal.
- Use liquorice root to help with bloating: dissolve 15g liquorice (from a health-food store) in about 250ml chamomile tea. Drink one cup a day.
- Let heat soothe the discomfort. Heat a spelt or cherry-stone cushion to about 40°C in the oven or microwave, and place it on your tummy.

NUTRITION

Flatulence can be aggravated by certain foods that are difficult to digest, while the problem can be aided by others. Also, people will react differently to certain foods so there are no hard and fast rules, but here are some things to try.

- Season food with spices that aid digestion, such as caraway seeds, anise, marjoram and ginger – they can reduce flatulence and bloating.
- Eat slowly and chew thoroughly – swallowing air is a major cause of wind.
- Avoid gassy foods such as carbonated drinks and beans.
- Cook legumes, leeks and cabbage thoroughly so they will be easier to digest.
- If you suffer from lactose intolerance or have difficulty digesting other foods, this could trigger a wind attack. Avoid the foods in question.
- Avoid sweets, slimming products and chewing gum sweetened with sorbitol, xylitol or mannitol. They are difficult to digest.

Use a pestle and mortar to prepare stomach-friendly spice mixtures.

> EVERY NOW and THEN GIVE FEET a break by WALKING BAREFOOT.

Foot care

Ill-fitting footwear, high-impact exercise and certain medical conditions are the main causes of painful feet. Thankfully, there are several traditional remedies that will help our feet carry us smoothly through life.

A foot rub is a treat for aching feet and the perfect way to relax and release stress – and there's no better treat for a friend or partner. But if there is no one around to rub your sore feet, here are some easy ways to relieve foot ailments or discomfort.

CORNS

Calluses can develop over time where a shoe pinches the skin or squeezes your toes together. Once the callus thickens and forms a hard core, it is a corn.
- Soften the callus before a corn develops with a footbath of chamomile or tea tree oil, or a daily rub with a little castor oil.
- Once softened, rub it off carefully with a pumice stone or special corn file.
- If a corn does develop, take the pressure off by putting a piece of gauze between the affected toes to reduce friction and rubbing, and to take pressure off the sore spot.

TO treat a corn
- Place a fresh, thin slice of onion on the corn and hold it in place with a gauze bandage until the core of the corn dissolves.
- Use a corn patch for stubborn cases. Pharmacies offer patches containing salicylic acid that soften the corn so it can be pulled off along with the patch.

BLISTERS

Blisters occur when the upper layers of skin have been damaged, often as a result of friction, and a clear fluid called serum fills the pocket. Most blisters heal naturally, but there are some traditional ways to prevent them forming – or to relieve the discomfort if they do.
- Prevent blisters from occurring by always wearing socks or tights with shoes.
- Going for a hike? Opt for two pairs of thin socks instead of one pair of thick ones. The socks will rub against each other instead of rubbing against bare skin.

Vinegar bread

one Pour a little wine vinegar over a few slices of white bread in a bowl. Let it stand for a few hours until it is soft and mushy.

two Spread the mixture onto painful corns or calluses, cover with a cloth and hold in place with a gauze bandage. Let it work overnight.

three Repeat the application as many times as is necessary to treat the area.

TAKE contrasting footbaths

Dip your feet in hot water for 5 minutes, followed by 10 seconds in cold water. This will combat athlete's foot effectively. These footbaths should also provide relief.

● Boil 3 tablespoons of chamomile or oak bark for 30 minutes in 1 litre of water, then strain and apply.

● Add a few drops of tea tree oil or apply it directly to the affected areas.

● Add 5 tablespoons of clary sage, which reduces sweating, to 2 litres of water. Alternatively, apply clary sage oil directly to affected skin with a cottonwool swab.

● Before a long walk, rub petroleum jelly into sensitive parts of the foot. It is always worth carrying blister plasters just in case – if you suffer from blisters regularly, apply them before you set out.

ATHLETE'S foot

Athlete's foot is a common condition caused by a fungal infection that thrives in moist, damp environments – therefore swimming pools or saunas pose the greatest risk of infection. Wearing flip flops, or socks and shoes made from natural materials helps to prevent this itchy, unpleasant ailment, but if you do catch it, try a few of these home remedies.

● To stop the itching, rub the affected area with a crushed garlic clove or fresh bear's garlic oil (however, do not use on broken skin).

● Alternatively, apply a thick paste of baking powder mixed with warm water. Rinse after 3 minutes and dry your feet thoroughly.

● Change socks daily and wash them in the hottest water possible.

● Leather shoes absorb sweat, and fungi will grow happily in moist footware. As it takes about a day for shoe leather to dry out, prevent the reinfection of your feet with fungus by not wearing the same pair of shoes two days in a row, and dust them twice a week with a talcum powder containing zinc oxide (available from pharmacies). This inhibits fungal growth.

FOOT pain

Walking or standing for long periods can leave feet tired and aching. Wearing high heels or other uncomfortable shoes can exacerbate the problem. Luckily, home remedies can produce quick relief.

● Try a relaxing warm footbath (at 38°C) with a few drops of eucalyptus, rosemary or juniper oil.

● Make a foot massage oil by slowly warming 50ml sesame or sunflower oil in a bain-marie and mixing in 5 drops of lemon balm oil. Gently massage dry feet thoroughly with the lukewarm oil mixture.

● Rubbing feet with ice cubes brings them back to life. Wrap the ice in a clean cloth first.

● Use surgical spirit to massage painful legs and feet.

SWEATY feet

Your feet are the natural habitat of millions of bacteria that thrive on sweat and skin cells. By-products produced by these bacteria are what make feet smell. Ensure shoes remain odourless with these traditional suggestions.

● Wash feet daily with warm, soapy water and dry well, especially between the toes. Change socks at least once a day.

● Take a 10 minute footbath using either spruce needles, cider vinegar or table salt (125ml of each per 1 litre of water), or black tea – pour 1 litre of boiling water over four teabags, steep and then add cold water to cool.

Gallstones

Your gall bladder acts as a kind of storage tank for bile – a substance the body needs to break down fatty food into digestible bits. But when there is too much cholesterol present, gallstones begin to form in tiny, hard globules that can grow to the size of an egg. A diet packed with rich, high-fat foods is a big contributor, as are alcohol and nicotine.

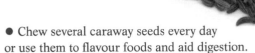

CHEW CARAWAY SEEDS DAILY as an aid to **DIGESTION.**

Bile fluid contains high levels of cholesterol and the pigment bilirubin, both of which precipitate as crystals and form stones. These may be as fine as beach sand or as coarse as gravel. Gallstones can develop in both sexes but are most common in overweight, middle-aged women.

HOME remedies

When your gall bladder goes on strike, it is a signal that your liver needs strengthening and the flow of bile must be restored.

● Artichoke juice can stimulate bile production. Mix 1 tablespoon of artichoke juice (from a health-food store) with a little water and take three times a day after meals.

● Eating bowls of boiled dandelions to counteract the effects of fatty meat is one ancient remedy. Today, dandelion extract or capsules can be bought at a health-food store. Alternatively, drink several cups of dandelion tea daily. Add 1 teaspoon of the leaves to 500ml water, boil then strain.

● Chew several caraway seeds every day or use them to flavour foods and aid digestion.

● Heat can ease the pain of a gall-bladder attack: apply a small, warm cherry-stone, spelt or flaxseed bag to the liver area, cover with a cloth and top with a wool blanket.

● Turmeric tea can help keep a gall-bladder in good health. Pour 250ml boiling water over ½ teaspoon powdered turmeric, let it steep for 5 minutes, strain and drink. Try to have at least 500-750ml a day.

● Avoid fatty processed and fast foods. These generally contain plenty of 'bad' saturated and trans fats (hydrogenated fat) that can lead to gallstones. If in doubt, read the ingredients label.

WHEN TO CONSULT A DOCTOR Frequent, severe distension, a feeling of fullness in the upper abdomen and pain in the liver area are early warning signs of gall-bladder ailments. Gallstones can form with time, interfering with bile flow and causing a great deal of pain. Get yourself to the doctor early and have the problem treated as quickly as possible.

Dandelion tea helps to counteract the effects of eating fatty meat.

Haemorrhoids

Haemorrhoids, or piles, are embarrassing and cause discomfort. In most cases lifestyle changes are the first step towards a solution, with these natural remedies to employ to alleviate unpleasant symptoms.

Haemorrhoids, caused by excessive straining during a bowel movement, are basically swollen veins in the anus accompanied by itching and sometimes sharp pain. They may be internal – that is, within the anal canal – or external, when they may be felt as little knobs or balls around the anal opening. Haemorrhoids are often the result of constipation, which also worsens those that already exist.

HOME remedies

● A cold wash after a bowel movement reduces itching and helps blood vessels contract. Soak a linen cloth with cold water (12-16°C), wring it out and wash the affected area, leaving a thin film of water on your skin. Do not dry it off but put on your underwear and sit still or lie under the covers in bed until your skin warms up.

● Take a sitz bath. Oak bark and horsetail contain tannins that have an anti-inflammatory effect. Boil two handfuls of oak bark or horsetail in 2 litres of water for 15 minutes, then strain. Pour the broth, cooled to body temperature, into a small basin and sit in it for 5 minutes twice daily.

● To alleviate itching place a cottonwool ball soaked in oil between your buttocks to prevent friction.

● Salves containing marigold or witch hazel reduce inflammation and bleeding effectively.

● An essential oil sitz bath also alleviates discomfort: put 2 drops each of cypress and chamomile oil and 1 drop of peppermint oil into a small basin filled with water at body temperature. Take a 5 minute sitz bath, then carefully dab the anal area dry and apply chamomile oil.

● Raspberries can aid digestion and soften stools to prevent further irritation to haemorrhoids. If you are susceptible to haemorrhoids, it is a good idea to stock up on raspberries when they are in season and freeze them in small portions.

● Dab on a little zinc oxide or petroleum jelly with a clean cottonwool ball. Tests prove that these everyday items are just as effective as pricier alternatives for soothing haemorrhoid pain.

PREVENTION

● A diet high in fibre produces softer stools, protecting against haemorrhoids.

● Adequate fluid intake prevents constipation. Drink at least six extra glasses of water daily to soften stools.

● Gentle and regular exercise encourages regular bowel movements.

● Use soft toilet paper or moisten the paper with a little water before use.

● Avoid prolonged reading sessions in the bathroom. Sitting on the toilet seat for a long time increases pressure on the rectum.

Steep a tablespoon of chamomile and yellow sweet clover in 250ml of boiling water, then strain after 10 minutes and add to a sitz bath.

Hair loss

Hormones or genes can be responsible for hair loss in both women and men. This can be difficult to combat. But to keep your hair in a generally healthy condition, there is no need to turn to expensive 'wonder cures'. Nature offers an array of effective remedies and applications that are much easier on the pocket.

Many hair problems respond well to home remedies. Hair loss that is not genetic could be a result of poor nutrition, ongoing stress, illness or incorrect or harsh treatment. In some cases, it may also be a side effect of medication.

HOME remedies

Anything that stimulates circulation to the scalp will aid hair growth and help you avoid hair loss.

● Rub your head with an onion. It may sound quirky, but rubbing your scalp for 10 minutes with the surface of a freshly cut onion is a tried-and-tested home remedy. The odorous vegetable contains plenty of sulphur that aids the formation of collagen, a substance that makes hair fuller and stronger. Follow up by washing your hair.

● To guard against hair loss, try a tonic consisting of about 50g each of nasturtiums and creeping thyme plus 1 litre of vodka. Let the ingredients steep in a closed container for ten days before straining. Massage the tonic vigorously into your scalp twice a day.

● Another traditional hair loss tonic consists of about 200g of stinging nettle root, 500ml wine vinegar and 1 litre of water. Boil the ingredients together for 30 minutes, then strain the liquid. Once it has cooled, pour it into a bottle. Use the tonic three times a week.

● Massage a few drops of pure tea tree oil into your scalp to stimulate hair growth.

● Try contrasting hot and cold water while shampooing every morning and evening. Always end with cold water, then carefully towel the scalp dry.

THE SULPHUR in nasturtiums HELPS TO FORTIFY HAIR.

PREVENTION

● To encourage blood circulation to your hair roots, massage your scalp with your fingertips for 5 minutes three times a day.

● Avoid overstyling. Washing with excessively hot water, extensive blow-drying, curling tongs and curlers can damage hair. Avoid perms and hair dyes, too.

● An unhealthy diet can contribute to hair loss. Avoid saturated fats and opt for foods high in hair-friendly vitamins and minerals, such as iron, zinc, protein and B vitamins.

Beer shampoo

one Wash your hair and rinse with warm water.

two Massage about 100ml beer into your scalp and let it sit for 15 minutes.

three Rinse hair with warm water and massage in another 100ml beer.

four Comb your hair and let the beer dry on your scalp. (It will soak in so well that no odour will remain.)

Hay fever

Springtime is greeted with mixed emotions by people with pollen allergies. The reawakening of nature is heralded by sneezing fits, red and watery eyes and, in severe cases, allergic asthma. The good news: there are many time-tested home remedies that offer relief.

Allergy symptoms are signs that your immune system is on the rampage, reacting to normally harmless substances such as pollen from grasses or plants. Here's how to wage war on the microscopic menaces that send your immune system into overdrive.

The reawakening of nature can be a nightmare.

BALANCE a twitchy IMMUNE system WITH BLACK cumin.

- Cider vinegar is a tried-and-tested treatment for hay fever. Put 1 teaspoon of cider vinegar into 125ml of water and sip the mixture slowly, preferably in the morning.
- Apply a moist cloth to itching eyes for fast relief. Use cold or warm water, whichever feels better, but make sure it has been freshly boiled and cooled.

BEHAVIOUR tips

The following tips will keep your exposure to allergy-causing pollen to a minimum.
- Keep windows closed during the day: that inviting breeze is bad news for an allergy sufferer, as it could potentially be carrying a load of pollen.
- Do not hang washing outdoors to dry. Pollen clings to moist surfaces.
- Change your clothes as soon as possible after outdoor activity.
- The middle of the day is peak pollen time, so stay indoors.
- Vacuum rugs frequently. Opt for a vacuum cleaner with an allergy filter and mop the floors regularly.
- Avoid having flowering plants and cut flowers in the house – regrettably, they spread pollen, too.
- Wash your hair before going to bed to avoid getting pollen on the pillow.
- Don't smoke. Nicotine increases susceptibility to allergies and heightens discomfort.
- Take shelter inside before a thunderstorm and for up to 3 hours afterwards. Storms are preceded by high humidity that makes pollen grains swell, burst and release their irritating starch.
- Wear wraparound sunglasses and/or a cyclist's mask to shield your face when you go out.

HOME remedies

- Black caraway (or black cumin), used for centuries to promote health and fight disease, can help to support a twitchy immune system. Over the course of three months, make a herbal tea by steeping seeds in 250ml boiling water for 10 minutes and drinking four cups a day.
- Watch what you eat. Nutrition plays an important role, especially in the lead-up to hay fever season. Vitamin C (fresh fruit, lettuce and vegetables) and magnesium (nuts, milk and grain products) strengthen your immune system. It may also help to eliminate meat from your diet.
- Use a simple soother. A nasal rinse with table salt may help clear a congested nose and remove trapped irritants. Dissolve 1 teaspoon of table salt in 250ml warm water. Put the salt solution into a container with a long, thin spout or a shallow bowl. Lean over the sink and sniff in the liquid one nostril at a time, allowing it to drain out through your nose or mouth. Nasal rinses are best used before going to bed.

Headaches

Headaches are a symptom, not an illness in themselves. There are a number of possible causes and most of these can be treated without reaching for the painkillers. Resourceful home remedies can provide gentle, quick and lasting relief.

Stress, over-exertion, sensitivity to weather, low blood sugar, colds, dental problems and psychological issues can all trigger a headache. Tension headaches involve a cramping of the neck and shoulder muscles, and respond well to acupressure treatments. Migraines, a specific type of headache, take the form of pulsating pain on one side of the head, often accompanied by sensitivity to light and noise, nausea, vision abnormalities and neurological problems. Possible triggers include alcohol, caffeine, cheese and the flavour enhancer monosodium glutamate (MSG), plus a lack of sleep, stress and hormonal influences.

HOME remedies

● Rest for a few minutes on the sofa and place some ice cubes wrapped in a cloth on your forehead.
● Rub a few drops of lemon balm, peppermint, clove or rosemary essential oil on your temples, forehead and neck (not suitable for people with neurodermatitis or children under two years).
● After removing the white inner skin, place the inside of a lemon peel on your temples for a few minutes.
● Sprinkle freshly squeezed plantain herb juice on a cotton cloth and place the cloth on your forehead like a headband.
● For a hot, relaxing neck compress wrap flaxseeds, chopped onions (both warmed in a bain-marie or microwave) or hot mashed potatoes in a cotton cloth and press onto your neck until the compress cools.
● Eat your greens. Studies suggest that migraine sufferers may have low blood levels of magnesium and could benefit from magnesium therapy. Dark-green, leafy vegetables, nuts and fruits are good sources of this mineral.
● Posture plays an important role in tension headaches, so pull those shoulders back and stand up straight.
● If a migraine strikes, head for a quiet, dark room for a little rest to help relieve the pain.

Juniper berry tea

3-4g juniper berries
250ml water

Crush the juniper berries and pour boiling water over them. Let the tea steep for 5 minutes, strain and then sip one cup three times a day. (Caution: do not take during pregnancy or if you have kidney disease.)

● Freshly boiled tea made from white willow bark contains salicin, a natural relative of the acetylsalicylic acid used in many common pain medications. It is easy to prepare: heat 1 teaspoon of white willow bark in 250ml of cold water and boil briefly. Leave to steep for 5 minutes, strain and sip one cup at a time, several times a day.

Neck compresses can soothe tension-related headaches.

Espresso flavoured with the juice from half a lemon can be a balm for headaches.

• If a headache is caused by nasal or sinus congestion, perhaps from a cold or hay fever, try a little bathwater aromatherapy. Put some eucalyptus or peppermint oil in the hot bathwater for inhalation and relaxation.

• Caffeine makes pain medications 40 per cent more efficient, so drinking small amounts may help hasten and increase relief, unless you are sensitive to it.

WHEN TO CONSULT A DOCTOR If unexplained headaches persist, make an appointment with a doctor as soon as possible.

PREVENTION

• If you take aspirin or ibuprofen frequently, stop. These drugs can cause 'rebound headaches' that start when a dose of medication begins to wear off.

• Avoid any form of nicotine as it constricts blood vessels.

• Limit alcohol consumption. Its toxic metabolic products increase the risk of a headache.

• Red wine and chocolate can trigger headaches in those with sensitivities.

• Get enough sleep and rest.

• Exercise such as jogging, walking and swimming encourages circulation and reduces stress.

• Feverfew doesn't just prevent fever, it reduces the frequency and intensity of migraines in those who take it regularly. Used for thousands of years by healers around the world, the herb can be grown in your garden, in a balcony pot or picked up in supplement form from a pharmacy or health-food store.

Acupressure

1 Use your index fingers to massage the depression just under the outer end of your eyebrows gently for 1 minute in a clockwise direction.

2 Massage the middle joint of the fourth finger on your right hand, on the side next to your little finger, for 1 minute in a clockwise direction.

Heart and circulation problems

The heart and circulatory system are the body's lifeline, so it is essential they run smoothly. A healthy diet, regular exercise and a smoke-free environment are all vital. But look to nature, too, as there are many natural ways to stimulate circulation and fortify our hearts.

Symptoms of circulatory disorders include tingling in the fingers or toes, pale skin and cold hands and feet. Don't ignore these signs; the first line of defence should be a visit to the doctor for a medical diagnosis and appropriate treatment. But cardiovascular disease is nothing new and the following traditional ideas may help. Just be sure to consult a doctor first before using a herbal remedy if you are taking prescription medicines.

HOME remedies

● Enjoy a gentle massage. Gentle, whole-body massages encourage circulation, particularly when a few drops of eucalyptus, pine needle or rosemary oil are added to the massage oil.
● To promote blood flow, sip tea made from equal parts of dried marigold flowers, daisies and heartsease (wild pansy). Pour 250ml of boiling water over 1 teaspoon of the mixture. Strain after 5 minutes and drink slowly.
● Encourage blood flow to your skin by rubbing it forcefully in the shower with a massage brush or a coarse washcloth.
● A mustard bath will increase circulation, open pores and stimulate sweat glands. Mix 200g of mustard powder with 2 litres of cold water. After a few minutes, strain the mustard water and pour it into a hot bath.

GOOD TO KNOW ☑

Versatile ginkgo

Ginkgo leaves yield an extract that contains a series of highly effective substances. Among other things, they encourage blood flow and prevent circulatory problems. Pharmacies sell commercial compounds in the form of drops and tablets.

● Take contrasting footbaths to boost circulation, especially in your legs. Soak your feet in a basin of hot water (38-42°C) for 5 minutes, then in a basin of cold water (18-20°C). Repeat. Dry your feet thoroughly and rest for an hour. The remedy is particularly effective when practised twice a week.
● Quite rightly, Chinese physicians have long recommended drinking green tea for health. Several clinical studies indicate the antioxidant-rich tea can reduce bad cholesterol and increase circulation.
● Health experts recommend 2,000mg of blood pressure-lowering potassium per day. Good sources include bananas, oranges and apricots.
● Reduce salt intake. The more salt blood contains the higher blood volume will be because sodium attracts and retains water. Spice up home-cooked meals with herbs instead, and stay away from packaged food.
● Eat celery. The crunchy vegetable is effective for controlling circulatory problems. Four stalks a day should do it.
● Eat chocolate. Dark chocolate is not just good for the soul, it has a proven ability to lower blood pressure. However, just 6g of chocolate a day will do the trick – and the darker the better.

The sulphur in raw garlic helps protect the heart. Cooked garlic is far less effective.

WEAK heart

The following are among the many heart remedies recommended for a weak heart in days gone by. Speak to a doctor before you try the herbal cures to avoid any adverse effects – especially if you are taking other medication.

● This low-cost home remedy is available in every household: freshly squeezed onion juice mixed with a little honey may strengthen the heart.
● Peppermint milk provides an economical drink that can help stimulate blood circulation. Pour boiling milk over some dried peppermint leaves and let it steep for 5 minutes, then strain and drink the milk in small sips.
● For six weeks, drink one cup of hawthorn tea twice a day, leaving at least 4 hours between drinks. The hawthorn shrub has been a stalwart of both European and Chinese herbal medicine since ancient times. In the 1800s, it became particularly renowned as a heart tonic, and now some clinical trials support this claim.
● Sprinkle cinnamon on cereal and include it in biscuit and cake mixes. Cinnamon may help to strengthen the cardiovascular system, shielding the heart from disorders. The spice also acts as a blood-thinning agent that increases circulation.
● Eat plenty of nuts. Nuts have many healthy effects on the heart. They help to lower the LDL (low-density lipoprotein or 'bad' cholesterol) levels in the blood – high LDL is one of the primary causes of heart disease. In addition, nut consumption

Menthol in peppermint and B vitamins in milk can help the heart.

reduces the risk of developing blood clots that can cause a fatal heart attack and improves the health of the lining of the arteries.

NERVOUS heart ailments

Anxious people have about a 25 per cent higher risk of developing coronary heart disease than their calmer counterparts, and are almost twice as likely to die of a heart attack over about ten years, according to researchers at Tilburg University in the Netherlands. So relax.
● Indulge in a daily cup of calming valerian tea, preferably in the evening: pour 250ml cold water over 2 teaspoons of minced valerian root; leave it to stand for a couple of hours, then strain. Warm up the tea and sip slowly.
● Caraway or lemon balm tea soothes nervous heart ailments, and they are quick and easy to prepare from fresh ingredients or teabags.
● Add essential oils (anise, lavender, wild mint, orange, rose) to bathwater or use in fragrant oil burners for a calming, relaxing effect.
● Get more rest. Lack of sleep has been linked to high blood pressure, atherosclerosis, heart attack and stroke. One theory is that poor sleep causes inflammation (the body's response to injury), infection, irritation or disease. That in turn revs up the sympathetic nervous system, which is activated by fright or stress.

WHEN TO CONSULT A DOCTOR If you feel exhausted at the slightest physical exertion, are constantly short of breath and retaining fluid in your legs, make an appointment to see the doctor.

Fortifying heart tea

40g dried hawthorn blossom and leaves
15g dried motherwort
15g dried lemon balm leaves

Pour 250ml boiling water over 2 teaspoons of the herbal mixture. Wait 10 minutes, then strain. Drink one cup three times a day.

Insect bites

The bane of beautiful summer days and balmy evenings, ticks, wasps and mosquitos have plagued humankind for centuries. As a result, many highly effective home remedies have evolved to relieve the pain and swelling and soothe that maddening itch.

Stings from bees, wasps or hornets can be dangerous. A few people are allergic to stings and may develop breathing difficulties with a dramatic fall in blood pressure (anaphylactic shock). In such cases, emergency treatment is vital. In Britain, mosquito bites are usually a harmless nuisance and you can rely on the wealth of experience from home medicine to treat these and other minor stings.

HOME remedies

● If you are stung by a bee, wasp or hornet, remove the stinger, scraping it out with a fingernail or hard object. Do not remove it with tweezers as this can squeeze more venom into the skin.
● To relieve the pain, soak a cloth in very cold water and wring it out, or make an icepack by placing ice cubes in a plastic bag and covering with a damp cloth.
● Cleanse the area. Stinging insects may have undesirable bacteria in their venom. Wash the sting well with soap and water or use an antiseptic wipe.
● To prevent swelling, place fresh slices of onion (a natural anti-inflammatory) or lemon on the sting.
● Stir together 2 drops of lemon essential oil and 1 teaspoon of honey and spread generously on the site of the sting to prevent inflammation.
● If you are away from home, coltsfoot and plantain herb can often be found growing by the side of a road. Crush a leaf between your fingers and press it onto the site of the sting.
● After removing a tick, disinfect the bite site with a few drops of tea tree oil, iodine or alcohol. (See 'Good to know', right, for further information.)

WHEN TO CONSULT A DOCTOR A sting on the mouth or throat carries a risk of suffocation and, for people with allergies, a sting anywhere can be life-threatening. In either case, go straight to hospital.

If you develop a rash around the bite site or a high temperature within three weeks of removing a tick, see a doctor.

GOOD TO KNOW ✓

Tick bite

If you get bitten by a tick, it is crucial to remove it quickly, including the head. Grasp it with tweezers as close to the skin as possible and slowly pull it out. If the head remains buried, consult a doctor. Be particularly wary of a circular redness spreading around the site of the bite some days or weeks afterward or if you develop a fever of 38°C or more – this can be the first indication of Lyme disease, a very serious condition.

PREVENTION

● Don't swat bees or wasps.
● Pack up leftovers quickly after a picnic so that the smell doesn't attract insects.
● Keep away from rubbish bins.
● Use perfumes and hairsprays sparingly – they frequently attract insects.
● Wear trousers and a long-sleeved shirt while hiking, if possible, and keep your skin covered up when in an area where insect bites are likely.
● Change out of sweaty clothing quickly, as sweat attracts insects.
● Don't walk barefoot over summer lawns.
● Avoid walking in tall grass or scrub land (especially where deer tend to graze) as ticks are likely to be present.
● Repel insects by adding five drops of citronella oil to 250ml water and dab it on exposed skin.

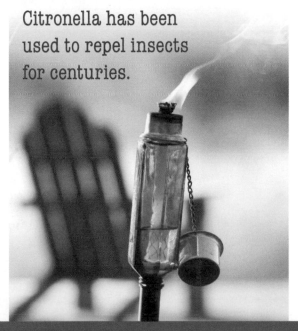

Citronella has been used to repel insects for centuries.

Joint treatments

Rheumatism was once a catch-all term for any ailment involving the joints. Nowadays, we distinguish between rheumatoid arthritis, osteoarthritis and gout, which have different causes but can all result in serious pain and reduced mobility. Thankfully, traditional remedies can be helpful.

Rheumatoid arthritis is an autoimmune disease that results in joint pain and deformity, osteoarthritis is a degenerative joint disease and gout is a metabolic disease that can result in joint damage. Try some of these home remedies to bring relief.

OSTEOARTHRITIS

- Apply cold moor mud or fuller's earth poultices to the affected joints once a day to soothe discomfort.
- Rub in arnica oil to combat joint inflammation and pain.
- A mud bath no hotter than body temperature (37°C) also eases arthritic pain.
- Drink a cup of stinging nettle tea three times a day to help with pain and inflammation. Use 1 tablespoon of the dried herb in 250ml of water and allow it to steep for 10 minutes.

- Try borage oil capsules or salves containing arnica, comfrey or capsaicin (the active ingredient in chilli peppers). The local health-food store or pharmacy can provide you with a range of compounds containing natural ingredients to reduce inflammation.
- The roots of devil's claw contain substances that soothe pain and inflammation. Originally from Africa, it has long been used around the world to treat painful joints.

RHEUMATOID arthritis

Whether you opt for a hot or cold treatment depends on the phase of the malady. If your joints are inflamed, hot and swollen, treat them with ice and cold packs made from mud, fuller's earth or clay (from a health-food store or pharmacy). When the condition is less acute, turn to heat to soothe discomfort and promote circulation.

Regular, gentle exercise helps to strengthen the joints.

Oil wrap for osteoarthritis

one Soak a cotton cloth in hot water and wring it out.

two Mix rosemary, marjoram and lavender oils in equal proportions, and put 10 drops of the mixture onto the cloth.

three Wrap the hot oil pack around your afflicted joints for about 10 minutes once a day.

four Repeat the application several times as needed.

● Relax. When your body is tense, pain will be more acute. Relaxation techniques may also ease pain.
● In addition to stinging nettle tea, teas made from meadowsweet or heartsease (1-2 teaspoons per 250ml hot water) can help to alleviate symptoms.
● A celery infusion can be a quick and effective remedy. Mince 1 heaped tablespoon of celery and pour 250ml water over it. Boil, steep briefly and strain. Sweeten with honey and drink 500ml a day.
Note: always prepare the infusion fresh and do not use it if you have a kidney infection.

ARTHRITIS prevention

● Oily fish such as mackerel, salmon and herring provide omega-3 fatty acids that ease swelling and pain. If you don't like fish, take fish oil supplements instead.
● Vitamin C has a positive influence on the course of the disease. It is particularly abundant in citrus fruits, kiwis and sweet peppers.
● Vitamin E (abundant in plant oils) intercepts so-called oxygen radicals, which form in greater quantities with acute inflammatory joint diseases.
● Avoid coffee, alcohol and nicotine.
● Some studies indicate that a diet rich in fruits, vegetables, salads, whole grains and reduced-fat milk products can help ease the symptoms of rheumatoid arthritis. Reduce saturated fat intake by cutting down on meat, fatty cheese, butter and cream.
● Slim down. Being overweight can increase joint damage.
● Take up tai chi or Chinese shadowboxing. When practised correctly, it relaxes all joints.

GOUT

Gout is a metabolic disease. When the kidneys become less effective at flushing away excess uric acid, the acid begins to crystallise in your joints, tendons and muscles, leading to swelling, pain and tenderness. To guard against gout, eat oily fish twice a week (but not eel), reduce the amount of meat you eat (no offal or consommés), and eat lentils, peas and red and white beans only in moderation. It is best to avoid alcohol and sweets entirely – both slow the body's ability to excrete uric acid considerably.
● Drink plenty of water or herbal tea to help kidneys flush out uric acid.
● Birch leaf tea acts as a diuretic, helping to flush out uric acid. It also contains salicylate – the same pain-relieving compound found in aspirin. Pour 250ml boiling water over 2 teaspoons of birch leaves, steep for 10 minutes and strain. Drink two cups a day.
Note: do not use if you have reduced heart, kidney or bladder function.
● Try mud and sulphur bath mixtures (available from a pharmacy or health-food store) added to body-temperature water to alleviate pain. Do this twice a week.
● Rub painful joints with camphor spirits (a solution of camphor and surgical spirit).

HORSE CHESTNUT SALVES can reduce INFLAMMATION.

Muscle pain

Over-use, tension and injury are the common causes of muscle pain, and intense but infrequent physical exertion often wreaks its revenge the following day. But sore, stiff muscles respond well to a bit of traditional wisdom.

Aching muscles are unpleasant but usually harmless and tend to ease after a few days. In the meantime, heat applications are a good way to alleviate the pain. However, muscle cramps can also be due to overload or other causes, such as circulatory disorders or a mineral deficiency. So you should consult a doctor if they are particularly painful or if the pain persists.

SORE muscles

● If you develop sore muscles, take it easy for the first 12-48 hours. When sore, muscles don't have full function and continued strenuous demands carry a heightened risk of injury.
● A hot bath can help you feel better. Add some moor mud, plus spruce needle or mountain pine extracts, for a soothing effect. The caveat: avoid heat for 2-3 hours after a tough workout, as it will promote circulation and increase inflammation.

Warm up and rub muscles with tea tree oil before intensive exercise.

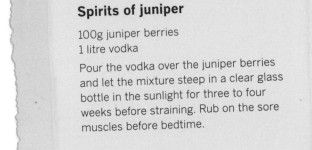

Spirits of juniper

100g juniper berries
1 litre vodka

Pour the vodka over the juniper berries and let the mixture steep in a clear glass bottle in the sunlight for three to four weeks before straining. Rub on the sore muscles before bedtime.

● A warm wrap with a few drops of arnica tincture can ease the pain.
● Massage can help ease sore, stiff muscles.
● Ample fluid intake flushes excess acids from the body and supplies it with important minerals. Opt for herbal teas and vegetable and fruit juices, diluted with mineral water containing little or no sodium.

MUSCLE cramps

● To relax a cramp in the calf, stretch the muscle carefully against the direction of the cramping, then walk back and forth. In stubborn cases, sit on the ground, pull your toes towards you and stretch your leg out fully. Then gently massage the muscle.
● Rubs containing extracts of menthol, camphor or horse chestnut can add extra power to a massage to loosen up cramps. So can essential oils containing eucalyptus, spruce needles or thyme.
● A lack of minerals, such as magnesium, potassium and calcium, is probably the biggest cause of nighttime leg cramps. These minerals are abundant in fennel, broccoli, bananas, dried fruits, oatmeal, nuts, milk, cream cheese and cheese.
● Cider vinegar provides your body with potassium. Drink 2 teaspoons of cider vinegar in 250ml water every evening for at least four weeks.
● If cramps are the result of a magnesium deficiency, taking magnesium in the form of effervescent tablets is a good idea. Check with a doctor first, however.

Muscle sprains

While a strain affects muscle or tendon tissue, a sprain is an injury to ligaments. The ankle is the most commonly sprained joint, usually the result of a single slip or stressful incident. Combine the correct first-aid rules with some home remedies and you will soon be up and running again.

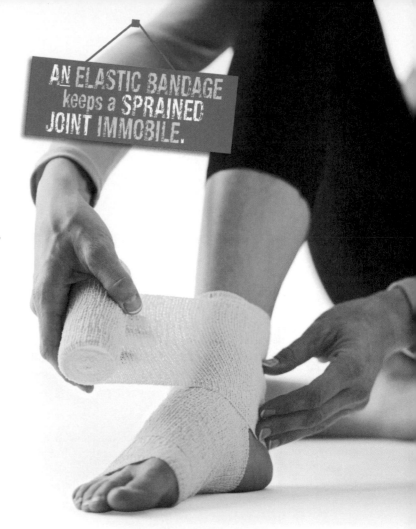

AN ELASTIC BANDAGE keeps a SPRAINED JOINT IMMOBILE.

FIRST aid

Act fast. A cold compress is the best and most effective aid for a sprain. It dulls the pain and decreases the blood flow, which lessens swelling. Ice the area immediately. Keep the ice in place for 15-20 minutes, then remove and leave off for an equal time. Do this four or five times daily for two days. After cooling, keep the sprained body part elevated to prevent further swelling. If you have sprained your ankle, place a pillow under the lower leg so that the leg is straight and slightly elevated. The easiest way to remember how to treat sprains and strains is the acronym RICE: rest, ice, compression and elevation.

HOME remedies

● Wrap the sprained area in an elastic bandage. The compression will help control the swelling. But don't make it so tight that you cut off circulation.
● Eat pineapple. Its active ingredient, bromelain, can help reduce swelling and speed your healing.

● Apply an ice-cold spelt or cherry-stone bag.
● Apply a little surgical spirit. This cools and combats swelling; cold wraps soaked in surgical spirit are a traditional treatment for sprains.
● When applied early, helichrysum oil can help combat inflammation.
● Carefully rub a few drops of tea tree oil onto the affected area to encourage healing.
● Salves and tinctures with a base of horse chestnut or comfrey accelerate the healing process.
● To make a tincture of arnica, let 100g of dried arnica flowers steep in 500ml surgical spirit for two weeks, then strain and store in a dark bottle. For a pain-relieving wrap, mix 1 teaspoon of the tincture with 250ml of cold water, moisten a cloth with it and apply to a sprained joint for 10 minutes. Keep the tincture stored in a cool, dark place.
● A liquid of oak bark (100g bark in 500ml water, boiled for 15 minutes) encourages healing.
● Get the correct shoes for your foot type.
● Spraining an ankle repeatedly can be a sign that your footwear is not giving you the support you need. Shoes designed specifically with an activity in mind can provide appropriate cushioning and traction.

Dill salve

Dill salve alleviates the pain caused by sprains.

2 tablespoons minced dill
1 tablespoon olive oil
A little beeswax

Mix the dill and olive oil, set aside for 24 hours then press through a strainer. Mix with warm beeswax to form a spreadable paste. Apply the salve to the affected area.

Nausea

Tainted food, too much alcohol, a virus, motion sickness, pregnancy or even stress can all trigger nausea and vomiting. Try some of the following suggestions, which can help to alleviate these unpleasant symptoms.

Travel or motion sickness is especially common in children, whose balance mechanisms are more sensitive than those of adults. A number of home remedies can effectively combat this and other types of nausea. However, as uncomfortable as it feels, a brief bout of vomiting can effectively rid your body of toxins. It is one of the body's most powerful defence mechanisms against harmful substances and, once it is over, nausea and stomach pain should disappear.

HOME remedies

- You can ward off rising nausea effectively by inhaling the aroma of a freshly cut apple. Breathe it in deeply.
- For sudden nausea, put a drop of peppermint oil onto your tongue, or drip the oil over the back of your hand and inhale the aroma.
- In days gone by, people kept scent bottles handy for such a crisis. Today, we are more likely to reach for essential oils. Put 2 drops each of lavender or sandalwood oil on a handkerchief and draw the scent into your lungs.
- To help calm your stomach, crush 1 teaspoon of fennel seeds in a mortar. Tip the crushed seeds into a pan and pour 250ml of water over them. Bring to the boil and leave to steep for 10 minutes before drinking.

Sucking on a freshly cut slice of lemon can help to nip rising nausea in the bud.

FOR motion sickness

- This tea will help prevent motion sickness. Prepare it in advance and take it with you on the journey in a small Thermos flask. Pour boiling water over one piece of cinnamon bark and let it steep for 10 minutes before straining.
- If nausea increases it sometimes helps to chew slowly on a little piece of ginger.

Chamomile wrap

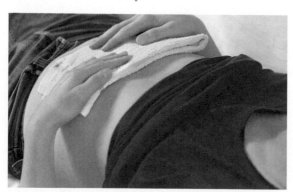

1 Fold a hand towel in two, then dip it in warm chamomile tea and wring it out.

2 Place the towel over your stomach area, spread a dry towel over it and hold in place with a wool scarf.

3 Let the wrap work for 10 minutes and replace it as needed.

Nervous problems

Overstimulation, conflict or a day packed with deadlines can jangle the nerves and leave you feeling tense. The result can be anxiety, nervous tension and sleep disorders. When it comes to re-establishing your equilibrium, these traditional teas, bath additives and scent mixtures can help.

When you feel overloaded, a little self-indulgence can work wonders – a relaxing bath by candlelight, perhaps, or a soothing massage. But if anxiety and nervous tension become overwhelming, you should seek professional advice.

HOME remedies

● Add a few drops of valerian and lemon balm to the bath water to stabilise emotional balance; lavender, bergamot, sandalwood or cedar will also do the trick.
● For a fragrant and relaxing herbal bath, boil up about 100g each of chamomile, lime and lavender flowers in 2 litres of water for 30 minutes. Strain and pour into hot bathwater. Add a little music and you have the perfect soak.
● Sandalwood helps to reduce anxiety and nervous tension. Mix 3 drops of sandalwood oil with 1 teaspoon of almond oil and gently massage your shoulders, neck, arms and legs. Better still, ask a partner or close friend to do it for you.

SIP LEMON BALM, HOP, passion flower or VALERIAN TEAS.

Breathing relaxation exercises

one Sit erect on a chair with your feet resting near each other on the floor. Let your arms hang down at your sides.

two As you breathe in through your nose, slowly spread your arms straight out to the sides and then bring your palms together over your head. Then stand up energetically.

three Rotate your hands so they are back to back. Lower your arms straight to your sides. As you do so, slowly sit back down and expel the air through your mouth.

four Wait a couple of seconds, then repeat the entire exercise. Do the whole exercise a total of ten times.

● To help calm your nerves and for a restful night's sleep, make a sachet of hop flowers and place it under the pillow.
● Three drops each of peppermint, basil, sage and lavender oils make a calming mixture for a scent burner.

PREVENTION

● Relaxation techniques, such as autogenic training, yoga, progressive muscle relaxation and meditation exercises, have a balancing and calming effect.
● Be sure to get plenty of sleep and rest breaks.
● Structure the day so there is some time for you.
● Taking a leisurely walk in the fresh air is good for the nerves.
● Lie in the sun for 10 minutes – it reduces stress and provides relaxation.
● Fish oils (omega-3 fatty acids) may elevate mood and calm anxiety. But, because oily fish can contain high levels of mercury, the NHS recommends pregnant women to eat no more than two portions a week and to avoid shark, marlin and swordfish.

Respiratory disorders

Pollution, stress, smoking, smog and office buildings with poor ventilation can all adversely affect your respiratory system. The good news: nature offers plenty of weapons to combat asthma, bronchitis and chronic coughs, or to complement medical treatments.

WARM elderberry, HONEY AND LEMON RELIEVE bronchitis.

The health of our lungs and respiratory system is affected by the quality of the air we breathe, and the effects of poor air quality can be far reaching. Thankfully, home remedies can help to alleviate the following conditions.

ASTHMA

Asthma is a serious condition, and without careful monitoring and ongoing treatment can rapidly become dangerous. The condition results in increased mucus production that narrows the airways in the lungs, causing a sudden feeling of breathlessness. Infections, allergies, stress, anxiety and stimuli such as dust, smoke and cold can all trigger life-threatening asthma attacks.

Asthma must be treated medically but, as scary as this condition can be, leading a healthy lifestyle can strengthen breathing passages and boost the body's natural defences. If you do have an attack, anything that breaks up mucus and relaxes your bronchial muscles will help. The caveat: if you have allergies or your asthma is allergy-related, be particularly cautious about using animal or plant-based remedies. These substances can trigger immune reactions and should be used only with medical supervision.

Thyme and lemon drink

Thyme can help to relieve coughing and make breathing easier.

15g dried thyme leaves
25g dried lemon tree leaves
1 litre water

Place the herbs in a heatproof bowl. Boil the water and pour it over them. Leave the brew to cool, then strain. Drink one cup three times a day.

● Try an airway-soothing mixture of honey and freshly grated horseradish in a 1:3 ratio. Take 1 tablespoon three times a day.
● To combat airway spasms and mucus, mix 2 teaspoons of cider vinegar and 1 teaspoon of honey in 250ml water and drink three times a day for three months.
● A warm chest compress breaks up mucus: soak a cotton cloth in water at about 50°C, wring it out and wrap it over your chest. Cover it with a dry cloth and rest in bed for 30 minutes.
● Excessively dry air irritates breathing passages. You can buy a humidifier, but bowls of water placed strategically around the house are a cheaper alternative. The water will evaporate into the air, providing moisture and some relief.

BRONCHITIS

Bronchitis is an inflammation of the bronchi, the tubes leading from the throat to the lungs, usually as a result of infection or irritation. To encourage healing, drink plenty of fluids. Teas and other hot drinks can soothe irritated airways and liquefy mucus, making it easier to cough up. In addition to traditional herb, clary sage or chamomile teas, home remedies that make use of common ingredients such as onions and sugar can be remarkably effective, and steam inhalations may soothe a persistent cough.

● Grind 1 tablespoon of anise seeds in a mortar and mix with 1 tablespoon of dried thyme. Pour 1 litre of water into the mixture. Let the tea steep for 5 minutes then strain, sweeten with honey and drink throughout the day.

● Peel and slice a large onion. Sprinkle the slices with a thick coating of brown sugar, layer them inside a glass jar, seal tightly and place in a warm spot overnight. Take 1 teaspoon of the mixture three times a day. Keep the remainder in the refrigerator.

● Chamomile soothes irritated respiratory passages. To inhale it, drop a handful of dried chamomile flowers in a bowl with boiling water, cover and leave to steep for 10 minutes before carefully inhaling the steam. As an alternative to chamomile flowers, add 1-2 drops of eucalyptus, thyme or cypress oil to the hot water before inhaling.

COUGH

Despite the huge sums spent on over-the-counter cough remedies every year, doctors say they are largely ineffective against a nagging cough. To relieve persistent symptoms, gentle, soothing home remedies can work wonders. Teas made from plantain, thyme or mallow can relieve a dry cough, and hot baths and cough remedies made from medicinal plants or ingredients may speed your recovery.

● To make your own cough syrup, combine 3 tablespoons of lemon juice and 250ml of honey in 50ml of warm water. Take 1 or 2 tablespoons every 3 hours. Note: honey should not be given to children under two years of age.

● For adults, warm beer can ease a dry cough and send you off to sleep. Before going to bed, warm up about 500ml of beer with 60ml of honey – don't let it boil. Sip slowly.

● A lemon wrap may help. Pour the juice from two lemons onto a cotton cloth and wrap it over your chest using a wool scarf to hold it in place. Give it an hour to work its magic before removing.
Note: if you have sensitive skin it could cause skin irritation. If so, remove the wrap immediately and wash your skin with warm water.

● Just 3 drops of eucalyptus oil and a little water in a fragrant oil burner can help to soothe troubled breathing passages.

Drink between two and five cups of fennel tea per day, which will have a soothing effect on respiratory ailments.

Shingles

About 20 per cent of people who have had chickenpox will later develop shingles, usually when they are age 50 or over. The infection causes a burning, blistering rash and mild to severe pain that can last for several weeks.

Most people have chickenpox in childhood. After the illness has gone, the virus remains dormant in the nervous system. But it can be reactivated many years later if the immune system is compromised by age, disease or stress, and cause shingles. This affects the nerves and the area of skin around them. You will need medical care to deal with the condition, but lots of rest and natural remedies can help speed the healing process.

HOME remedies

● Try various essential oils to alleviate pain. Mix 2 drops each of bergamot and eucalyptus oils and 2 tablespoons of almond oil. Use a cottonwool ball to dab the mixture onto the blisters.

A herbal wrap

This encourages the rash to heal.
What you need:

25g oak bark
200g lady's mantle
20g chamomile
20g clary sage
10g sweet yellow clover
1 litre cold water

Pour the water over the ingredients and bring to the boil. Steep for 5 minutes and strain. Dab the warm liquid onto the affected skin with a cottonwool ball or soak a cloth in it and cover the area.

Used in a herbal wrap, lady's mantle can be a useful ally.

● A fuller's earth poultice can also alleviate pain and dry up blisters. Stir the fuller's earth with a little water to form a thick paste, spread it finger-thick on a cloth, cover with gauze and apply, cloth-side down, to the rash. Replace the pack as soon as it warms up. Apply twice daily.
● Apply linen cloths soaked in soothing chamomile or yarrow tea to the rash.
● Blisters heal more quickly when they are sprinkled with a little diluted homemade tincture of marigold (one part tincture, four parts water).

PREVENTION

● Strengthen your body's defences to make sure viruses don't have a chance. Regular exercise, adequate rest and a vitamin-rich diet will help.
● Avoid extended sunbathing, as ultraviolet rays are stressful to the body.

WHEN TO CONSULT A DOCTOR Shingles is the product of the herpes zoster virus. If you have a rash on your forehead or anywhere near your eyes, see a doctor immediately to avoid the risk of damaging your corneas. Although you can expect some pain – unfortunately it comes with the condition – if the pain is unbearable, it could indicate the presence of nerve damage (post-zoster neuralgia), so seek medical help. Home remedies can support a doctor's treatment, but it is equally important to get plenty of rest. Physical exertion can exacerbate the problem.

Skin treatments

Skin acts as a barrier protecting us from the outside world. Crucial to our survival and well-being, we need to be kind to it – especially as its health and appearance can play an important role in our sense of identity.

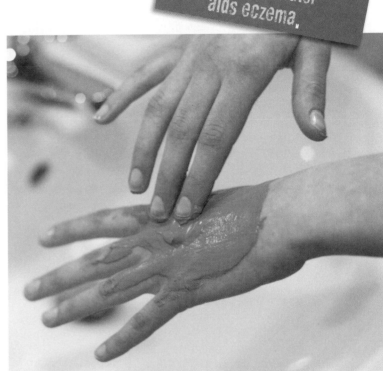

A PASTE OF FULLER'S EARTH AND water aids eczema.

Despite scientific advances in treatments for skin problems, many people still prefer to rely on natural ingredients and home remedies.

ECZEMA

This chronic skin condition, characterised by an itchy, red, scaly rash, can be relieved with natural remedies. Cleansers, salves and plant-based poultices can help to soothe the itching and inflammation, and moisturise dry skin.

BASIC skin care

If you are prone to eczema, your skin needs oil and moisture to restore its natural balance.
Note: if you have allergies, animal and plant ingredients can trigger an allergic reaction, so use them with caution.
● Soap-free cleansers are gentle and therefore far less irritating to the skin.
● Plant-based salves with jojoba or evening primrose oil are good moisturisers.

TREATING eczema

Check with a doctor before you use a home remedy, as eczema varies from person to person. The remedy you use must be right for you.

● Press raw cabbage leaves with a rolling pin until the juice comes out. Warm the leaves in a strainer held above steam and apply twice daily.
● Pour 250ml of cold water over 2 teaspoons of walnut leaves. Boil for 5 minutes, strain and cool to lukewarm. Soak a cloth in the liquid, wring it out and apply to the rash for 15 minutes.
● Stir together 3 tablespoons of fuller's earth and an equal amount of cold water to form a thick paste. Apply to eczema for 20 minutes. Rinse with cool water, then treat your skin with an anti-inflammatory salve containing vitamin E.
● Chamomile or marigold ointments can moisturise skin and soothe the relentless itching. The healing effect is strengthened when you keep products in the refrigerator and use them cold.
● Oil baths (maximum bathing time of 10 minutes, at no higher than 35°C) with chamomile or rosemary oil return moisture to the skin.
● An oatmeal bath leaves skin feeling soft and supple, and calms the itch: pour about 500g of oatmeal into an old stocking, knot the end and add to the bathwater.

THINGS to avoid

- Alkaline soaps, cosmetics containing alcohol and synthetic grooming products dry your skin out even more and make it susceptible to secondary infections from bacteria, viruses or funguses.
- Frequent contact with water and baths over 35°C.
- Intense sunbathing.
- Handling chemicals without protective gloves.

SKIN & the sun

Sunlight has many beneficial qualities but we must also protect ourselves from health problems caused by its rays, including skin cancer.

Since sunburn is a first-degree burn, anything that cools the skin will help relieve the pain – from a wet T-shirt to a cold linen cloth soaked in saline solution and placed over the burn. To make the solution, mix 1 teaspoon of salt to 2 litres of distilled water.

HOME remedies

You could try one of the following, but never use any remedies on skin that is blistered or open.
- Wraps with milk, yoghurt or buttermilk soothe pain and cool skin. Apply at least twice a day for 30 minutes. Thin slices of cucumber, potato, onion or apple also cool hot skin.
- Rub the burn with the cut surface of a lemon or a tomato – the vitamin C encourages healing.

A goosegrass bath

A bath with goosegrass soothes the burning and tightness of stressed skin.

100g goosegrass
2 litres water

Boil the water and goosegrass for 5 minutes, then strain and squeeze out the plant. Add the mixture to a lukewarm bath and soak for 20 minutes.

- Place a wrap soaked in cooled black tea or witch hazel tea (1 tablespoon of witch hazel in 375ml water) on reddened skin several times a day.
- Mix a few drops of evening primrose and lemon oil in equal proportions and apply daily.
- Try gels or creams containing aloe vera or arnica.
- A lukewarm bath with 8 tablespoons of fuller's earth added to it cools and encourages healing.
- Drink plenty of fluids, preferably cool herbal teas, mineral water or diluted fruit juices.

WHEN TO CONSULT A DOCTOR For severe headaches, sunburn in babies and little children, severe pain and burn blisters, consult a doctor. Never pierce or burst the blisters.

PREVENTION

- **0-2 (low) UV Index:** minimal protection required. Wear sunglasses and use sunscreen with a SPF factor of 15+. Cover up if outside for more than an hour.
- **3-5 (moderate) UV Index:** wear a hat, sunglasses and a sunscreen with a SPF factor of 30+. Cover up if outside for 30 minutes or more. Reduce time in the sun between 11am and 4pm.
- **6-7 (high) UV Index:** protection is a must. Wear a hat, sunglasses and apply a sunscreen with a SPF factor of 45. Try to avoid direct sunlight between 11am and 4pm.
- **8-10 (very high) UV Index:** use extra precautions – unprotected skin will burn quickly and could suffer long-term damage. Avoid going outside between 11am and 4pm. If you must, stay in the shade, cover up and apply a sunscreen with a SPF factor of 45+.
- **11+ (extreme) UV Index:** take exceptional precautions. Stay inside, or remain in the shade. Cover up and apply SPF 60 sunscreen.

Only expose yourself to midday sunlight for a few minutes.

CUTS & scrapes

Whether it is a finger cut during food preparation or a grazed knee following a fall from a bicycle, there are plenty of traditional remedies that can treat minor skin damage effectively.

FIRST aid

The most crucial aspect of wound care is to clean and disinfect thoroughly.
1 Carefully remove any foreign objects from a cut, scratch or abrasion with disinfected tweezers.
2 Disinfect with an antiseptic solution (from a pharmacy).
3 To stop bleeding, wrap a clean cloth or towel around the affected area and apply pressure.

HOME remedies

A day after an injury occurred, the wound – which should now be closed – can be treated to aid healing. Only use these remedies on unbroken skin. You will find any number of commercial antibacterial ointments at the pharmacist, but why not try one of Mother Nature's simple, low-cost solutions?

● Tea tree oil compresses have proven their worth to many generations of healers. Add 5-8 drops of oil to a clean cloth and cover the wound for 24 hours. Repeat regularly.

● For a yarrow wrap, add 100ml of boiling water to 1 tablespoon of dried yarrow flowers and strain. Moisten a cloth with the solution, gently wring it out and apply to the injured skin.

● To speed up healing, scald a lavender teabag, leave it to cool then place it on a wound.

● Vitamin C helps wounds heal. Treat yourself to an extra serving of strawberries or a juicy orange.

Bandage with chamomile tincture

one Pour about 100ml surgical spirit over 15g chamomile flowers. Leave it to steep for ten days.

two Carefully strain the solution and thoroughly squeeze the juice out of the chamomile flowers. Pour the mixture into a clean bottle.

three Dilute the tincture with water in a 1:4 ratio and apply to a piece of muslin. Leave on your injured skin for at least 30 minutes.

CABBAGE wrap

Not only is cabbage packed with important nutrients but its leaves also have many healing properties. A cabbage wrap alleviates pain and promotes healing. Use twice a day.
To prepare:
1 Rinse a few inside leaves from a head of cabbage and remove the central rib.
2 Soften the leaves with a rolling pin, then apply to your wound for several hours at a time using a bandage or clingfilm to hold them in place.

PREVENTION

● Minor accidents are inevitable, but taking sensible precautions such as wearing protective gloves, storing knives correctly or wearing kneepads when cycling will minimise injuries.

● Check your tetanus shot is up to date.

Try boiling some cabbage, then cleansing your skin with the water.

Sleep

Your body needs at least 7 to 8 hours of sleep a night to regenerate fully. But anxiety, a disturbed sleeping environment or an underlying health problem can all result in insomnia. Here are some tried-and-tested ideas that will help you to get a good night's sleep

Stress, worries or hormonal changes during menopause are often the root cause of an inability to fall asleep or a tendency to wake up repeatedly. Sleep disturbances are frequently temporary, but sometimes they drag on for weeks, months or even years and become a nightmare for the victim. You may not be able to control all the factors that are interfering with your sleep, but go back to having that warm milk before bed you were given as a child. Sometimes a calming drink and a change to your bedtime routine can be enough to solve the problem.

HOME remedies

● Slowly sip a cup of calming valerian or hop tea before going to bed. If you wish, you can sweeten the beverage with a little honey.
● Have a tiny tipple. Hops produce their calming effect not only in the form of teas and scent bundles, but also in beer. In moderation, alcohol adds to hops' soporific qualities A small glass (about 185ml in the evening) can work wonders – but drinking more could have the opposite effect.
● An hour before bedtime, sip a glass of warm milk containing 20g finely ground almonds.
● Take a dip. It is difficult to fall asleep if your feet are cold. Warm socks can help, as can contrasting footbaths before going to bed. Dip your feet into warm water (38°C) for 5 minutes, then into cold water (12-16°C) for 20 seconds. Repeat, ending with another dip in the warm water.
● A sleep-inducing mixture for a scent burner consists of 4 drops of chamomile oil and 2 drops each of lavender, sandalwood and neroli oil. Place the scent burner in the bedroom an hour before bedtime or add the drops to a small bowl of warm water to disperse the scent.
● Drip a few drops of lavender oil onto a spelt cushion and lay your head on it. Your body heat will cause the oil to evaporate gradually, bringing on drowsiness.

Hang fragrant, dried lavender clippings – cut just before flowering – in the bedroom to aid sleep.

● Take a warm, relaxing bath containing lime flowers about half an hour before going to bed.
● Here's another trick you can use: lie down on the bed in complete darkness, keep your eyes open and force yourself to stay awake – usually the reverse happens and you will quickly fall asleep.
● Modern methods such as autogenic training help to induce sleep through autosuggestion.

A slumber drink

This tea will make it easy to get to sleep and then help to keep you asleep through the night.

40g valerian root
20g hop cones
15g lemon balm
15g peppermint leaves
10g bitter orange peel

Mix the ingredients, then use 1 teaspoon of the mixture per 250ml tea.

TIPS for a good night's sleep

● Keep your bedroom quiet, dark and not too warm (no more than 17°C).

● Before going to bed, make sure the bedroom has been thoroughly aired.

● There should be no television in a bedroom – it belongs in the living room.

● Going to bed and getting up at regular times provides a healthy sleep rhythm and synchronises the body's biological clock.

● Make sure the mattress is of good quality. Test it out by lying down on it before making a purchase.

● Use sheets made from natural materials (such as cotton or linen) to avoid night sweats.

● Get lots of exercise during the day, preferably in the fresh air, to ensure that you are tired at bedtime. But avoid exercise within 4 hours of going to bed.

● Take a short walk before going to bed.

● Avoid fatty foods or eating a heavy meal in the evening, as they take a long time for your stomach to digest.

● Don't consume alcohol in excess, or coffee, black tea and cola containing caffeine. They are stimulants and interfere with sleep. And an excess of any drink in the evening could mean you will have to get up in the night.

● Avoid wrestling with problems in the evening – worrying spoils sleep. Make a to-do list of all the things you need to address the following day, then relax.

● Establish a routine. Children need to be run off their feet during the day to be tired at night, but the hour before bedtime should be low-key. Aim for a calm bedtime routine that might include a bath and a book.

Sleep sachets

A sleep sachet provides a pleasant, aromatic route to a refreshing slumber.

1 handful lavender flowers
1 handful hop cones
1 handful lemon balm
1 small pillowcase

Mix the dried herbs and enclose them in the pillowcase. Use as a pillow or place underneath your pillow.

You may need to make some lifestyle changes in order to sleep well.

Stomach complaints

A frantic lifestyle, poor nutrition, stress and too little exercise can result in an upset stomach. But popping a pill as soon as you start to feel uncomfortable could make things worse. A sensitised stomach is likely to react just as well to nature's much gentler medications.

STOMACH ache

It's an unfortunate fact of life that many of the foods we most enjoy are the ones that our stomachs like the least. And eating too much at once can leave your stomach with too much to handle. However, if you drink plenty of mineral water and unsweetened herbal teas (preferably at least 2 litres per day), eat a healthy diet and don't feast too often, your stomach will thank you.

● Sip fennel, chamomile, lemon balm or peppermint tea with and in between meals to aid digestion.

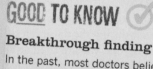

GOOD TO KNOW ✓

Breakthrough finding

In the past, most doctors believed stress and dietary factors caused stomach ulcers (peptic ulcers), with their symptoms of bloating, pain and nausea. The prescription: get rest, reduce anxiety, eat bland food and eliminate coffee and alcohol. Recent research, however, has revealed that the culprit in one in every five cases of peptic ulcer is actually a bacteria called *Helicobacter pylori*. A simple test is enough to confirm the diagnosis and, after treatment with a course of antibiotics, only a small proportion of patients relapse.

● Eat a small piece of fresh ginger if you are feeling nauseous. Or, if you prefer, grate roughly 2 teaspoons of fresh ginger and leave it to steep in 250ml of water for about 10 minutes before drinking.
● Some experts suggest that those cultures that value the after-meal burp have been right all along. Sparkling water or soda can help to activate beneficial wind.
● Prepare an anti-inflammatory tea from 1 teaspoon of liquorice root and 1 teaspoon of valerian: pour 250ml of boiling water over the ingredients, cover and leave to steep for 10 minutes. Strain and sip one cup slowly at mealtimes.
Note: do not use during pregnancy.
● Chamomile tea can help ease cramps.

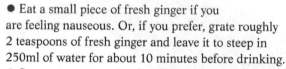

A BOWL OF PORRIDGE can HELP to SETTLE an UPSET STOMACH.

- Vitamin A helps to rebuild damaged mucous membranes in the stomach lining. Good sources include carrots, green cabbage, spinach, sweet peppers, apricots and honeydew melon.
- The traditional use of liquorice dates back several thousand years. There is some evidence to suggest that it can soothe your stomach and help if you have an ulcer.
- Massage your stomach in gentle, circular motions using 1 tablespoon of almond oil mixed with 3-4 drops of chamomile oil.

Replacing fatty foods with fresh grains and fibres is a major step towards calming an unruly stomach.

HEARTBURN

Sometimes your stomach will rebel against rich, heavily spiced or fried foods, especially if you eat them quickly or shortly before going to bed. The result can be heartburn, which causes a burning sensation just under your ribcage, often accompanied by a feeling of fullness, nausea and stomach pain.

In addition to traditional home remedies such as peppermint and fennel tea, a change in your eating habits and other gentle, natural solutions may help to put out the fire.
- Nux vomica is a homeopathic remedy for relieving heartburn. You will find it at a local health-food store.
- Sip a cup of tea made from elderflowers, lime flowers and peppermint (mixed in equal proportions) to relieve cramps and calm your stomach. To prepare, pour 250ml of boiling water over 1 teaspoon of the mixture.
- Black tea with a pinch of salt and a cream cracker can help to calm the stomach.
- A gentle stomach massage in a circular motion with chamomile or lavender essential oil (3-4 drops mixed with 1 tablespoon of almond oil) alleviates pain.
- Coffee, alcohol and spicy or fried foods can upset your digestion. Watch your diet and avoid food triggers.
- Eat smaller meals more frequently, rather than one or two big feasts a day, and consume your last meal at least 4 hours before bedtime.

IRRITABLE bowel syndrome (IBS)

Doctors don't fully understand the mechanism behind irritable bowel syndrome. What they do know is that it is marked by continuing problems with constipation, bloating, diarrhoea, heartburn and nausea. It is likely that stress can exacerbate IBS and certain foods also seem to make it worse.

On the positive side, it won't kill you and with a little help from age-old wisdom you may be able to regulate and soothe your troubled digestive system.
- Lingering over a meal allows enough time to digest it properly. When you are relaxed, you are less likely to swallow air that can increase abdominal discomfort.
- Avoid excessively rich foods and divide your food intake into several small meals a day.
- The aptly named chamomile roll cure – which involves tea and a rolling movement – can provide relief when used in the morning over the course of several days (see box, below).
- Alcohol and cigarettes are poison for a nervous stomach, so avoid them.
- Also avoid coffee and black tea. They stimulate digestion and that is not good if you are prone to diarrhoea.
- Avoid strong spices, as well as sweets, smoked foods, doughnuts and fried foods.

Chamomile roll cure

one In the morning make a tea using 2-3 teaspoons of dried chamomile and 250ml water.

two Go back to bed and then sip the tea on an empty stomach.

three Lie on your back for 5-10 minutes, then for an equal time on your left side, stomach and right side.

four Keep drinking chamomile tea throughout the day.

STRENGTHEN YOUR DEFENCES

A strong immune system can fight off most viruses, but when it's weakened they can get the upper hand. A healthy diet and lifestyle can strengthen your body's defences, aiding prevention or assisting if a virus does take hold.

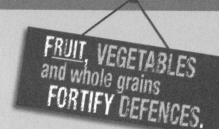

FRUIT, VEGETABLES and whole grains FORTIFY DEFENCES.

Reducing stress

Stress, conflict and worry are poison not only for your mental well-being but for your body's defences. Try to reduce them.

- Build in regular breaks during the day for a little 'me time'.
- Learn a relaxation technique, such as yoga, progressive muscle relaxation or Qigong.
- Get adequate rest, always going to sleep and rising at the same time each day.

Healthy eating

- Vitamin C is essential for your immune system. You'll find it in many fruits and vegetables, but particularly good choices are citrus fruits, kiwis, sweet peppers and fruit juices (especially orange, guava and cranberry).
- Eat a varied diet with plenty of fruits, vegetables, whole grains, plant oils, milk products, fish and lean meats. You will be providing your body with a full supply of substances to fortify its defences: vitamin E, beta-carotene, zinc and selenium.
- Avoid nicotine and alcohol: both weaken the immune system, and tobacco smoke is a Class A carcinogen.

Relaxation techniques such as yoga help beat stress

Echinacea tea is a useful ally.

Help from the health food store

Compounds made from coneflower (*Echinacea purpurea*) activate your immune system and strengthen the body's defences. Take echinacea daily for a period of about eight weeks. If you take it for longer than this, it could have the opposite effect as you may overstimulate and weaken your immune system.

Note: Don't take echinacea if you are allergic to coneflowers.

Toughening up

- Even a good, old-fashioned barefoot stroll through the grass can perk up your immune system.
- Start each morning with a contrast shower: turn on the warm water then, after 2 minutes, switch to a cold shower for 15 seconds. Repeat the process three times. End on cold water.
- Avoid car exhaust (it is toxic) and don't spend too much time in poorly ventilated indoor places where chemicals are being used (such as beauty salons and petrol stations), or where new materials, such as carpets, have recently been installed.
- Drink at least six 250ml glasses of water every day, which will boost your immune system and lessen feelings of fatigue.

Plenty of exercise

The fastest way to feel energised is to exercise – you'll feel the effects right away. A simple 10 minute walk will decrease tension, banish fatigue and boost mental alertness for hours afterwards. Make it part of your daily routine and pretty soon you'll be toning muscles, strengthening your heart and improving the functioning of most organs and bodily systems. Exercise immediately lightens the workload of the immune system, speeding up the elimination of germs and other threats by stimulating circulation, making you breathe deeply, accelerating perspiration and increasing muscle activity. Try cycling, swimming or taking a brisk daily walk.

Exercise aids the immune system by stimulating circulation.

Teeth

Sugary foods produce acids that attack enam[e]
and can result in tooth decay, especially in th[e]
young. As we get older, gum and periodontal
diseases are the major threat. Fortunately,
there is a lot you can do to keep teeth and
gums healthy and looking good.

GUM inflammation & periodontal disease

Pressure sores and injuries from dentures, plus
plaque and tartar, can lead to inflammation of the
gums accompanied by redness, pain, bleeding and
even receding gums if left untreated.
● Gargle with clary sage oil, which disinfects and
alleviates pain. Put 4 drops into 125ml of warm
water and gargle with it several times a day.
● For a simple mouthwash, dissolve 1 teaspoon of
table salt in a glass of water.

Eat an apple a day to
strengthen your gums

● Put 150g of quince seeds in 1 litre of water, boil
for 15 minutes, strain and leave to cool. Rinse your
mouth with the liquid three times a day.

TOOTH pain

Home remedies can alleviate the pain of a cavity but
they can't eliminate the hole in your tooth – see a
dentist for help with that.
● Tooth pain improves quickly when you pour a little
clove oil onto a cottonwool ball and hold it to the
painful tooth.
● If you don't have clove oil, place a clove on an
aching tooth and bite down on it carefully.
● Roll a savoy cabbage leaf with a rolling pin until
soft and then press it onto the outside of the
appropriate cheek.
● Rub the gums surrounding a painful tooth with a
crushed garlic clove.
● Rinse your mouth three to five times a day with an
analgesic mixture of arnica, clary sage and
chamomile. Mix together 10g arnica, 30g sage and
40g chamomile. Pour a cup of hot water over
1 tablespoon of the mixture, then strain and cool.
● Ice cools and soothes pain: press an ice pack onto
your cheek or suck an ice cube.
● Willow bark and meadowsweet contain substances
related to the painkiller acetylsalicylic acid (a
component of many synthetic pain medications). You
can make a tea from these herbs that helps relieve
tooth pain.

A COLD PACK provides
quick relief from
TOOTH PAIN.

PREVENTION

● Have your teeth cleaned and checked by a dentist and a dental hygienist at least once a year, even if you are not experiencing problems. Go every six months if possible.

● To cleanse your mouth of bacteria, rinse it thoroughly several times a day with tea tree oil (3-4 drops of tea tree oil in a glass of warm water) after brushing your teeth.

● Regularly use mouthwash containing clary sage or chamomile to clean and disinfect.

● Massage gums regularly with your fingers to strengthen them. Apply gentle pressure and work in circular motions.

● Keep your diet low in sugar, which bacteria thrive on, and your teeth and gums will benefit.

● Brush your teeth two to three times daily, after each meal.

● When you brush, use toothpaste containing fluoride to harden your teeth.

● Replace your toothbrush regularly every two to three months, because the bristles soften with use and become inefficient at cleaning.

● If you are on the go during the day and have no chance to brush, chew some special teeth-cleaning or sugar-free gum to stimulate saliva production and help remove food particles.

● Eat tooth-friendly foods. After meals, eat nuts or cheese, which counteract acidity, or a fibrous food such as celery that removes plaque.

● Don't smoke. Smoking is a major cause of periodontal disease, when pockets form between the gums and teeth causing foul breath from the constant discharge of pus.

GOOD TO KNOW ☑

Night-time trick

Avoid grinding your teeth at night by using this simple trick: chew a hard crust of bread or a carrot before going to bed to tire out your chewing muscles.

GRINDING your teeth

Teeth aren't built to withstand constant grinding. Anger, worry and poor alignment are the most frequent causes of tooth grinding, which can lead to tooth wear and gum problems.

● Go to the dentist to find out if tooth misalignment, fillings or crowns that don't fit properly are the cause of the problem.

● Try relaxation techniques if psychological strain is causing the grinding. They may provide relief.

● Use a mouth guard, fitted by a dentist, to reduce the impact of tooth grinding at night.

Cleaning teeth properly

one Use dental floss or a dental water jet to remove food particles from between your teeth.

two Use the red-to-white technique when brushing: always clean from the gum towards the tip of the tooth.

three Brush the chewing surfaces of your teeth.

four Thoroughly rinse your teeth and gums with mouthwash.

Throat problems

A sore throat – also known as pharyngitis – is unpleasant and typically the companion of a cold or flu. Most cases are not serious and will pass within a week, but natural remedies can often provide fast relief until they do.

A sore throat is a common complaint normally caused by a viral or bacterial infection, allergy, dry air or by inhaling smoke or another airborne pollutant. It is often the first sign of a cold and can be accompanied by fever and congestion.

SORE throat

Symptoms can vary from a mild scratchiness to severely inflamed mucous membranes in the throat and pharynx. Fortunately, there's a range of proven remedies for this common ailment within the kitchen cupboards.

● Drink lots of fluids to keep mucous membranes moist. Hot herbal or fruit teas (especially anti-inflammatory sage tea) are ideal drinks. Sweeten to taste with honey.

POUR BOILING WATER OVER chamomile flowers AND inhale.

Onion milk with thyme and cloves

Onion milk alleviates pain and helps combat inflammation.

1 litre milk
1 onion
1 sprig of thyme
2 cloves

Cut the onion into large pieces and add to a pan with the milk, sprig of thyme and cloves. Bring mixture to the boil, then remove from the hob and leave to steep for 10 minutes before straining. Drink onion milk three times a day.

● Honey has a natural antibacterial effect that makes it a particularly effective home remedy. Allow 1 teaspoon of honey to run slowly down your throat several times a day to chase the soreness away.
● Heat helps support the body's natural defences. Wear a scarf when you feel a sore throat coming on and avoid draughts and cold.
● Dry air further irritates mucous membranes. To keep humidity levels in your house high, place moist cloths on the radiators and bowls of water on the windowsills. Add a couple of drops of essential oils for a soothing effect.
● Gargling soothes pain and swelling in the throat and pharynx. Mix 1 teaspoon of salt with 500ml water to make a solution for gargling. Or try an old home remedy: 25ml of cider vinegar in 250ml water. Sweeten with honey.

- A potato wrap speeds healing. Boil, peel and mash 2-3 potatoes. Spread the warm – but not hot – spuds onto a cloth and wrap around your throat with a wool scarf to hold them in place. Wear until the potatoes cool. Repeat three times a day.
- Make a refreshing throat compress by soaking a linen cloth in warm thyme, sage or mullein tea. Gently wring out the cloth and wrap around your throat, covered by a wool scarf.
- Add a few drops of either lavender or eucalyptus essential oil to 250ml of boiling water and inhale to ease hoarseness.
- Ice cream is also a favourite pain relief remedy, and not just for children. Or try a hot drink flavoured with lemon and honey.

- Garlic has antiviral and antifungal properties, so it can be particularly effective at relieving a sore throat caused by a virus. Add to food or blend a few cloves of garlic, fresh or dried, into a vegetable cocktail.
- An effective home recipe for sore throats is a mixture that tastes better than the ingredients might suggest: stir ½ teaspoon freshly grated horseradish into a glass of warm water along with 1 teaspoon of honey and a pinch of ground cloves. Drink slowly.
- Get a new toothbrush. Bacteria trapped on a toothbrush may cause a sore throat to linger.

HOARSENESS

There are many possible causes for that frog in your throat: strained vocal chords, excessively dry ambient air or a viral infection, to name just a few. Resting your vocal chords may be all that is needed, so speak softly and as little as possible. In addition, an array of simple remedies can speed up the healing process. If there is no improvement, you should consult a doctor.
- Hot tea of any kind can soothe a sore throat, especially sage tea, which should ideally be made fresh every time from dried sage leaves. Slightly warmed mulberry juice (from a health-food store) also alleviates discomfort.
- If your mother treated your sore throat with a soothing mug of hot milk and honey, she was spot on. Drinking this beverage several times a day will ease hoarseness quickly.
- Tasty, economical and effective: cut an apple into slices, lightly fry it in a little butter or oil, sprinkle with plenty of sugar and eat.

Honey has natural antibacterial properties that make it a particularly effective home remedy.

Varicose veins

As varicose veins cannot be undone without medical intervention, prevention is as important as treatment. A little old-fashioned advice can help prevent varicose veins from getting any worse.

Those swollen, darkened veins in your legs can be traced to the faulty functioning of the veins' valves, which causes blood to accumulate in the legs.

HOME remedies

● Keep legs elevated as frequently as possible to encourage blood backed up in the veins to flow out.
● Stir together 250ml of warm water, 1 tablespoon of cream and 5 drops of lemon essential oil. Soak linen cloths with the mixture, wring them out slightly and apply to your calves.

Walking up steps helps stimulate blood circulation.

A cold knee shower

one Direct a cold stream of water (with the shower head removed) along the outside of the right leg from the back of the foot to a hand's breadth above the knee. Hold it there for 10 seconds.

two Move the stream downward on the inside of your leg.

three Repeat the process with your left leg.

four Finally, briefly rinse the soles of your feet.

five Dry off and put on warm wool socks. Rest for 20 minutes.

● Massage your legs from bottom to top using a mixture of 5 drops of cypress, lavender and juniper essential oil with about 50ml olive oil.
● Get a chair that fits your body. When you sit in a chair that is too deep for you, the edge of the seat presses into your legs constricting blood flow.
● Go low. Low-heeled shoes require calves to do more work, greatly aiding circulation.
● Rub your legs twice a day with marigold salve.
● Researchers have found that supplements containing horse chestnut extract (available from a pharmacy or health-food store) can combat leg pain and swelling as efficiently as compression stockings.
● Raise the foot of the bed or elevate your feet with a pillow to encourage blood flow.

PREVENTION

● Avoid standing or sitting for long periods.
● Don't cross your legs. It can slow circulation to and from your lower legs.
● Wear support stockings prescribed by a doctor.
● Sit on a chair, stretch out your legs and repeatedly bend and stretch your feet to activate the pumps in the veins.
● Get moving. Activities such as hiking, swimming, cycling and cross-country skiing promote circulation and prevent blood from pooling.
● Climbing stairs is good training for calf muscles – avoid lifts and escalators.
● Lose weight. Carrying around extra pounds puts additional stress on the circulatory system.

Warts

MILD MARIGOLD SALVE IS ideal for warts ON THE FACE.

Paediatricians used to advise children to draw their wart on a postcard and send it far away – a psychological boost that might have helped. These days, medical responses to a wart consist mainly of burning, scraping, cutting or freezing. These are painful techniques that can leave scars. In most cases, warts respond to natural remedies.

Warts occur when viruses, usually the papilloma type, invade skin cells. They may be unsightly and can itch. A verruca on the heel or ball of the foot can cause pain when walking, because the weight on the foot presses it inward. Warts often disappear within six months to two years without treatment.

HOME remedies

Treating warts requires patience – unfortunately there is no cure-all. Resist scraping or picking them off. If you have painful verrucas, you may require medical treatment.
● Marigold salve works well for warts on the face as it contains relatively mild substances. However, your face is delicate, so check with a doctor before trying this remedy.

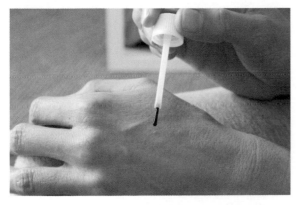

Apply oils and mixtures with a small brush to avoid spreading the warts through touch.

GOOD TO KNOW ✓

Prevention

Warts, which occur most frequently on hands and feet, are not only unattractive and annoying, they are also contagious. For that reason, never share creams, handkerchiefs or facecloths. To avoid getting warts in the first place, don't walk barefoot at swimming pools, saunas, gyms or in hotel rooms.

● Apply a few drops of tea tree, lavender or clove oil to the wart.
● Brush pure castor oil onto the affected areas of the body several times a day to prevent the viruses from reproducing.
● Mix together Epsom salts and cider vinegar in a 1:4 ratio and dab onto the wart several times a day to promote healing.

HELP from the pantry

● Rub warts with a peeled raw potato.
● Alternatively, brush the skin growths with the juice of an unripe fig.
● Apply the inside of a banana peel to a wart on the sole of the foot and secure it with a bandage.
● Cut a lemon into slices and put them into a glass. Add cider vinegar to cover. Let the lemon slices steep for two weeks, then rub the wart with them.

Women's reproductive health

While women have the exciting ability to create new life, the hormones that control the reproductive cycle can trigger discomfort. Fortunately, there are a number of traditional ways to combat these pains.

From adolescence on, a woman's reproductive system has a major impact on her life. Her complex hormonal balance may be disrupted as she progresses from the onset of periods to childbearing to menopause. Here are some time-tested remedies.

MENSTRUAL discomfort

Breast pain, bloating, acne, cramps and feelings of intense irritability are a few of the symptoms that may signal the start of your period. It may be preceded by uncomfortable PMS (premenstrual syndrome). More intense cramping and pains in the head, back and stomach are among the problems that can be experienced when the period begins.

● Monk's pepper (chasteberry) has been used for gynaecological conditions since the time of Hippocrates. It helps regulate hormonal balance, soothes the discomforts of PMS (such as breast tenderness and itchy skin), and even inhibits severe bleeding. This is available as pills from a pharmacy. Ask for agnus castus.

● Warm baths increase comfort and reduce cramping. Additives might include relaxation-inducing lemon balm or lavender, chamomile or yarrow to counter severe bleeding, or thyme to bring on menstruation.

● Massages can help ease pains and cramps: gently massage your lower abdomen and back with evening primrose or lavender oil.

● A tea made from valerian root, chamomile and peppermint, mixed in equal amounts, may help to soothe severe bleeding accompanied by cramps.

● Tea made from tansy and marigold may help stop cramping.

● Tea made from rue and yarrow, mixed in equal amounts, may help regulate menstruation and eliminate pain.

PREGNANCY

Although a natural condition, pregnancy can be difficult for some women.

● In the first trimester of pregnancy, many women are plagued with morning sickness. This simple trick can help: take a packet of dry biscuits or cream

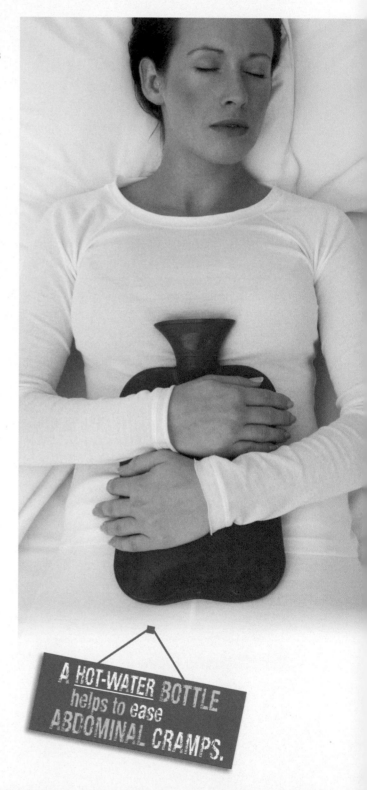

A HOT-WATER BOTTLE helps to ease ABDOMINAL CRAMPS.

GOOD TO KNOW ✓

Black cohosh

The roots of black cohosh contain triterpane glycosides, a popular treatment for premenstrual discomfort, menstrual irregularities and painful menstruation, as well as hot flushes during menopause. Women around the world have been using the herbal supplement for generations and a wealth of modern evidence backs up their choice. In some studies, black cohosh outperformed conventional therapies such as hormone pills and antidepressants in combating menopause symptoms. However, it takes four to six weeks from the time you begin taking it for the effects to become apparent.

crackers to bed and nibble a few right after waking up in the morning. Then rise, have a leisurely breakfast and sip some peppermint tea.

● Stretch marks may appear on your breasts and expanding belly. Gentle massages with jojoba or evening primrose oil keep your skin moist and elastic.

● A warm wrap with lavender oil helps soothe breast tenderness.

● Swimming or a gentle back massage can soothe pelvic pain.

● Magnesium, found in whole-grain products and dried fruits, helps eliminate cramping in your calves.

THE menopause

An inevitable part of ageing, menopause generally affects women some time between the age of 45 and 55. The drop in hormones that is its hallmark can also trigger a series of uncomfortable symptoms, including hot flushes, vaginal dryness, insomnia and mood swings.

● Various tea preparations help people get through menopause. Valerian tea has a calming, balancing effect, sage tea reduces sweating attacks and lemon balm tea can help with sleep disturbances.

● Contrast footbaths may help fend off those upsetting hot flushes. Fill two foot bowls with hop flowers and water: one hot (38°C); the other cold (10°C). Alternate placing your feet in the warm water for 5 minutes and the cold for 10 seconds.

● Nutrition plays a role in hot flushes. Eat plenty of fresh food and abstain from coffee, alcohol and nicotine, which can further lower your oestrogen level.

● Soya and red clover contain phytoestrogens – plant-based compounds that bear a chemical resemblance to oestrogen. Asian women have long enjoyed the benefits of soya, making them far less prone to hot flushes. Try incorporating soya milk, soya flour and tofu into your diet, or pick up some red clover pills from your local pharmacy or health-food store.

Ginger tea can help combat nausea during pregnancy.

Beauty and body care

Keeping your body in good condition is important to help you look better, feel better and stay youthful. Nature can play an important role, with plant-based eye creams, tempting fragrances, luxurious massage oils, natural shampoos and much more …

Age spots

These little brown marks usually appear on the back of the hands but sometimes on the arms or face as well. You can lighten and possibly reduce them using natural care products – but these treatments require plenty of patience.

Age spots, which are sometimes referred to as 'liver spots', should really be called 'sun spots' as they are actually the result of excessive exposure to the sun.

HOME remedies

It is possible to conceal age spots with make-up, or there are costly medical procedures such as laser removal. But home remedies are still worth a try.

Test them first on less noticeable places, such as the hands, before using them on your face. Wait a few days to see if you get a reaction and to get an idea of what effect it has on your skin – we're all different.

● Try applying pure lemon juice to age spots twice a day. Or, try buttermilk for a similar effect.

● For an easy to make paste to lighten age spots, mix 10g each of powdered ginger and rose petals. Stir 2 teaspoons of the mix into a small amount of warm water to make a spreadable paste. Apply to the spots, cover with a cloth and leave for about 30 minutes before rinsing your skin clean.

AGE SPOTS ARE A RESULT OF EXPOSURE TO THE sun's ultraviolet rays.

Lemon-based lightening agent

1 egg white
2 teaspoons lemon juice
½ teaspoon vitamin E oil

Beat ingredients together and spread onto age spots. Leave to work for 20 minutes, then rinse thoroughly. This daily regimen should be repeated over the course of a few weeks to lighten age spots.

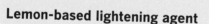

● Make a natural lightening agent by mixing together about 1 teaspoon each of honey and plain yoghurt. Spread the mixture on age spots once a day and leave for 30 minutes before rinsing well.

● Rub age spots several times a day with a piece of papaya, or use gels containing aloe vera (from a pharmacy). They can be helpful because they contain substances that stimulate the growth of healthy cells.

PREVENTION

● Free radicals contribute to the occurrence of age spots. Ultraviolet radiation plays a role in this, so it is important to avoid intensive sunlight.

● Add certain vitamins to your diet and avoid alcohol and nicotine that encourage the formation of free radicals. A well-balanced diet provides the body with nutrients that can combat their effects – such as vitamin C and pro-vitamin A (in fruit and vegetables), vitamin E (in plant oils and grains) and selenium (in nuts, legumes and grains).

Blemishes

The ancient Greeks used vegetable-based treatments while the Romans preferred mineral baths to help heal blemishes. Many traditional methods can still prove effective when it comes to fighting pimples, spots and blackheads.

Natural ingredients are at the forefront of today's skincare research and can be found in many of the latest beauty products. Achieve similar results using these simple methods.

HOME remedies

Nature has a few useful weapons we can employ in the battle against pimples, spots and blackheads.

● To calm reddened, irritated skin, use the soothing properties of mallow. To make a healing mask, beat 1 egg white until stiff and infuse with 3 tablespoons of mallow tea. Apply with a brush, leave it to work for 15 minutes then rinse off.

● To combat acne, create a yeast mask. Crumble 20g of brewer's yeast and combine it with 2 tablespoons of plain yoghurt, then mix in 1 teaspoon of honey. Apply the mask to your moistened face and then rinse thoroughly after 20 minutes. But remember: you will need to be patient – it could take several months before you notice any results.

● Before the manufacture of over-the-counter and prescription products to combat acne, people used natural remedies. The best known include thyme oil, raw potato slices, lemon juice and freshly cut garlic cloves.

PREVENTING pimples

Follow these tips to help prevent the outbreak of pimples.

● Refresh skin daily using mild cleansers.

● Thoroughly remove make-up every night, and avoid products that contain oil.

● Limit your skin's exposure to the sun, and use the appropriate sunscreen when you are out in it (*see Skin Treatments*, page 56).

● Monitor your diet. Determine whether alcohol, nicotine, coffee, black tea or fatty foods fuel your breakouts.

A facial steam bath

A weekly deep cleansing of the pores can soothe blemished skin.

2 tablespoons chamomile
2 tablespoons peppermint
2 tablespoons thyme
3 litres water

Place the herbs in a large bowl and pour boiling water over them. Leave to steep briefly. Cover your head with a towel and carefully hold it over the bowl for 10 minutes to let the steam open up your pores.

Be careful with facial steam baths – too much heat could scald your skin.

Cellulite

Those unattractive dimples on the hips, thighs and buttocks have been plaguing women for centuries. Consequently, over the years women have found several natural ways to deal with it.

Cellulite appears when a pocket of fat beneath the skin pushes on the connective tissue, creating a cottage-cheese effect. Women are particularly prone to cellulite because their connective tissue is softer than men's.

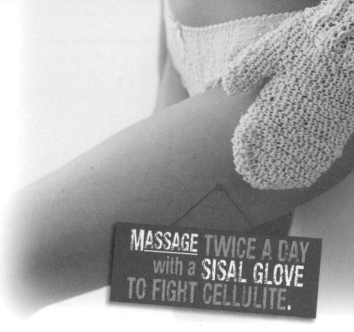

MASSAGE TWICE A DAY with a SISAL GLOVE TO FIGHT CELLULITE.

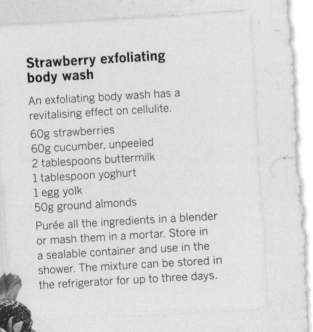

Strawberry exfoliating body wash

An exfoliating body wash has a revitalising effect on cellulite.

60g strawberries
60g cucumber, unpeeled
2 tablespoons buttermilk
1 tablespoon yoghurt
1 egg yolk
50g ground almonds

Purée all the ingredients in a blender or mash them in a mortar. Store in a sealable container and use in the shower. The mixture can be stored in the refrigerator for up to three days.

HOME remedies

There are a range of methods to combat cellulite, involving a number of effective substances including potatoes, sea salt and massage oils. But home remedies are not enough. It is also essential to eat a low-fat diet and exercise regularly.

● To make an effective massage oil, pour 200ml of wheatgerm oil over a handful of fresh ivy leaves and leave to steep for two weeks in a sealed container in a warm location. Strain and mix in 2 drops of rosemary oil. Rub into cellulite daily using circular motions.

● Other oils such as cinnamon, lavender, chamomile, clove, rosemary and sandalwood can be mixed with a base oil (such as almond oil) to create a massage oil for cellulite patches. (*For more on essential oils, see page 90.*)

● Potatoes possess remarkable healing powers, including the ability to tighten connective tissue. To take advantage of this, peel a potato, slice it thinly and spread raw slices onto the skin affected by cellulite. Cover with a cotton cloth and let it work for 15 minutes.

● Create your own mixture to tighten the subcutaneous tissue. Boil about 3 tablespoons of ivy leaves in 2 litres of water for 2 minutes, then strain. Moisten cotton cloths with the cooled liquid and apply to the affected areas once a day for about 20 minutes.

● Massaging sea salt into moist skin can also be effective. Do it after taking a shower, then rinse it off with warm water.

PREVENTION

Diet and exercise are key to preventing cellulite. First and foremost, people carrying excess pounds must slim down – being overweight is a risk factor.

● Exercise promotes circulation and keeps skin and connective tissue taut. Weight training is particularly helpful, and specific exercises can be designed to target trouble spots.

● Cut back on hydrogenated and saturated fat, sugar and salt as they encourage fat and water build-up, leading to the dimpled skin effect.

● Drink plenty of fluids to flush out waste and provide skin with moisture from the inside. Eat a low-fat, high-fibre diet containing plenty of green vegetables and fresh fruits.

Deodorants

Using a homemade blend of natural ingredients to refresh your skin gently and prevent odours is both healthier and less expensive than buying deodorants.

Underarm perspiration is caused by secretions from the apocrine and the eccrine sweat glands. The amount you perspire is linked to the body's efforts to regulate temperature. The hotter the temperature, the more you sweat. This is how the body cools itself.

MIX 2-3 DROPS OF rose oil WITH a little water AND DAB ON UNDERARMS.

BASIC formula

Essential oils are an important element of many deodorant formulas. The essential oils of citrus plants, such as lemon, bergamot, lemongrass, lime, neroli and grapefruit, are pleasantly cooling. Rose or lavender oils, on the other hand, lend a delicate, feminine aroma to a deodorant, while sage and cypress oil have an astringent and contracting effect that helps reduce sweating. They can keep you smelling sweet all day.

● To make a basic but fragrant deodorant, mix about 3 tablespoons of vodka and 3 tablespoons of either witch hazel, cornflower or rose water. Add to the mix 40 drops of essential oil. For best results, steep the deodorants for about a week before using.

VARIATIONS

● Another option is to use deodorant stones or crystals, which are made from crystallised natural mineral salts that kill odour-causing bacteria. Deodorant stones are inexpensive, but may clog pores if applied too liberally.

● To make a homemade deodorant with a delightful fresh summer scent, mix 3 tablespoons each of dried sage and dried lavender, the juice and grated rind from half a lemon and 3 drops each of lemon and neroli oil in a sealable container. Pour in 250ml of witch hazel and steep for a week. Strain and add 2 tablespoons of cider vinegar.

● Cider vinegar by itself is also useful to counter excessive sweat. Sprinkle it onto a cotton cloth and rub the underarms with it. Drinking sage tea regularly also nips sweat production in the bud.

● Homemade deodorants are best stored in pump spray bottles and shaken before each use.

Dried lavender can be used to add a delightful fresh summer scent to a homemade deodorant.

Eye care

The skin around our eyes is delicate, so requires special care. It is also the first part of our face to show signs of ageing, so it is important to ensure it is kept properly hydrated. Thankfully, natural eye creams can be used to care for this sensitive area.

Almond oil can help prevent wrinkles.

Eye creams moisturise the delicate skin around the eyes and treat dark circles and wrinkles.

CARE products

● Plant-based eye creams are available in most pharmacies. These cosmetics contain eyebright, marigold, aloe vera and oils such as wheatgerm, avocado and macadamia. Apply them morning and night.
● For a homemade fragrant skin oil that will leave you looking fresh, mix about 100ml of avocado oil and 3 drops of rose oil.
● Using easily spreadable plant oils, such as almond, apricot, jojoba and coconut oil, is an excellent way to moisturise the areas around the eyes and keep skin smooth. Apply a few drops directly onto the skin or moisten a gauze bandage or cottonwool pad with oil and place it on closed eyes and the surrounding skin.

CIRCLES under the eyes

Whether the cause is lack of sleep, allergies or just ageing, no one looks their best with dark circles under their eyes. They appear when blood vessels show through the delicate skin under the eyes.
● Eye creams or gels containing horse chestnut or butcher's broom may help to reduce puffy eyes.
● Mix 1 tablespoon of egg white with 3 tablespoons of plain yoghurt to make a refreshing eye mask. Dab onto the circles under your eyes and let the mask dry, then rinse with warm water.
● To make an invigorating eye compress, pour 250ml of boiling water over 1 teaspoon each of eyebright and lime flowers. Steep for 5 minutes and strain. Moisten two cottonwool pads with the cooled liquid and place onto closed eyes for a few minutes.
● One easy way to eliminate circles under the eyes is to place fresh, cold cucumber slices onto the eyelids. If you are in a hurry, use concealer to cover circles and provide a little visual first aid.

Slices of cool, fresh cucumber can soothe eyes and help to eliminate dark circles.

CROW'S-FEET

As we grow older, our skin becomes dryer and wrinkles form around our eyes. It helps to moisturise regularly, but crying, laughing, blinking and winking all leave traces. And don't try to get by without wearing your glasses – continued squinting also produces wrinkles.

● To delay developing crow's-feet for as long as possible, keep your skin smooth with a little daily care. Pure avocado and almond oil and aloe vera gels moisturise and prevent wrinkles.

● For a homemade wrinkle cream, stir together a little almond oil and lanolin, then apply to the skin under your eyes before bedtime.

● Once a week, treat the skin around your eyes to a nourishing skin-firming mask. Mix together 1 egg yolk, 3 drops of lemon juice and ½ teaspoon of olive oil, then apply the mixture with a brush. Leave on for 15 minutes, then rinse with warm water and refresh with cold water.

Doing eye exercises twice a day helps prevent wrinkles.

EYELASH care

Thick, dark, long lashes are easy to achieve by applying mascara.

1 Use mascara with almond oil to strengthen your eyelashes.

2 Carefully remove eye make-up in the evening. Dab a little water onto a soft cloth and wipe off mascara by rubbing gently towards your nose.

3 To ensure eyelashes look healthy, carefully apply a little olive or castor oil with a cottonwool swab for some deep moisturising – but be careful not to get the oil in your eyes.

DRINKING WATER KEEPS SKIN MOISTURISED and LOOKING HEALTHY.

Eye exercises

one To prevent wrinkles, close your eyes then open them after 1 minute and let them circle slowly in a clockwise direction. Stop and relax for a moment.

two With your eyes open, use your fingertips to pull your lower eyelids down slightly. Lift lids against this resistance and close your eyes. Relax for 1 minute with closed eyes.

three Now pull your eyebrows up firmly. Slowly close your eyes against this resistance. Again rest for a full minute with eyes closed. Repeat twice a day.

Facial cleansers

Beautiful skin won't be achieved without a good skin care regimen. Always removing sweat and make-up is a good first step towards this, while nourishing facial masks have been used for years to promote a youthful appearance.

Harsh cleansers such as soaps or toners with a high alcohol content strip skin of natural oils as well as removing the dirt and grease. Opt instead for mild products that match your skin type. All these can be applied using a cottonwool ball.

Note: always keep lotions you have made at the recommended temperature and never keep them for longer than the time advised. Bacteria can start to grow if they are not stored carefully.

Lady's mantle can be used in a cleanser for oily skin.

CLEANSING normal skin

● Stir together half a pot plain yoghurt, 1 tablespoon of olive oil and 2 teaspoons of lemon juice. Apply, then rinse off with warm water after 2 minutes. It can be kept in the refrigerator for two to three days.
● Warm 50g buttermilk and dissolve 1 teaspoon of honey in it. Stir in 1 teaspoon of lemon juice. Dab on and cleanse your face with gentle, circular motions. After 2 minutes rinse off with warm water.

CLEANSING greasy skin

● Pour 250ml of boiling water over 1 tablespoon each of lady's mantle, chamomile and sage. Strain after 5 minutes and allow to cool. Add 200ml of buttermilk and bottle. Shake before using. Leave on briefly before rinsing with warm water. The mixture can be kept for up to three days in the refrigerator.
● A stinging nettle milk promotes circulation and treats oily skin. Heat a handful of stinging nettle leaves in 300ml of milk. Strain the milk just before it comes to the boil and let it cool before applying.

Wash cleanser residue off using plenty of warm water.

Calming lotion

This flower lotion is cooling and refreshing for both normal and sensitive skin.

4 tablespoons witch hazel
4 tablespoons rose water
2 tablespoons lemon juice
2 drops lavender oil
2 drops rose geranium oil

Add ingredients to a bottle and shake vigorously. Apply a small amount of the lotion after cleansing.

• Fennel lotion has long been used to improve skin's appearance. Steep 3 tablespoons of fennel seeds in 150ml of buttermilk over a very low heat for 30 minutes. Allow to cool, strain and apply.
• Alternatively, stir together 1 tablespoon of finely crushed flaxseeds, 1 tablespoon of oat bran, 1 drop of lemon oil and a little hot water.

CLEANSING mature & dry skin

High-quality oils cleanse and return moisture to skin.
• For dry skin, add 100ml of aloe vera gel, 3 tablespoons of almond oil and 3 drops of lavender oil to a bottle and shake. Apply, leave for several minutes then rinse off the excess with warm water.
• For mature skin, melt 10g of beeswax, 50g of lanolin and 10g of cocoa butter in a bain-maire. Add 80g of almond oil and heat to 60°C. While stirring, add 80ml of rose water heated to 60°C. Stir until cool, then apply to skin.

CLEANSING sensitive skin

• Mix about 150ml of warm water with 50g oatmeal and 1 tablespoon of glycerin.
• Heat a handful of fresh violet flowers in 300ml milk. Just before it comes to the boil, remove from the heat and strain. Allow to cool before applying.

TONER

A toner removes leftover cleansing milk and helps clarify skin. Dab a small amount onto skin. Toners are generally left on the skin to dry, not washed off.

• For normal skin, add 5 tablespoons of witch hazel and 3 tablespoons each of rose water and orange blossom water to a bottle and shake. Alternatively, rub your skin with a piece of fresh cucumber.
• For an astringent or anti-inflammatory toner for oily skin, add 50ml sage water, 1 teaspoon of surgical spirit and 1 drop each of tea tree and sage oil to a bottle and shake. Alternatively, use dandelion tea.
• For dry skin, add 6 tablespoons of orange blossom water and 1 drop of lavender oil to a bottle and shake.

TONERS WITH ALOE VERA ARE GENTLE on sensitive skin.

Facial masks

Masks and exfoliators cleanse and remove dead skin cells, revealing more lustrous skin beneath.

NORMAL skin

● For a tea-based mask, mix together 1 tablespoon of plain yoghurt, 1 teaspoon black tea and ½ teaspoon lemon juice. Apply the mask, cover with a warm, moist cloth and let it work for 15-30 minutes.
● Mix together and apply 1 tablespoon of plain yoghurt, 1 tablespoon of sunflower oil, 1 teaspoon of plain honey and 1 teaspoon of lemon juice.
● Dissolve 1 tablespoon of gelatine in 125ml apple juice in a bain-marie. Cool until it forms a gel. Apply the mask and let dry, then rinse off with warm water. Pear or peach juice work just as well.
● Mix an egg yolk with 1 teaspoon of honey and a few drops of olive oil and apply to the face. After 15 minutes, rinse with warm water.

OILY, blemished skin

● Beat one egg white until stiff and mix with half a finely grated carrot. Stir in 1 teaspoon of chamomile and anti-inflammatory yarrow teas and apply.
● Stir together 3 tablespoons of oat flour with 3 drops of antiseptic tea tree oil, 2 drops of sage oil and an equal amount of warm water to ward off pimples, blackheads and inflammation. Apply thickly.
● Mix 2 tablespoons of fuller's earth with water to create a spreadable paste to calm inflamed skin.
● Stir together about 20g of brewer's yeast, 20g of yoghurt and 1 teaspoon of honey.
● Mix 2 tablespoons of honey, 3 tablespoons of ground almonds and the juice of half an orange for a gentle, exfoliating scrub. When applying avoid the eyes. The scrub can be kept for a week in the fridge.

MATURE skin

Fruit masks can help to firm and smooth saggy skin.
● Peel and purée a ripe peach and mix with a little facial cream to make a thick paste.
● Grate an unpeeled apple and stir together with 1 tablespoon of cider vinegar and 1 teaspoon of cornflour for a wrinkle-defying mask.
● To care for mature or dry skin, purée half a banana and mix with 1 teaspoon of honey, an egg yolk, 1 tablespoon of yoghurt and 1 tablespoon of wheat flakes. Leave to work for about 15 minutes.

The right way to apply a mask

one Apply a facial mask only to freshly cleaned skin, leaving space around the eyes and mouth.

two Leave the mask on for 15-30 minutes. During that time, relax.

three Carefully wash off the mask with warm water before applying a moisturising cream, gel or skin oil.

DRY skin

● Soak 2 tablespoons of fine oat flakes in 125ml of buttermilk for 30 minutes. Apply a thick layer.
● Mix together 2 teaspoons of honey, 2 tablespoons of plain yoghurt and 1 teaspoon of wheat flakes.
● Spread on a thick mask made from 2 tablespoons of fuller's earth, 5 drops of wild rose oil and an equal amount of rose water to improve skin tone.

SENSITIVE skin

● To promote circulation and clear up skin, stir together 2 tablespoons of finely ground oatmeal with 3 drops of peppermint oil and an equal amount of hot water. Spread the mask on thickly.
● Stir together 2 tablespoons of fuller's earth and an equal amount of lukewarm chamomile tea to make a spreadable paste.

Foot pampering

During an average lifetime, a person's feet will carry them around the world three times – or at least an equivalent distance. So it's important to ensure they remain in good shape. Natural footbaths, scrubs and lotions are the solution.

Corns, calluses, bunions and ingrowing toenails are not only unsightly and uncomfortable but they can also make it difficult to buy shoes that fit. These problems are often preventable and many natural home remedies can ease them. By caring for hard-working feet, walking and other forms of exercise remain pain free and pleasurable.

Moisturise tired feet after a pedicure.

Herbal spray

Refresh tired feet with this traditional herbal mixture.

2 drops rosemary oil
10 drops peppermint oil
10 drops lemon oil
10 drops cypress oil
100ml witch hazel

Add all the ingredients to a bottle with a spray top and shake vigorously. Spray feet several times a day, as needed.

DAILY care

Rub your feet with this nurturing foot lotion morning and evening.
1 To make the lotion, bring 250ml of milk, a handful of peppermint leaves and 2 tablespoons of fresh rosemary to the boil.
2 Leave the mixture to steep for 15 minutes, then cool and strain.
3 Mix 3 drops of peppermint oil into the herbal milk and pour into a sealable bottle. It will keep in the refrigerator for about four weeks.

REGULAR care

● A weekly foot scrub will remove scaly skin and make feet smooth and supple again. Mix together 1 tablespoon of almond oil, 1 teaspoon of sea salt and 3 drops of eucalyptus oil. Massage into your feet for a few minutes, then rinse them thoroughly. It's a good idea to have a foot scrub before a pedicure.
● Calluses can be painful. To avoid them, remove hardened skin occasionally with a pumice stone, preferably after a shower or bath. (To further soften a callus, tape a cottonwool ball soaked in vinegar to it overnight.) After gently rubbing down callused skin, massage in olive oil to keep the skin supple.

TIRED, stressed & sore feet

● A foot massage soothes aching feet. To make a simple massage oil, mix together 2 tablespoons of almond oil and 3 drops of lavender oil and massage feet using gentle, circular motions. Begin with the sole of the foot and proceed from the toes towards the heel. A massage brush or foot roller can also be used.

● To revive tired, swollen feet try an elderflower footbath. Boil two handfuls of elder flowers in 1 litre of water with a handful of peppermint leaves. Cool, strain and pour liquid into a small basin of warm water. Soak feet for 10 minutes.

● Another therapeutic footbath can be made by filling a bowl with warm water and adding a few drops of chamomile or lavender oil. Or add 1 litre of warm milk to a footbath. This will help to rehydrate cracked skin on your feet.

● To make stimulating bath salts, dissolve 2 teaspoons each of rosemary and spruce needles in 2 tablespoons of surgical spirit, then add 250g of sea salt. Mix the ingredients well. Dissolve 2 tablespoons of the bath salts in about 2 litres of warm water and soak feet for 10 minutes. Store the remainder of the salts in a tightly closed container.

● A sand and sea salt scrub removes callused and scaly skin on the soles. To make it, mix 250ml of fine sand, 2 tablespoons of sea salt, about 175ml of olive oil and 2 drops each of rosemary, peppermint and lemon oil. Rub into the soles of the feet with circular motions before rinsing with warm water and rubbing dry.

A foot roller provides relief for tired soles.

Perfect pedicures

one Take a 10 minute footbath with warm water and a few drops of the essential oil of your choice.

two Gently remove callused skin. Pumice stones have been used for skin care since the time of the ancient Greeks and Romans, and still work well. Or use a special callus knife, available from pharmacies.

three If necessary, soften cuticles with a little olive oil then push them back with a cuticle stick.

four Trim nails regularly. Always cut or file them straight to keep them from becoming ingrowing and painful.

● Thin, brittle skin can be painful and, if left untreated, can lead to further problems. To make a soothing herbal oil, heat 125ml of olive oil, add 10g each of marigold and lavender flowers, and leave to steep for 3 minutes over a low heat. Strain the cooled oil and squeeze out the flowers thoroughly. While still warm, apply a thin coat to the feet.

● To help swollen feet, try eating bananas. They are a natural source of potassium, which helps relieve fluid retention.

FOOT conditioning

Feet are like any other part of the body – they require regular exercise. Walking barefoot is one way to stimulate circulation and toughen them up.

● To strengthen feet, spread some marbles out on the floor and roll your feet back and forth on them while sitting. Then try to grasp the marbles with your toes.

● Wearing properly fitting shoes at all times is crucial.

Hair care

There are many gentle shampoos and rinses, nurturing conditioners and fortifiers based on traditional formulas that will keep hair clean, shiny and glowing with vitality.

NURTURING shampoos

A mild, neutral shampoo from the pharmacy is a good starting point for many nourishing formulas. After washing, let hair air dry, If you must use a hair dryer, avoid the hot setting.

Remember diet is also important when it comes to achieving healthy looking hair. Eat salmon, dark leafy vegetables, nuts and legumes to ensure you get plenty of vitamin B_2, vitamin B_7 (biotin) and zinc.

Oily hair

If your scalp produces too much oil, hair will soon become limp, lank and greasy. Strong shampoos dry out the scalp encouraging more oil production, but pH-neutral shampoos control oil levels. Since oil can form an ideal breeding ground for bacteria, it is best to wash greasy hair at least every two days.

> A MILD, NEUTRAL SHAMPOO is a good STARTING POINT.

Soapwort shampoo

Soapwort (or Wild Sweet William) has been used as a cleanser for generations.

10 soapwort stalks with leaves
500ml water
2 handfuls of fresh herbs

Chop soapwort finely and bring to the boil in water. Strain after 15 minutes. For greasy hair, add peppermint. For dull hair: parsley or rosemary. For dry hair: rosemary. For normal and fine hair: parsley. Allow to cool while covered. Strain again and bottle.

● To make a mild herbal shampoo, pour 1 litre of boiling water over a handful each of sage, rosemary and peppermint (fresh or dried). Steep for 30 minutes, strain then mix with about 150ml neutral shampoo before bottling the mixture.
● For a fragrant shampoo, mix 100ml of neutral shampoo with 3 drops each of cedar, bergamot and lavender oil.

Dull, lifeless hair

Using chemical treatments can render hair limp and lank.
● Make a nurturing shampoo that gives hair lustre by mixing 2 teaspoons of wheatgerm oil and 2 teaspoons of honey. Grate a little olive oil soap to make 1 tablespoon of soap flakes and dissolve them in 100ml of hot water, then mix in 1 teaspoon of lemon juice. Pour into a bottle and shake before use.

Normal hair

● To make a mild, fragrant shampoo mix 100ml neutral shampoo with 2 drops each of neroli, lemon and ylang-ylang oil, and leave to steep for a week.
● To make a traditional fortifying shampoo, beat 1 egg yolk with a fork until foamy and mix with 2 tablespoons of beer, 1 tablespoon of cognac and 5 drops of lemon oil.

Dry, over-processed hair

If the glands in your scalp secrete too little oil, hair will become dry and brittle. Too much sun, salt water, chlorine, frequent blow-drying and chemical hair treatments can increase the problem. Substances such as egg yolk or oils will boost hair's moisture level and bring back the shine.

● Make an egg shampoo by beating together 2 egg yolks with 2 shot glasses of rum, 1 tablespoon of olive oil and the juice of half a lemon. Massage the shampoo into wet hair and scalp, let it work briefly then rinse thoroughly.

● Rub a little olive oil into your scalp before washing hair to prevent the shampoo from drying it out. Leave it on for 30 minutes, rinse with warm water then wash hair with an easy-to-make moisturising shampoo of 2 eggs and 1 tablespoon of avocado oil.

● Make dry hair soft and silky by soaking a facecloth in whole milk and rubbing it into washed, damp hair. Rinse thoroughly after 15 minutes.

● Treat dandruff with an infusion of rosemary and thyme. Place 2 tablespoons of dried rosemary and 2 tablespoons of dried thyme in a bowl and add 150ml of boiling water. Cover the bowl and allow to steep for 15-20 minutes. Strain into a 300ml clean plastic bottle with a tight-fitting lid. Add 150ml cider vinegar and shake before each shampoo.

● Make a dry shampoo with 1 tablespoon cornflour or finely ground oatmeal. Apply to hair and rub it through to absorb as much oil as possible. Comb hair to remove tangles, then brush thoroughly until all excess cornflour or oatmeal has been removed.

RINSES & conditioners

Rinses, poultices, hair tonics and other conditioners can make dull hair shine again.

Oily hair

● Make a lemon rinse by mixing 1 litre of boiling water with the peel from 2 lemons and steep, covered, for 20 minutes. Strain and squeeze out the peel. Let the lemon water cool before rinsing.

● A conditioner containing essential oils may help. Mix 3 tablespoons of cider vinegar with 2 drops of sage, juniper and lemon oil and 200ml of warm water. Rinse hair with it, then rinse with tap water.

● To rid hair of oil and make it smell wonderful, make a vanilla rinse. Mix 100ml each of white rum and beer, 2 eggs, 1 teaspoon of lemon juice and ½ teaspoon of vanilla pulp (slit open a few pods and scrape them out). Massage into hair then rinse thoroughly after 10 minutes.

● A tea tree oil poultice will reduce oily build-up. Mix 50ml of almond oil with 3 tablespoons of lemon juice and 10 drops of tea tree oil. Comb through wet hair, leave on for 30 minutes then rinse and wash.

● Reduce oil production by applying an oil cure every ten days. Mix 2 tablespoons of jojoba oil and 10 drops each of juniper and tea tree oil, and apply to dry, unwashed hair and scalp. Comb through, cover hair with a plastic wrap and a towel, and leave on for 30 minutes. Wash hair with a mild shampoo.

Fine hair

● Make a weekly egg and beer poultice that can add body to lank locks by stirring together 3 tablespoons of beer and 1 egg. Massage into hair, cover with a plastic wrap and a towel, and leave on for 30 minutes. Wash hair gently with a mild shampoo.

Herbal conditioner

Use this conditioner for oily hair after every shampoo.

1 teaspoon burdock root
1 teaspoon chamomile flowers
1 teaspoon sage
1 teaspoon peppermint
1 teaspoon lavender flowers
1 teaspoon rosemary
500ml water
1 tablespoon lemon juice

Mix the herbs and the boiling water, and steep for 30 minutes. Strain and add the lemon juice. Use warm and do not rinse out.

Use lavender and chamomile flowers for treating oily hair.

1 teaspoon of cider vinegar and 3 drops of lemon oil. Spread through hair after shampooing and let it dry. Do not rinse it out.

Dry, stressed hair

● Avocados are packed with nutritious oils. To make a deep conditioner, purée the flesh of 1 avocado and mix with 2 egg yolks and 1 tablespoon of molasses. Knead the paste evenly into wet hair and rinse thoroughly after 30 minutes. Apply once a week.

● Bananas also make a nurturing conditioner. Mix one crushed banana with 1 tablespoon of avocado and massage the paste into hair and scalp. Cover hair with a plastic wrap and a towel, leave on for 20 minutes then rinse hair with warm water and wash with a mild shampoo.

● Thyme protects your hair from drying out. To make a thyme rinse, pour 1 litre of boiling water over 3 tablespoons of thyme (fresh or dried). Steep for 10 minutes, then strain and cool to lukewarm. Apply after every shampoo.

Lustreless, dull hair

● For shiny hair, mix 1 tablespoon of cider vinegar, 5 drops of lavender oil and 500ml of warm water and rinse washed hair with it. Do not wash out.

● Alternatively, mix 500ml each of chamomile and orange blossom tea and apply while still warm.

Normal hair

● A henna poultice will leave hair with a silky shine. Stir together 1 tablespoon of neutral henna (which strengthens and conditions but does not colour hair) with 1 egg yolk, 1 teaspoon of olive oil and enough warm water to make a spreadable paste. Massage into washed hair and leave in for 15 minutes. Rinse with warm water.

● For a healthy scalp and shiny, smooth hair mix 3 drops each of neroli, chamomile, lavender and sage oil in the palm of your hand and spread through freshly washed hair. Do not rinse off.

● To make a natural hair fortifier, mix together 500ml of warm water, 1 teaspoon of honey,

GOOD TO KNOW ✓

Split ends

If the ends of your hair are dry and brittle, apply a moisturising poultice or oil cure once a week after shampooing. It is also a good idea to massage in some olive or jojoba oil every evening and thoroughly rinse it out the next morning.

DANDRUFF

Those annoying flakes of dandruff are usually the result of an excessively dry or oily scalp and/or inappropriate hair-care products.

● To make a dandruff-fighting rinse, mix 1 teaspoon each of comfrey, rosemary and stinging nettle (fresh or dried) with 200ml witch hazel and steep for five days. Massage into the scalp after every shampoo and do not rinse out.

● Make an essential oil remedy by mixing 1 tablespoon of almond oil with 2 drops each of cedar, rosemary and lemon oil. Massage into the scalp. Rinse after 2 hours.

Hair colouring

Natural substances such as henna are enjoying renewed popularity as concerns grow over the safety of harsh chemical hair dyes. They are easy on your hair and scalp – and provide a wide variety of glossy shades.

Always colour test a lock of hair first and stick to the specified exposure time.

BRIGHTEN up blonde hair

● To lighten and add shine, mix 150ml boiling water with the juice and grated peel of a lemon and steep for 30 minutes. Strain and stir in 1 teaspoon of cider vinegar. Apply evenly to washed hair and rinse with warm water after 10-15 minutes. Repeat once a week.
● Pour a little boiling water over 150g of chamomile flowers and steep for 30 minutes. Strain and put the flowers into a bowl with 100g of dried, finely ground rhubarb. Add 1 tablespoon of olive oil and hot water and mix. Apply with a brush and cover hair with cling film for 30 minutes. Rinse out and wash hair.

BRUNETTE hair dye

● Blend 125ml natural henna with 250ml ground walnut shells. Make a paste with boiling water. Add 1 teaspoon lemon juice and 1 tablespoon olive oil, stir and apply with a brush. Cover hair with cling film, leave for 20-30 minutes then rinse and wash.
● Create a medium brown dye by mixing 100g crushed onion peel, 50g finely ground sandalwood

Henna powder comes from the dried and ground leaves of the henna bush.

and 1 tablespoon olive oil. Create a spreadable paste with boiling water, then apply and rinse as above.

DYEING hair dark brown or black

● Mix 3 tablespoons black tea, 90g black henna powder, 1 tablespoon olive oil and 1 teaspoon cider vinegar. Add boiling water to make a paste. Apply as for brunette dye, above. Rinse after 30-40 minutes.
● For a darker shade, steep 4 tablespoons dried sage, 2 tablespoons dried rosemary and 1 tablespoon lemon juice in 400ml water over a low heat for 30 minutes. Strain, cool, then use as a rinse after washing hair.

COLOURING & darkening grey hair

● Mix 100ml of boiling water with 4 tablespoons of dried sage and 1 teaspoon of black tea. Let steep for 30 minutes, then strain. Moisten hair with the lukewarm solution and rinse after 30 minutes. Repeat this for several weeks – so be patient.

A rich, red natural dye

Henna turns dark brown hair a rich, attractive red.

180g red henna powder
250ml black tea
1 tablespoon olive oil

Mix the henna, warm tea and oil and apply immediately. Cover your head with cling film and a large towel and leave on for 2 hours. Rinse hair with warm water and wash.

Hand care

We use our hands constantly, so they need good protection. Avoiding harsh soaps and extremes of temperature, and wearing protective gloves for housework and gardening will help, while nature offers a range of natural moisturisers.

Well-maintained hands remain supple, move gracefully and stay attractive for longer. So pamper your hands with these soothing home remedies.

Healing mask

For smooth, soft hands, apply this mask once a week.

½ cucumber
1 egg white
1 tablespoon yoghurt
1 tablespoon avocado oil
1 teaspoon lemon juice
2 drops peppermint oil

Peel and purée the cucumber, then beat the egg whites until stiff. Mix all the ingredients together and spread on your hands. Leave for 15 minutes before rinsing off with warm water.

TREATING rough hands

● To make chapped hands soft, pour 300ml boiling water over about 125ml fennel seeds, steep for 10 minutes, strain and cool. Soak your hands in the infusion for 2 minutes every time you wash them.
● To rejuvenate dry skin, use a yoghurt and egg poultice once a week. Mix together 2 tablespoons yoghurt, 1 tablespoon cream cheese, 1 egg yolk, 1 tablespoon honey and 2 tablespoons lemon juice. Apply a layer 3mm thick to hands. Wash off with warm water after 15 minutes or so.
● For an exfoliating scrub that will smooth rough skin, stir 1 teaspoon of sugar into a little lemon or grapefruit juice and rub hands with it before rinsing with warm water.

● To make an oil and honey massage that leaves skin noticeably softer, mix 1 teaspoon of honey and 2 teaspoons of almond oil. Massage into dry, chapped hands. Let it work overnight (wear cotton gloves) and rinse with warm water in the morning.
● For a moisturising emulsion that will cleanse, stir 1 teaspoon honey and the juice of half a lemon into 100ml warm milk.
● Make a daily hand cream by pouring the juice of 2 lemons through a muslin. Mix the clear juice with an equal amount of almond oil. Melt 2 tablespoons of beeswax in a bain-marie, add the lemon-almond oil and stir until the liquid cools. Add 5 drops of citrus oil (lemon, grapefruit, neroli or lemongrass). Pour into a clean jar and store in the refrigerator. It will keep for three days.
● Soaking hands once a week in warm olive oil will also restore their softness.

To keep hands soft, use a daily hand cream.

HOMEMADE BEAUTY PRODUCTS

Some cosmetics contain harsh chemicals. Homemade beauty products take advantage of plant-based oils for moisture and essential oils for delightful scents. Here are some of the most popular natural ingredients.

Plant-based oils

- Apricot kernel oil makes an excellent massage oil and has a faint scent of almonds. It is appropriate for all skin types but is particularly good for dry, sensitive skin.
- Avocado oil provides moisture, nourishment and hydration for the skin.
- Jojoba oil, a liquid wax, controls moisture and doesn't leave skin with an oily sheen. It is suitable for every skin type.
- Macadamia nut oil is rich in fatty acids, making it an ideal ingredient for creams.
- Use almond oil for massages and to care for all skin types.
- Olive oil nurtures every skin type, but is especially suitable for dry, rough skin.
- Wheatgerm oil is rich in vitamin E, which combats the ageing of dry, mature skin.
- Evening primrose oil stimulates and improves the appearance of skin. It is also sometimes used in conditioners for brittle, overprocessed hair. It is made from the seeds of the evening primrose. Only use it in small amounts as it is extremely potent.
- Wild rose oil comes from rose hips and provides ample moisture for rough skin. Only use it in small amounts – and only for scent, not as a base oil – as it is very strong.

Essential oils

These common oils, found in most health food stores, provide the finishing touch – a delightful scent – to many homemade beauty products. Which one you choose depends on personal preference.

Bergamot oil • The finest of all citrus oils with a sweet, citrus-fresh scent

Geranium oil • A soft, flowery, feminine touch

Lavender oil • A pure, fresh, flowery scent used in many products

Neroli oil • A fresh, flowery scent with a touch of bittersweet orange

Rose oil • A sweet, flowery scent; a very feminine oil

Rosemary oil • An intense scent similar to camphor; use sparingly

Sandalwood oil • A warm, heavy and long-lasting scent used by men and women

Incense oil • A heavy oil with a touch of lemon and camphor

Ylang-ylang oil • An exotic, sensual scent, ideal for perfume, deodorants and bath additives; use sparingly

Lemon oil • A pure, fresh scent with a subtle and sweet touch; ideal for cleansing

Natural beauty products are kind to the environment as well as to your hair and skin.

MOST INGREDIENTS can be found in HEALTH-FOOD stores

Useful ingredients

- Beeswax gives creams, salves, lotions and lipsticks a thicker, more solid consistency.
- Fruit or cider vinegar reduces itching, cools and refreshes. It regulates the pH value of skin, acts as a natural antiseptic and promotes blood circulation. The vitamins and minerals it contains also make organic cider or other fruit vinegar an ideal additive for cleansers and baths. (But don't use the malt vinegar you put on chips.)
- Glycerin is a clear, syrupy alcohol used as a lubricant in creams and lotions.
- Cocoa butter melts at a low temperature and makes a good base for soaps and creams.
- Lanolin – the pure oil from sheep's wool – is a moisturising skin care all-star, in large part because of its water-repellent properties.

Mineral ingredients

- Fuller's earth, found in health-food stores, is a fine-grained soil consisting of white and red clay, loam and aluminium silicates. The sterilised powder is mined from Ice Age loess deposits and can be used both internally and externally to absorb toxins. In the latter case it can be used in wraps, compresses, face masks, baths and hair-care products.
- Rhassoul and kaolin, found in some health-food stores or online, are natural clay cleansers that provide a gentle, non-sudsy alternative to shampoos and soaps.

BEES WAX

BEESWAX WAX

Lip care

Our lips have just three layers of skin cells, compared to the 16 layers on most of the face. As they have neither sweat nor oil glands, they dry out easily so it is worth using some natural remedies to keep them soft.

Lips are often exposed to sun, wind and other irritants, but there are a number of easy, natural ways to protect them.

Coloured lip gloss

2 tablespoons coconut oil
1 tablespoon almond oil
1 tablespoon beeswax
1 tablespoon cocoa butter
1-2 drops red food colouring

Melt the oils, wax and cocoa butter together in a bain-marie and stir. Next, stir in the food colouring. Put into a small container to cool. Remember to test first before applying all over your lips.

CARE & protection

● Massage lips carefully with a soft toothbrush every morning. Adding a little honey to the toothbrush will make lips softer.
● The active ingredients of papaya also work well. Make a balm by puréeing quarter of a papaya. Apply generously to the lips and surrounding skin. Rinse with warm water after 10 minutes and apply regular lip balm.
● Make a moisturising lip balm by melting 2 tablespoons of wheatgerm oil, 1 tablespoon of beeswax and 1 tablespoon of honey in a bain-marie. Add 3 drops of peppermint oil and 2 drops of chamomile. Stir until thick, then set aside to cool.
● For an everyday balm, melt 100ml of olive oil and 25g of beeswax in a bain-marie; cool until lukewarm. Mix in 1½ teaspoons of honey, 20 drops of chamomile tincture and 1 teaspoon tincture of propolis. Stir until cool and store in a cool place.
● A simple, weekly exfoliating lip scrub consisting of 1 teaspoon of sugar and a little olive oil will remove flaky skin and promote circulation.

● Always protect lips from ultraviolet rays by using lip balm with a built-in sun protection factor.
● Drink plenty of fluids to keep the sensitive skin of your lips tender and smooth.
● Avoid licking your lips in cold weather – the combination of wet and cold robs them of even more moisture, rendering them dry and rough.

ROUGH, cracked lips

● Smooth chapped lips by applying some cocoa butter or carrot juice several times a day, or by placing slices of cucumber on them.
● Alternatively, mix 1 teaspoon each of yoghurt and honey and apply to lips. Wait 10 minutes and rinse with warm water.
● To make a warm compress, mix equal amounts of thyme and chamomile. Pour a cup of boiling water over 1 teaspoon of the mixture, steep for 10 minutes then strain. Dip a soft, sterile cloth into the warm liquid and apply to lips for 20 minutes.

Apply all-natural lip gloss to help protect your lips.

Moisturisers

Advertisements from the cosmetics industry promise consumers a flawless complexion and perpetual youth if they use their (often pricey) products. However, quick, inexpensive and easy-to-prepare natural formulas are often just as effective.

The skin is our largest sensory organ. It excretes sweat and oil, stores fat and moisture and, because it interacts with the environment, helps to protect us from heat, cold and pathogens. There are five basic skin types, and each one needs special care.

NORMAL skin

This is largely free from blemishes and has adequate moisture. Clean it with a mild, pH-neutral soap or cleansing lotion, apply a moistening floral water such as rose or orange-blossom water, and finish with a thin application of a not-too-rich cream or gel.

● For a tried-and-tested skin cream, beat 1 egg white until stiff and add 1 teaspoon of honey and 3 drops of almond oil. Beat the mix to produce a thick, smooth cream that will keep in the refrigerator for three to four days.

● Alternatively, melt 50g of beeswax in a hot bain-marie and add 100ml of wheatgerm oil and about 4 tablespoons of elderflower water. Stir the ingredients, put the cream into a small porcelain container and store in the fridge.

GOOD TO KNOW

Day and night cream

After washing in the morning and evening, treat your skin with a cream appropriate for your skin type. A day cream can help to protect against the environmental effects of cold, wind and sun. A night cream or oil may help your skin to eliminate toxic substances and absorb nutrients. All-purpose creams can be used in the morning or evening.

SENSITIVE skin

This may be affected by the sun, detergents and some make-up products. Apply:
● Almond or jojoba oil as a moisturiser.
● Orange peel facial scrubs.

OILY skin

This is characterised by large pores, an oily sheen, pimples and blackheads. But despite overactive oil glands, it still needs moisture. A consistent care

ONLY KEEP homemade products FOR THE RECOMMENDED TIME.

programme includes pH-neutral cleansing with a soap or lotion, a clarifying, alcohol-free facial toner, an oil, a light cream or a gel with soothing properties, and a weekly exfoliating scrub and follow-up facial mask.

● Make a soothing cream by pouring 125ml of boiling, distilled water over 20g of marigold flowers. Cover and let it cool, then strain through a fine sieve, thoroughly squeezing out the flowers. Stir in 20ml of almond oil and blend the mixture with rhassoul or kaolin until it is the consistency of a cream. Store in a porcelain jar in the fridge.

COMBINATION skin

If you have enlarged pores, an oily sheen and perhaps blemishes on the forehead, nose and chin but the rest of your skin is often dry, you have combination skin. In addition to using a mild, pH-neutral cleanser, you should:

● Use an exfoliating scrub made from natural ingredients once a week.

● Apply a moisturising cream on dry skin in the cheek area and a cream for oily skin on the forehead, nose and chin, preferably in combination with astringent herbal waters such as witch hazel or sage water.

Oil from the orange blossom has long been an ingredient in traditional cosmetics.

DRY skin

This is characterised by small pores, wrinkles, scaly areas and a feeling of tautness. Gently cleanse skin with moisturising lotions or plant oils. Use floral water such as orange-blossom water and rich facial creams, gels or oils. Try an exfoliating scrub once or twice a month.

● Mix 110g cocoa butter, 2 tablespoons of beeswax (grated), 2 teaspoons of distilled water, 3 tablespoons of sesame oil, 2 tablespoons of coconut oil and 1 tablespoon of olive oil. Combine beeswax with water and melt over a low heat, then add cocoa butter and blend. Gradually add coconut, sesame and olive oil. Pour into a glass jar. The mix will thicken as it cools down.

● Alternatively, stir together 20g each of lanolin and petroleum jelly and add 2 drops each of rose and lavender oil. Stir until the cream is smooth. Store in a porcelain container in the refrigerator for up to two months.

MATURE skin

● Rose gel may help to plump up mature skin. Heat about 150ml of distilled water and mix in 1 teaspoon of powdered gelatine. Stir in 1 teaspoon of rose oil and 4 drops of lavender oil, plus 1 tablespoon of glycerin. Let the gel cool and store in a porcelain jar.

● A beeswax-based cream will preserve skin's elasticity. Melt 10g of beeswax in in a bain-marie at 60-70°C. Stir in 50ml of almond oil, 1 teaspoon of jojoba oil, 15g of aloe vera and 5 drops of rosewood oil.

EGG WHITE, HONEY and ALMOND oil make a good SKIN CREAM.

FACIAL oils for night use

Facial oils, when applied sparingly to clean skin, soak in almost completely. All oils are made by adding ingredients to a small, dark bottle and shaking vigorously. When stored in the refrigerator, homemade oils last for about two weeks.
● For normal skin: blend about 50ml each of jojoba and almond oil with 5 drops each of geranium, rose, lavender, chamomile and incense oil.
● For mature skin: pour 50ml of almond oil, 1 teaspoon of wheatgerm oil and 5 drops each of incense, lavender and chamomile oil into a dark bottle. Shake well.
● For dry skin: mix 5 tablespoons of olive oil, 2 teaspoons of avocado oil and 10 drops of lemon oil.
● For oily skin: stir 50g of powdered aloe into 40ml of distilled water and add 2 teaspoons of orange blossom water. Set aside. Combine 1 teaspoon of honey with 100ml of apricot kernel oil in a bain-marie. Mix all ingredients together.
● For sensitive skin: mix 75ml of almond oil and 10 drops of neroli oil.

NECKLINE & throat

A woman's décolletage has long been a symbol of beauty and femininity. But skin on the neck and throat is significantly thinner and more sensitive than on other parts of the body, so daily care is required to keep it firm. Try these natural cosmetics, but remember you won't see instant results. They will take time to have an effect.

HOME remedies

● Combat loose skin and ageing with an oil wrap. Mix 1 tablespoon of almond oil with 1 tablespoon of honey, spread the mixture onto a cloth and apply to your throat for 1 hour. Apply weekly.
● For neck wrinkles, use parsley milk. Heat up 500ml of whole milk and add a bunch of chopped parsley. Let it steep for a few minutes then strain the milk and let it cool to lukewarm. Moisten a soft, clean cloth with the parsley milk, place it on your throat and let it work for 15 minutes.
● To help tighten skin, make a lemon cure by beating an egg white until stiff, then stirring in the juice of a lemon. Apply and wait 30 minutes. Rinse with warm water and apply moisturising cream.
● To prepare a skin balm for nightly skin regeneration, mix 2 tablespoons apricot kernel oil, 15 drops of jojoba oil, 3 drops neroli oil and 2 drops each of almond, rose and lavender oil with

To refresh skin, finish up a morning shower with a blast of cold water.

2 teaspoons of glycerin. Pour into a dark, sealable container. Gently rub in a few drops before bedtime.
● Purée about 100g of raspberries and mix with 2 teaspoons each of almond bran and honey, plus an egg yolk. Apply to your neckline as a mask and rinse after 30 minutes with warm water.
● Throat exercises help to keep skin firm. Reach over your head to grasp your left ear with your right hand. Carefully turn your head to the left, hold the tension for a few seconds and then relax. Do the same with the other side.

SUN protection

The sensitive skin of the neckline and throat should be sheltered from the sun. To prevent slackening of the tissue and the formation of wrinkles, always use sunscreen on your face, neck and throat.

Nail care

Attractive nails are a sign of good health and they complement our appearance. When they are well manicured, they help us to feel more attractive and better groomed. Make your nails shine with regular care and a diet rich in vitamins.

Nails are made of a hard protein called keratin, which helps to protect fingers and toes from injury. Excessive hand washing and some household cleaners can rob them of vital oil and moisture, sometimes resulting in split or peeling nails. Natural remedies can do much to help keep nails looking strong and healthy. But see a doctor if you suspect you have a nail disease or other nail disorder.

THE essentials of nail care

Ideally, you should give yourself a manicure every other week. Start with a gentle exfoliating scrub to remove scaly skin and dirt, then push your cuticles back. Shape your nails with scissors or a file (filing from the edge of the nail and towards the centre). Follow up by massaging your hands and nails with a moisturising gel, such as aloe vera.

Nail oil

For flexible, strong nails.

2 tablespoons almond oil
1 teaspoon jojoba oil
5 drops lavender oil
5 drops lemongrass oil

Mix all the ingredients and massage into nails morning and evening.

1 To begin with, basic daily cleaning is necessary and simple. Scrub your nails with a soft brush and coat them with a little olive or almond oil.
2 Use a basic hand and nail scrub before a manicure. Heat up 2 tablespoons of cocoa butter and mix it with 2 tablespoons of ground almonds and 5 drops of lemon oil. While it is still warm, massage it into your hands and nails, then rinse with warm water.
3 Push cuticles back carefully – don't cut them. Soften them first with a little olive oil.
● Alternatively, soften cuticles by mixing together 1 egg yolk, 1 tablespoon of fresh pineapple juice, 1 teaspoon of lemon juice and 2 drops of lemon oil. Apply with a brush or cottonwool swab. Rinse after 15 minutes.
4 Use an emery board or a glass nail file (or even a diamond file for very strong nails) to file your nails; it is a matter of personal preference which you use. Metal files can make nails brittle and fingernail clippers can cause nails to become grooved and frayed.
5 Finish by shining the nails with a chamois nail buffer or a polishing file, or apply clear nail polish.
● Give fingernails a therapeutic oil bath from time to time. Warm up 50ml of macadamia, almond or olive oil and add 1 drop each of geranium, sea-buckthorn and lavender oil (from a pharmacy or health-food store). Dip your fingers in the oil bath for 2 minutes, then massage your hands with the remaining oil.

Use only high-quality scissors for manicures, or your nails could split.

2 Apply the nail polish in two coats: start with a stroke in the centre of the nail and then do the sides in two or three strokes.
3 Apply a protective top coat so that the nail polish lasts for longer.
4 When the time comes to replace it, use nail polish remover that is acetone-free and contains oil to prevent nails from drying out too much.

● To make a nourishing poultice that should be applied to nails once a week, combine thoroughly 1 egg yolk and 2 tablespoons of wheatgerm oil, then stir in 1 tablespoon each of grated carrots and lemon zest. Apply the poultice to nails and the backs of the hands and cover with a cloth. Rinse with warm water after 30 minutes.

BRITTLE nails & soft nails

Soft or brittle nails that crack or split easily can be frustrating. There are, however, simple ways to strengthen them.
● Coat nails with a little warm olive oil mixed with a couple of drops of lemon juice every evening. Wear soft cotton gloves in bed and let the mixture work overnight.
● Vitamin B_7 (biotin) also strengthens nails. It is abundant in brewer's yeast, soy products, offal and egg yolk.
● Avoid growing long nails, which are naturally weaker, and don't use fast-drying nail polish, as both can make fingernails brittle.
● Soft nails can be toughened up by rubbing them once a day with a mixture of cider vinegar and lemon juice.

THE right way to apply nail polish

Look for nail polish that contains as few chemical ingredients as possible. Thousands of years ago, the Chinese painted their nails with a mixture of egg white, gum arabic, gelatine and beeswax.
1 Before applying nail polish, it is a good idea to prime your nails with a protective, nurturing base coat.

When painting your toe nails it helps to use foam separators to keep your toes apart.

Natural perfumes

Since the earliest civilisations, perfume has been produced to make our homes or bodies smell more fragrant – and scent plays an important role in human attraction. Fortunately flowers, fruits, woods and spices can all be employed to make us smell more appealing.

Succulent fragrances stimulate the mind and any of these traditional formulas can help to make us feel refreshed and revitalised. Smells sway emotions, modify perception and are the ultimate wake-up call for the senses.

BASIC formulas

● Mix together about 50ml of vodka, 1 teaspoon of distilled water and about 25 drops of the essential oil of your choice, then shake vigorously. The resulting fragrance is ready for immediate use.
● Sandalwood and cedar are favourites for masculine scent mixtures. For a flowery aroma, use patchouli, geranium or lavender oil. Opt for orange or bergamot oil to obtain a charming fresh scent, and use violet or rose oil to achieve heavy, sensual notes.

For a masculine scent, try chamomile oil.

● Oxygen, heat and light destroy fragrances. Keep homemade perfume in a tightly sealed bottle in a cool, dark location and it will last for about three months.

HOMEMADE fragrances

● Eau de cologne, a time-tested classic, can be made at home. Mix 300ml of water with 12 drops each of bergamot and lemon oil, 10 drops each of orange and geranium oil, 6 drops of rosemary oil and 3 drops of neroli oil. Shake vigorously and set mixture aside for two days. Then, add about 75ml of distilled water, shake once again and leave to steep for about a week before using.
● To produce a feminine fragrance with a subtle vanilla aroma, slit open 2 vanilla pods and soak them in about 100ml of vodka. Remove the vanilla pods after three days and add 250ml of distilled water.
● To make a classic summer perfume that relies on the pure, fresh scent of citrus fruits, mix together 1 tablespoon of vodka, 10 drops each of orange, neroli, lemon, mandarin orange and rose oils, plus 5 drops of bergamot oil, then add about 100ml of orange flower water. Let it steep for at least a week to develop fully.

Fragrance for him

1 tablespoon vodka (40 per cent proof)
10 drops chamomile oil
10 drops geranium oil
10 drops clary sage oil
10 drops bergamot oil
10 drops neroli oil
5 drops coriander oil
100ml witch hazel

Put the vodka and the essential oils into a glass bottle and shake well to dissolve the oils. Add the witch hazel and shake once again. Leave to steep for a week then shake before each use.

● For a distinctive violet perfume, pour about 40ml of surgical spirit and 50ml of distilled water over 100g of violet flowers and leave to steep for a full week. Strain the liquid and mix with 100ml of distilled water. Alternatively, use two handfuls of lavender or rose flowers instead of the violets.

Lavender water

Scenting handkerchiefs is a tradition well worth reviving – especially when the perfume is calming lavender:

2 teaspoons dried lavender flowers
20 drops lavender oil
250ml water

Bring the water to the boil, add the lavender flowers and oil. Mix well and set aside, covered, for 24 hours. Strain through fine muslin, then transfer to sterilised bottles or jars. Seal tightly. Keep in a cool, dark place.

● For a fragrant rose water, pour 1 litre of boiling water over 100g of fresh rose petals and steep for an hour, then strain. Bring the rose water back to the boil and pour it over another 100g of rose petals. Store the cooled rose water in a dark bottle.

USING fragrances

Consider these guidelines when using fragrances.
● In contrast to heavy evening perfumes, the subtle aroma of less intense fragrances can be used from head to toe.
● Never use too much perfume or the scent will become overpowering. A couple of drops from a phial or two or three squirts from a pump bottle will be sufficient.
● In order to make the most of a scent, apply perfume to clean skin.

MAKE A SOFT, tantalising PERFUME FROM VIOLETS.

NUTRITION FOR BEAUTY

A healthy, balanced diet is good for more than just keeping off weight. Vitamins and minerals make your skin smooth and supple, strengthen your fingernails, teeth and gums, and keep hair looking shiny and silky.

Vitamins

- A good supply of vitamin A is the basis for healthy skin and a fresh face. Vitamin A keeps skin looking smooth and young by stimulating cell regeneration. You can get it in the form of retinol (liver, whole eggs, milk and some fortified food products) or from the beta-carotene contained in many fruits and vegetables (carrots, cantaloupe melons, sweet potatoes and spinach). You can make it easier for your body to absorb beta-carotene by boiling or steaming foods containing the substance and adding a little butter or other fat.
- B vitamins are key for the growth of skin, hair and nails. Vitamin B_6, in particular, helps form collagen for firm connective tissue. Biotin, which is often referred to as vitamin B_7, is another of the B-group vitamins that protects skin and hair cells.
- Vitamin C accelerates collagen production in your cells and helps keep gums healthy. Since vitamin C is quickly destroyed by heat, foods that are high in vitamin C are best eaten raw.

MINERALS ARE CRUCIAL FOR many body functions.

Minerals

- Iron facilitates oxygen transport in the blood. Pale, brittle fingernails and hair loss can be signs of an iron deficiency – one of the most common nutrient deficiencies in the world.
- Potassium helps the body to eliminate harmful substances and plays an important role in keeping skin supple.
- Calcium (which is contained in milk products) makes teeth strong and healthy.
- Magnesium keeps cell walls stable, helping to stop skin from wrinkling.
- Selenium, found in whole-grain products, may help damaged skin.
- Zinc works with vitamin A in the body to produce and maintain a strong immune system.

Fluids

If your skin is to remain supple, clear and fresh, it needs plenty of fluid. Adequate fluid intake also helps your internal organs filter out harmful substances. A rule of thumb: drink at least 1.5 litres of fluid per day – ideally water, fruit juice spritzers or herbal and fruit teas. Freshly squeezed organic fruit and vegetable juices will boost your appearance and your health. Drink them in any combination, but don't underestimate the calorie content of fruit juices.

Dairy foods are rich in calcium.

Sources of vitamins and minerals

Beta-carotene • orange and yellow fruits and vegetables, dark leafy vegetables

Biotin • eggs, legumes, offal

B vitamins • grains, soya beans, nuts, legumes

Iron • meat, whole-grain products

Potassium • dried fruits, legumes, nuts, soy products, vegetables, mushrooms, avocados, bananas

Calcium • milk and milk products

Magnesium • whole-grain products, legumes, mineral water

Selenium • brewer's yeast, whole-grain products, seafood, mushrooms, brown rice

Vitamin C • citrus fruits, kiwi, rose hips, blackberries, sweet peppers, papaya, broccoli

Vitamin E • cold-pressed oils, nuts, cereal germ

Zinc • legumes, cereals, nuts, poultry, seafood

Vitamins are classified according to how they are absorbed and stored in the body:

vitamins A, D, E and K are soluble only in fats;

vitamin C and the B vitamins are soluble only in water.

Your best source of vitamins and minerals is food, but a multivitamin supplement can offer insurance against gaps in your diet.

Fruit and vegetables should account for about one-third of your diet.

Antioxidants

Environmental pollution, cigarette smoke, ultraviolet radiation and stress contribute to the formation of what are known as free radicals in the body. These aggressive oxygen compounds can damage tissue and cause your skin to age prematurely when present in excessive amounts. Natural antioxidants such as vitamin C, vitamin E, beta-carotene and selenium stabilise free radicals and keep them from causing damage in the body. A diet high in antioxidants also reduces your risk of developing many serious diseases. Eating a range of whole foods can help to ward off damaging toxins and free radicals. The following foods are among the top picks.

- Berries are full of healing antioxidants. Blueberries, cranberries and blackberries contain proanthocyanidins that have strong antioxidant properties. Strawberries, raspberries and blackberries contain carcinogen-fighting ellagic acid. Eat them on their own, sprinkle a handful in your morning cereal, make a smoothie or enjoy them for dessert with a little low-fat whipped cream.
- With more vitamin C than an orange and more calcium than a glass of milk, broccoli gets a gold star for nutrition. On top of that, it is packed with healthy phyto-nutrients. Boil it, steam it or use it to make a delicious soup.
- Garlic, also known as 'the stinking rose', acts as a natural antibiotic, killing off some strains of harmful bacteria. Roast it whole and spread it on a baguette, or sauté it with a little butter and toss with hot pasta.

Oral hygiene

Brushing and rinsing, combined with a well-balanced diet, help to ensure you have a healthy smile and pleasant breath. They may also improve your long-term health, as many illnesses start in the mouth.

People have believed for centuries that bad teeth signify bad health. It was only comparatively recently that gum inflammation was identified as the problem. The theory is that bacteria from dental plaque seep into the bloodstream via inflamed gums and produce enzymes that make blood platelets stickier and more likely to clot, contributing to the hardening of arteries. The good news is that this risk factor can be easily controlled.

Healthy teeth and strong gums are the product of conscientious care and good nutrition. Sugar is the number one enemy as it damages teeth in two ways: by interfering with the absorption of calcium; and, by causing tooth decay. Milk products, however, contain a healthy amount of calcium, which hardens teeth. To keep gums healthy and strong, eat plenty of fresh fruit and vegetables that contain vitamin C.

TOOTH-CARE tips

● Good tooth care does not end with brushing. Cleaning between teeth at least once a day with dental floss is essential. Guide the floss between teeth and wrap it in a 'C' shape at the base of the tooth, slightly under the gum line. Slide the floss up to the top of the tooth several times. Finish off by rinsing your mouth.

● Recent findings indicate that there's no difference between a regular or electric toothbrush when it comes to thorough cleaning. Because of their rapid, rotating motion, electric toothbrushes clean effectively in a much shorter time, but brushing carefully by hand for 2 to 3 minutes is equally effective. In either case, replace your toothbrush every two months.

Moderately sized heads reach into corners and angles effectively.

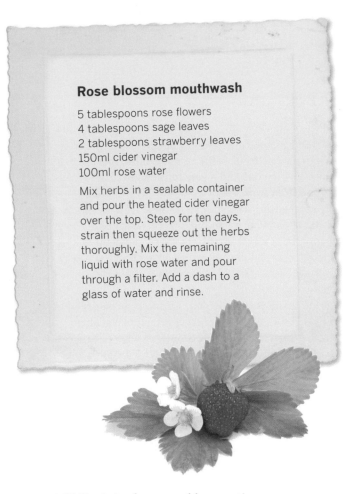

Rose blossom mouthwash

5 tablespoons rose flowers
4 tablespoons sage leaves
2 tablespoons strawberry leaves
150ml cider vinegar
100ml rose water

Mix herbs in a sealable container and pour the heated cider vinegar over the top. Steep for ten days, strain then squeeze out the herbs thoroughly. Mix the remaining liquid with rose water and pour through a filter. Add a dash to a glass of water and rinse.

● Halitosis is often caused by a coating on your tongue, but this is simple to remedy with a tongue scraper. Just slide the scraper over your tongue three to four times, twice a day.

HOME remedies for tooth brushing

Fresh breath is practically guaranteed with a toothpaste made mainly from sea salt and fuller's earth. Mix together 100g of fine fuller's earth and ½ teaspoon sea salt and drizzle in enough boiled water to make a creamy liquid. Add 2 drops each of peppermint and tea tree oil.

● For a gum-refreshing tooth powder, finely grate 40g dried orange peel and mix with about 30g dried peppermint leaves and 10g sea salt. Store in a screw-top container. When brushing your teeth, just sprinkle a little powder onto your moistened toothbrush.

● For a lightening tooth powder, mix 1 small container of bicarbonate of soda with 2-3 drops of caraway oil. But don't use it too often as the abrasive action of the bicarbonate of soda may damage weak tooth enamel.

● You can also keep your gums healthy by rubbing them with the inside of a lemon skin. Alternatively, brush your teeth occasionally with warm sage tea.

MOUTHWASHES

Using mouthwash after brushing leaves your mouth fresh and clean.

● To make a pleasant smelling, refreshing mouthwash mix 50ml each of water and vodka and 3 drops each of eucalyptus, anise and clove oil in a small bottle. Add 1 teaspoon of the mouthwash to a glass of water and gargle.

● For another refreshing rinse, mix 500ml of vodka, 2 teaspoons of peppermint oil, about ½ teaspoon of cinnamon oil and ¼ teaspoon of anise oil. Add a dash of the mixture to a glass of water and rinse.

● To make a mouthwash that will help to keep gums healthy, mix 2 teaspoons each of arnica, propolis and sage tinctures. Add 10 drops of the mixture to a glass of water and rinse.

Flossing will clean your gums and the spaces between your teeth.

Relaxing baths

Take advantage of the power of a long soak to soothe, moisturise and heal your skin. Once you have stepped out of the bath, be sure to apply a moisturiser.

First, determine your skin type (*see page 93 for more on skin types*). Dry and oily skin require different treatments and even normal skin requires special care.

NORMAL skin

- Open your pores with a yoghurt-based bath additive. Purée 1 serving of plain yoghurt, 1 tablespoon of honey, 2 tablespoons of almond oil and 1 vanilla bean first pulped in a blender. Add 10 drops of orange oil and swirl into the bathwater.
- Invigorate skin with a spruce needle bath additive. Fill a jam jar two-thirds full with fresh spruce needles, add enough sea salt to fill the jar, then pour in water to the top. Leave to steep in a warm place for two weeks, shaking vigorously every day. Put 2 tablespoons of this mixture into a cloth bag and add it to the bathwater.
- Pamper yourself with a vanilla shower gel. Mix 100ml of neutral, unscented shampoo, 50ml of warm water, a pinch of salt and 15 drops of vanilla oil.

OILY skin

- Make a buttermilk bath by adding to the water a purée made from 1 litre of buttermilk, the juice of 4 lemons and 4 generous handfuls of peppermint leaves.
- Soothe your skin with an oatmeal bath. Combine 250g of oatmeal, a handful of fresh sage and peppermint leaves, and 10 drops of lemon oil in a cloth bag or old pair of clean tights. Drop the bag in your bathwater, squeezing it out occasionally while you soak.
- Stimulate your body with a firming shower gel. Stir together 250ml of shower gel base (available from beauty supply stores or online) with 10 drops each of geranium and lemon oil and 5 drops each of rosemary, juniper and sage oil.

MATURE & dry skin

- Help to make skin silky-soft with a bath oil made from 50ml of almond oil, 10 drops of grapefruit oil and 5 drops each of lemon and orange oil.
- Soak in a lavender bath. Cover a handful of dried lavender flowers with water, boil for 5 minutes and strain. Add 2 tablespoons of honey and 1 tablespoon each of cream, buttermilk and olive oil, and add to your bathwater.

TO RELAX, SPEND 20 minutes IN A BATH below 36-38°C.

Scrubs

Treat your face and body to exfoliating scrubs regularly to remove dead cells, improving the skin's appearance and making it feel smoother. There are many natural ingredients you can use for this.

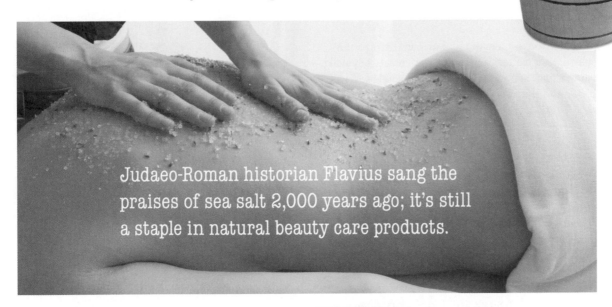

Judaeo-Roman historian Flavius sang the praises of sea salt 2,000 years ago; it's still a staple in natural beauty care products.

Oranges, lemons, sugar, olive oil and almond bran are ideally suited for facial scrubs. Essential oils and salt are key ingredients for body scrubs.

EXFOLIATING facial scrubs

● Treat sensitive skin with orange peel. Grate the zest from an orange and dry it carefully, then grind it in a coffee grinder or in a mortar using a pestle. Add a little warm water and rub onto skin. Rinse thoroughly.

● Rejuvenate your skin with a scrub that is made from two common cooking ingredients. Stir together 1 teaspoon of sugar and 1 tablespoon of olive oil. Massage the scrub onto skin, then rinse thoroughly.

● Cleanse normal and oily skin until it is glowing by mixing together 1 tablespoon each of almond bran and oatmeal with 1 teaspoon of grated lemon peel. Add a little warm water, then massage the scrub onto your skin. Rinse thoroughly with warm water after 2 minutes.

● Smooth dry skin with a fragrant oil scrub. Mix 2 tablespoons of almond bran with an equal amount of rose water to produce a thick paste. Add 2 drops of rose oil and 1 drop of lavender oil. Gently massage into the skin and rinse with warm water after 2 minutes.

EXFOLIATING body scrubs

● Stimulate dry skin with a sea-salt scrub. Mix 1 tablespoon of sea salt, 2 tablespoons of almond oil and 10 drops of lemon oil. Apply to a facecloth and massage skin while still wet. Rinse thoroughly.

● Make a sweetly fragrant scrub by mixing together 2 tablespoons of sea salt, 1 tablespoon of walnut oil, 5 drops of patchouli oil and 2 drops of jasmine oil. Apply the scrub, wrap yourself in a towel and let it work for 30 minutes before rinsing off thoroughly.

GOOD TO KNOW ☑

How often?

Since exfoliating scrubs are abrasive, they shouldn't be used frequently. Exfoliating facial scrubs and body scrubs should be used on dry skin once a month. But you can treat normal facial skin to a scrub every two weeks, and oily facial skin weekly. After a scrub, always apply a moisturising skin-care product made specifically for your skin type.

Shaving

It's the most common way to remove excess body hair but shaving can damage your skin. Luckily, there are plenty of natural creams, salves and lotions available to soothe it. Home cures will also provide relief from shaving cuts and ingrowing hairs.

Whether you use a manual or an electric razor for shaving is a matter of personal preference, but avoid using a manual razor on your face if you have blemished skin. It may open pustules and could lead to infection.

SOOTHING skin irritations

● Soothe and cleanse your skin with a pleasantly fragrant aftershave lotion. Pour 1½ tablespoons of surgical spirit, 5 tablespoons of witch hazel, 1½ tablespoons of rose water and 2 drops each of clary sage, sandalwood, cedar, lemon and cypress oil into a small bottle. Shake the mixture vigorously and leave to steep for four weeks. Apply straight or pour a little aftershave into your hand, rub in a few drops of jojoba oil and smooth it onto your skin.
● Make a balanced aftershave lotion by combining thoroughly 75g of pH-neutral moisturiser (from a pharmacy) and 1 teaspoon each of jojoba, apricot kernel and almond oil, and 1 teaspoon of aloe vera gel, plus 2 drops each of chamomile, mint and geranium oil. Store the aftershave in a jar in a cool place.
● Relieve and soothe irritated skin with a witch hazel salve (available at health-food stores). The effect is intensified by mixing in a few drops of tea tree oil.
● Combat irritation and soothe your skin after shaving with creams and salves containing marigold, chamomile or aloe vera (available from health-food stores and pharmacies).

TREATING small cuts

If you use a blade razor, you are bound to cut yourself occasionally. To stop the bleeding:
● Dab small cuts with a cottonwool ball moistened with surgical spirit.
● Use a moist styptic pencil, which is designed to promote coagulation.
● Grind dried yarrow flowers to powder in a mortar using a pestle. Apply to the cut and press lightly with a damp cloth.

PLACE A RAZOR HIGH ON THE CHEEK and move it IN SHORT STROKES.

REMOVING ingrowing hairs

● Apply a hot compress (for example, a damp cloth with 2-3 drops of tea tree oil on it), then pull out the hair with clean tweezers and disinfect the skin with a drop of tea tree oil. To prevent ingrowing hairs, use only clean, sharp blades and always shave in the direction of the hair growth.

Skin care

Cleopatra, legend has it, indulged in daily milk and honey baths. Natural ingredients certainly provide nutrients, moisture and subtle fragrances to soothe and refresh skin, and they also help protect it from the sun's harmful ultraviolet rays.

Much time and attention is given to facial moisturising, but our skin from the neck down requires attention, too. There are many traditional treatments to help keep skin in good condition – just choose those best suited to your skin type (*turn to page 93 for information on skin types*).

NORMAL skin

● Make a body-nurturing cream by mixing ½ teaspoon of lanolin and 1 teaspoon of cocoa butter melted in a hot bain-marie. Then add 50ml of almond oil and stir frequently until the mixture cools. Add fragrance by blending in 5 drops of rose oil (for women) or sandalwood oil (for men).

● Make your skin soft and silky with a poultice of 100g Dead Sea mud (available at health-food stores or online). Mix with 2 teaspoons of aloe vera gel and apply to clean, dry skin. Wrap your body in a large towel and let it work for 30 minutes before showering. Use weekly.

● Refresh your skin with body powder. Mix 75g each of arrowroot and polenta. Stir in 5 drops each of ylang-ylang and neroli oil. Store the powder in an opaque, sealable container and apply with a powder puff to clean, dry skin.

OILY skin

● Moisturise skin with a nurturing body oil of 5 tablespoons of jojoba oil with 1 tablespoon of almond oil and 5 drops each of lavender and geranium oil. Apply sparingly to skin after bathing.

● Fight blemishes on your back and neckline with a lavender-honey mask. Stir together 2 tablespoons of wholemeal flour, 1 tablespoon of honey and 2 tablespoons of orange-blossom water in a bain-marie to form a paste. Add 5 drops of lavender oil and spread on problem areas. Shower off after 20 minutes. Apply once or twice a week.

Squash poultice

Provides moisture that is appropriate for every skin type.

100g squash, skin removed
1 peach
50g cucumber, unpeeled
1 tablespoon cocoa butter
1 tablespoon honey
1 egg white

Purée ingredients in a blender. Apply to skin, wrap your body in a linen cloth or large towel and let it work for 20 minutes before showering.

Body lotions made from natural plant ingredients can help to refresh and regenerate skin.

hot bain-marie and stir in 1 tablespoon each of sesame, avocado and coconut oil. Place cream in a container and keep refrigerated for up to two months.
● Soak up the moisture from an exotic poultice. Purée a banana and mix with 2 tablespoons of buttermilk, 300ml coconut milk, 2 tablespoons of yoghurt and 1 tablespoon of honey. Massage gently into skin and cover with a large towel. Rinse off after 30 minutes.

ELBOW care

Elbows may be hidden from view for much of the time, but that doesn't mean they should be ignored.
● Treat rough elbows to a poultice. Apply a mixture of 2 tablespoons of warmed honey and 1 tablespoon of lemon juice and leave on for 30 minutes. Rinse and liberally apply moisturising cream.
● Alternatively rub them with warm almond oil, wait 5 minutes and wipe off with a clean cloth. Then treat your skin with a mixture of equal parts lemon juice and glycerin.

SUN protection

Sunlight may brighten our lives, but its ultraviolet rays can damage skin and cause premature ageing.
   UVA rays, which have the longest wavelength, are dangerous because they penetrate deeply into the skin, causing damage to collagen and cells. UVB rays are shorter but more powerful, and are most intense during the summer months. They cause sunburn, ageing and wrinkling. Research shows repeated exposure to UVB rays can affect the immune system and lead to skin cancer. (*For more information on treating and preventing sunburn, see Skin treatments, page 56.*)

FOLLOW A SKIN-CARE regimen suitable FOR YOUR SKIN TYPE.

● Remove shine from skin with body powder. Mix 15g of talcum powder, a pinch of zinc oxide and 2 drops of lemon oil.

MATURE & dry skin

● Moisturise skin after bathing or showering with a fragrant oil mix. Combine 50ml of jojoba oil and 10 drops of essential oil. Good choices include patchouli, mandarin, neroli, lemon, myrrh, rose or lavender oil.
● Treat skin with a body oil with a floral fragrance. Mix 3 tablespoons each of avocado, almond and apricot kernel oil, 15 drops of rose oil and 5 drops of lavender oil. Rub in sparingly after a bath or shower.
● Moisturise dry skin with a body cream. Melt 1 tablespoon each of cocoa butter and beeswax in a

Skin test

one Cleanse your face with a mild soap. Apply no other skin-care products; wait roughly 2 hours.

two Press a sheet of tissue paper against your face.

three Check the imprint. Oily skin shows imprints of your forehead, nose, chin and cheeks. Normal skin should leave just a glimmer of oil. Combination skin leaves imprints of only the forehead, nose and chin. No imprint will be visible with dry skin.

BEHAVIOURAL tips

To protect against sunburn:

- Drink lots of fluids, especially water, to keep skin from drying out.
- Introduce skin to the sun slowly – it's best to begin in spring.
- Opt for the shade (or indoors) during the hottest parts of the day.
- Use sunscreen even on cloudy summer days – although the weather appears dull, ultraviolet rays still pass through.

- Apply sunscreen 15 to 20 minutes before going outdoors and let it dry thoroughly before getting dressed. Use creams for normal skin and gels for those with sensitive skin or if you have an allergy to the sun.
- Reapply cream immediately after swimming, showering or drying off, even when using so-called waterproof sunscreens.

Note: natural sunscreens such as avocado or sesame oil and lemon juice do not provide much protection and are inadequate for children or adults with sensitive skin.

GOOD TO KNOW ☑

Sun protection factor (SPF)

Without protection, you can stay in the sun for only a short time before skin begins to burn. The SPF factor is a laboratory measurement of the effectiveness of a sunscreen. Use sunscreen with an SPF of 30, for example, and it will take 30 times as long for skin exposed to the sun to burn. But how long it takes to burn is dependent on many factors, including skin tone, time of day, geographic location and weather conditions.

- Wear lip balm with built-in sun protection.
- Protect children with sunglasses and a sunhat, as well as keeping their skin protected.
- Wear clothing made from natural fibres – ultraviolet rays can penetrate artificial fabrics.
- The closer you get to the Equator, the higher the SPF you need. The same applies to altitudes: the sun's rays are stronger in thinner air. Reflective surfaces (water and snow) also produce intense radiation.

PROPER application of sunscreen

- Apply sunscreen protection several times a day. Be sure to apply plenty to your nose, cheeks, ears, neckline and shoulders.

Choose a sunhat with a wide brim to protect your neck and ears as well as your head.

Good housekeeping

In today's hectic world, we are often so busy with the challenges of work and family that there's little time left for household chores. The good news? There are many easy traditional ways to clean and care for your home that still produce amazing results.

Bathrooms

Keeping a bathroom clean is an essential chore, but you can avoid using harsh chemical products. Try these traditional cleaners and cleaning methods to ensure your sinks, tiles and fixtures are spotless.

For hygiene as well as appearances, your bathroom needs to be cleaned regularly. Limescale and soap especially will leave traces in the bath and sinks where bacteria and mould can build up. However, these sure-fire tricks will keep things sparkling clean.

SINKS, baths & showers

● Clean sinks daily with a mild bathroom cleaner, then rub them down with a fabric softener sheet – the water will run off easily, leaving no residue.

● Rub off stubborn dirt or limescale with a slice of lemon peel dipped in salt.

● Scrub baths, sinks and tiles with a paste made from salt and white spirit to renew lustre.

● Wipe glass-fibre baths with a damp cloth and a little bicarbonate of soda.

● Acrylic baths, which are more delicate than steel baths, have a reputation for being hard to clean. But as long as you shine them up regularly, all you should need is a little washing-up liquid and water.

● Help keep shower enclosures looking crystal-clear with a solution of washing-up liquid and a soft cloth. Minor soap scum build-up can be rubbed off with vinegar, while serious build-up should be treated with a paste of salt and vinegar applied in a circular motion with a bath cleaning brush. Rinse and dry when you've finished. Use a squeegee daily to prevent lime spots and rub off chalky streaks with vinegar.

● Apply a little petroleum jelly to the shower-curtain rod so the curtain glides more easily. Soak new or freshly washed curtains in salt water to prevent mould growth.

● If mould appears anywhere in the shower, treat it with vinegar, lemon juice or bicarbonate of soda.

● Make old tiles gleam by rubbing them gently with a piece of newspaper or a chamois leather moistened with ammonia solution (1 teaspoon ammonia in about 500ml water). Polish with a little cooking oil to add shine and protect against moisture, but take care to avoid getting oil on the grouting as this will attract unwanted dirt.

● Rub away stubborn stains with a neat solution of household ammonia.

● Remove rust spots by rubbing them with a mixture of water and vinegar.

WINDOWLESS BATHROOMS SHOULD INCLUDE extractor fans

Use an old toothbrush to clean hard-to-reach places.

- Scour discoloured grout between tiles with a solution of household ammonia or bicarbonate of soda. Dab it on with a moist cloth or use an old toothbrush, leave it to work, then rinse it off.
- Whiten grout by scrubbing it with a little toothpaste on an old, soft-bristled toothbrush.
- Use very fine sandpaper to rub severely discoloured grout, but take great care not to damage the glaze on the tiles while doing so.

DRAIN

- Use a small, flat strainer to prevent hair from going down the plughole and clogging the drain.
- Use an old-fashioned rubber plunger – the suction should release the blockage.
- Sprinkle bicarbonate of soda into the drain and rinse it down with white vinegar followed by boiling water to unclog blocked pipes.
- Eliminate unpleasant smells by sprinkling bicarbonate of soda directly into the drain and leaving it to work overnight.

PLUMBING fixtures

- Achieve a scent-free shine by rubbing plumbing fixtures with a mixture of 1-2 teaspoons lemon juice and about 500ml water.

- To remove limescale on fixtures, wrap an old rag moistened with vinegar or lemon juice around the fixture and leave it to work for a couple of hours or overnight. Remove the softened limescale with a toothbrush.
- Remove limescale build-up in tap aerators and showerheads by leaving them to soak in a solution of vinegar and salt. Then clean the holes in the showerhead with a nailbrush or toothbrush.

TOILET

- De-scale the toilet bowl by placing a layer of toilet paper soaked in vinegar over the lime deposits and rubbing them off the following morning.
- Avoid urine scale by putting a dash of vinegar in the bottom of the toilet bowl once a week and leaving it overnight.

Eliminate most odours with bicarbonate of soda.

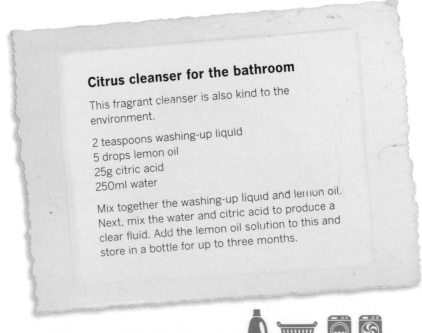

Citrus cleanser for the bathroom

This fragrant cleanser is also kind to the environment.

2 teaspoons washing-up liquid
5 drops lemon oil
25g citric acid
250ml water

Mix together the washing-up liquid and lemon oil. Next, mix the water and citric acid to produce a clear fluid. Add the lemon oil solution to this and store in a bottle for up to three months.

Beds and bedding

We spend about a third of our lives in bed, so it's worth investing in a good-quality mattress and ensuring that blankets, pillows and other bedding are clean and aired.

A good night's sleep gives the body a chance to rejuvenate. It gives muscles and soft tissue time to recover and repair, and the mind an opportunity to process memories. Poor bedding can disturb sleep, adversely affecting health, mood and quality of life.

BLANKETS, pillows & sheets

● Air and shake out your bedding regularly. This distributes the filling evenly and combats dust mites and other pests that like warm, dark places.
● Hang blankets and pillows on a clothes-line when the air outside is fresh and dry. However, don't hang out anything with a feather filling in intense sunlight as this can make it brittle and porous.
● Wash bedding in a washing machine only if the washing drum is big enough to handle it. If not, have it cleaned commercially.
● Use a gentle detergent or hair shampoo to wash feather or down pillows, then put them in the dryer at a low temperature along with a clean tennis ball, which prevents the fillings from clumping. Do not vacuum down or feather quilts or eiderdowns as you risk thinning out the filling.
● If you suffer from allergies, wash your bedding more frequently and use pillows and blankets made from synthetic materials or rayon, not feathers. These are easier to wash and dry to get rid of mites.

GOOD TO KNOW ✓

Mattress fillers
Good mattresses come in many varieties. They may have a core of synthetic material, such as foam, latex or interior springs, or the core may be made from natural materials such as coconut, horsehair, straw or seaweed. The covering usually consists of linen, cotton or raw silk. Mattresses with a natural core should be refurbished after about five years, as they wear out more quickly than latex or spring mattresses. If you're plagued by allergies, opt for a dust-free, antibacterial mattress of foam or latex.

MATTRESSES

● Prevent lumps from forming by turning a mattress every three months and flipping it twice a year unless it is a memory foam mattress as these should never be turned.
● A mattress must be aired and cleaned regularly. Take it out of the bedroom and let it breathe. Then, place a damp sheet on the mattress and beat it. The cloth will pick up the dust and can be laundered.
● Dust a mattress using the upholstery nozzle on the vacuum cleaner whenever you change the sheets.
● Buy a special mattress protector if you have severe dust-mite allergies. It will protect the mattress against mites and can be washed at a high temperature.
● Consider mattress covers made of cotton with a polyurethane backing. They are easy to remove and wash, and protect mattresses against stains.
● Rub off fresh bloodstains with cold water, then rub in a little shampoo. Lather with a stiff brush then rinse with cold water. Leave to dry naturally.

Ensure that sheets and blankets are clean and aired.

Carpeting

Rugs and carpets in the main areas of the home tend to get a lot of wear. Keep them maintained on a regular basis to ensure that they feel soft under foot, add warmth to a room, look good and last a long time.

Rugs and carpets need to be cleaned and vacuumed regularly as footprints, food, grease or other stains will spoil their appearance and may also leave smells. Follow these tips to keep rugs and carpets spotless.

DUSTING

- Vacuum a carpet regularly and thoroughly to keep it in good condition, rather than waiting until the dirt becomes visible.
- For the first six months gently brush new rugs to loosen and pick up material.
- Authentic Oriental rugs should only be vacuumed occasionally. Keep the suction level low to avoid damaging the fine fibres.

RUG cleaning

- Wet-clean every rug or carpet from time to time. Put small, washable rugs into the washing machine and clean larger ones by hand with a mild detergent, using a soft-bristled brush or sponge to release the dirt and grime. Rinse afterwards with water and, whenever possible, hang rugs outdoors to dry.
- Use a homemade cleaning powder for rugs. Mix 3 tablespoons of soap flakes with 550g cornflour. Sprinkle the mixture on to the rug, work it in with an old hairbrush or scrubbing brush, then vacuum.
- Clean and freshen a rug by sprinkling it with moist salt. Leave the salt to work for a few minutes and then vacuum.
- Grate fresh potatoes and scald them with boiling water. After 3 hours, strain them and brush the rug with the potato water. Leave it to dry, then vacuum.
- Freshen up a rug's colours by rubbing it with vinegar and water.

TREATING stains

The occasional accident is unavoidable, but prompt action will help prevent a permanent mark.
- Soak up rug spills immediately with clean, dry cloths. Weigh them down and leave them for several

Vacuum a carpet regularly to keep it in good condition.

hours before treating the stain. When removing stains work from the edge towards the centre.
- Soften dried stains with glycerin to help loosen them before further treatment.
- Treat coffee stains immediately by dabbing the soiled spots with a solution of 25ml vinegar and 60ml water, or a solution of mild detergent. Coffee stains should always be treated as quickly as possible.
- Use cold mineral water on bloodstains, blotting with a clean towel as you go.
- A minor burn spot that hasn't destroyed the pile of a carpet or rug can be removed with an onion-based mix. Boil up 250ml vinegar, 50g talcum powder and two coarsely chopped onions. Allow to cool, then apply to the spot and leave to dry. Once dry, brush off the residue then rub the fibres gently until the burned bits disappear.
- To repair burn holes, use a razor blade to cut some fibres from a section of the carpet that's hidden. Put some all-purpose glue into the burn hole and arrange the fibres in it. Weigh down until the glue dries.
- Eliminate grass stains with a solution of 15ml ammonia and 250ml water, then dab with tap water.
- Soak up grease or oil stains by sprinkling them with plain flour or cornflour. Let it sit for an hour, brush off and remove the residue with tap water.
- Blot up wine stains with paper towels, then treat them with a solution made from quarter of a teaspoon mild detergent and 1 litre of water. Blot up any excess liquid.

TREAT WINE STAINS with DETERGENT, NOT SALT

● Never put salt on fresh red wine spills because it will set the stain and could leave an indelible blue mark on the pile. Also, it will cause the carpet to remain damp.

● Remove hardened chocolate with a knife, then dab away the residue with cold water followed by warm water.

● Horrified to find a big wad of chewing gum stuck to the rug? Don't worry. Place a plastic bag filled with ice cubes over it. This will make the chewing gum brittle and it can then be chipped away with a spatula or spoon.

WHAT else you need to know

● To prevent furniture from leaving imprints in a carpet, buy little round plastic coasters just slightly larger in diameter than the feet of the furniture to place underneath.

● Remove imprints by placing a damp cloth on them and ironing it carefully. Then brush the fibres in the opposite direction.

● Don't let rugs slide. Place rubber rings from preserving jars under the four corners of rugs to keep them in place and use anti-slip underlay to stop larger rugs from moving.

● Avoid electric shocks by spraying synthetic-fibre carpets with a mixture of 250ml liquid fabric softener and about 2.5 litres of water.

● Straighten rug fringes by spraying them with laundry starch then smoothing them with a comb.

● Spray straw rugs occasionally with a little salt water to keep them flexible.

● Use bicarbonate of soda to remove bad smells from rugs. Simply sprinkle it on the surface, leave it to work for half an hour, then vacuum it up. For stubborn odours, rub the carpet first with vinegar and water, then apply the bicarbonate of soda.

Removing furniture impressions

1 Move the item of furniture to one side so you can reach the impression.

2 Place an ice cube on the compressed fibres to make them swell up, then vacuum to make the pile stand up again.

Curtains

Nets, curtains and blinds shield your family from prying eyes, keep the cold out on winter nights and protect you from the summer's heat. By maintaining them carefully, you can extend their life and keep them looking great.

There is a tendency to ignore curtains but, like every other item in your house, they will collect dust, hair and pollution. Regular maintenance will ensure they stay clean and remain an attractive feature of your interior decor.

DUSTING

● Dust curtains and drapes regularly with the upholstery nozzle on the vacuum cleaner to ensure they don't have to be washed or cleaned quite so often.
● Curtains made from synthetic fabrics attract dust. Use cold salt water to dissolve it.

WASHING

● Soak coloured curtains in salt water to prevent fading and help dissolve dirt.
● Wash delicate curtains by hand in the bath with plenty of hand soap or use the detergent below.
● Protect net curtains in the washing machine by putting them in a pillowcase.
● Always wash curtains using the machine's delicate cycle, with plenty of water. That should help to reduce the amount of creasing.
● Allow hand-washed curtains to drip dry; never wring them out. To ensure that they hang properly, weight the bottom edges with clothes pegs before hanging them.
● To prevent shrinking, stretch out cotton curtains while they are still wet.
● Stiffen net curtains with a 1:3 sugar-water solution added to the last rinse, or put them in water used for boiling rice.

Hang newly washed fine fabric curtains while still damp so they don't have to be ironed.

● Give your curtains a wonderful fragrance by adding a few drops of perfume or essential oil to the wash cycle.

COMBATING yellowing

● Yellowed net curtains recover their gleaming white colour even without chemical bleaches if you soak them in a bicarbonate of soda solution (about 180g bicarbonate of soda in 10 litres of water) before washing them. You can get the same effect by adding a glass of cola or 2-3 denture cleaning tablets to the curtains' wash cycle. Dry in the sun.
● Salt, the cure-all, also helps with this. Simply soak smoke-yellowed curtains overnight in a salt solution (about 550g salt in 10 litres of water) and then wash as usual.
● If older nets lose their whiteness, give them an attractive cream colour instead. Just add an infusion of weak tea to the rinse cycle. The intensity of the colour depends on how long you let the tea steep.

Detergent for curtains

50g soap
10 litres water
4 tablespoons ammonia
4 tablespoons white spirit

Dissolve the soap in hot water and stir in the ammonia or white spirit. Pour the solution over the curtains laid flat in the bath and leave for an hour before rinsing. This is particularly well suited to delicate curtains.

Dishes

Always hand-wash valuable silverware, fragile glassware and delicate porcelain, which might be damaged by a dishwasher. To remove stubborn stains and baked-on residue from other dishes, pans and utensils, old-fashioned elbow grease and a bit of traditional ingenuity are often still more efficient than modern appliances.

If you are using a dishwasher, check that the dishes you are putting in are designated dishwasher safe by the manufacturer. Wooden chopping boards and plates should not be machine washed as they can't stand the heat and will lose their lustre or even crack and split. You're also better off washing dirty pots and pans by hand. They take up too much space in the dishwasher and require a special, less energy-efficient wash cycle to get them clean.

BEFORE doing the dishes

- Soak dried-on food remains to soften them before rinsing. You'll find that grease rinses off better with hot water, but proteins and carbohydrates are best removed with cold.
- Protect delicate porcelain by lining your sink with a towel.
- Rubbing lipstick marks with salt makes them much easier to wash off.

WASHING by hand

- Make up some soft soap (see panel, right) from odds and ends of old soap. As well as being a tried and trusted cleaner, it avoids waste.
- To kill germs, use water heated to at least 60°C when washing by hand, and make sure you change your dishwashing sponge and tea towels frequently.
- As a general rule of thumb, wash non-greasy items first. The proper sequence should be: glasses, cutlery; plates, bowls and other dishes; and, finally, pots, pans and baking sheets. If possible, fill a second sink or basin with hot, clear water for rinsing.
- Wash glazed and unglazed earthenware pottery by hand, without detergent if possible. Remember that the glaze on earthenware pottery is heat-sensitive.
- Take special care when cleaning cutlery with wood, bone and ivory handles. Rinse the metal parts

Making soft soap

Soap making is an age-old tradition, and this recipe has been used for generations. A teaspoonful of the gel will clean the dishes, though don't use it in the dishwasher.

90g good-quality, unscented household soap
1 litre cold water
A heatproof jar with tight-fitting lid

Using a fine grater, grate the soap into an old saucepan.

Add the water and place the saucepan over a low heat. Bring just to the boil, stirring occasionally, then simmer gently for 15 minutes.

Pour into the jar, leave to cool then seal and leave for a day to form a gel..

with a damp sponge but don't soak or dip the handles into water. Place the cutlery into the drainer basket with the handles up.
- Scrape food remnants from valuable porcelain immediately, then wash each piece separately in warm water.
- Never wash gold-rimmed porcelain in the dishwasher or with bicarbonate of soda or other harsh products that could remove the finish.
- Remove hairline cracks in fine china by soaking it overnight in a large bowl of warm milk (no warmer than milk you would give to a baby). Then, gently hand-wash as usual – the tiny lines should disappear.

REMOVING stains from porcelain

● Remove tea stains or residue from porcelain cups by mixing hot water with 1 teaspoon of bicarbonate of soda in the cup, leave it to soak then wash it out thoroughly. Or, mix 2 tablespoons of chlorine bleach and 1 litre of water. Soak the cup in the solution for no more than 2 minutes, then rinse immediately.
● Minor limescale is easy to wipe off with a damp sponge and vinegar.
● Wash off stubborn limescale by pouring a dash of citric acid and hot water into the container to be cleaned and leaving it to sit for 1 hour. Repeat as needed until residue is dissolved, then wash and rinse thoroughly.
● Wipe brown stains off a teapot with a paste of vinegar and salt.
● Scrub away stains with a mixture of salt and vinegar or lemon juice.
● Remove nicotine stains from porcelain by using a cork dipped in salt.

CLEANING teapots & coffeepots

● Never wash teapots with washing-up liquid or in the dishwasher; just use hot water. A layer of tannin residue actually enhances the aroma of the tea.
● However, if you don't like the look of the tannin, remove it gently by adding vinegar to the teapot, then leave it to steep before rinsing it out. Another option is to dip a damp cloth in bicarbonate of soda and then use this to wipe out the pot before rinsing it.

Gluing porcelain

one Lay out the porcelain, quick-drying glue and dressmakers' pins or Blu-Tack for attaching the porcelain pieces to a work surface (protected with newspaper or an old towel).

two Clean the broken pieces of porcelain and the areas to be glued with a lint-free rag. Leave to dry.

three Apply a very thin layer of glue to the pieces.

four Carefully fit the pieces together and let them dry. If necessary, hold them in place with pins or Blu-Tack. Immediately wipe off any glue that squeezes out.

● Clean a glass coffeepot by adding a handful of uncooked rice and filling it with dishwater. Put the lid on and shake until the stains are gone.
● Dissolve denture-cleaning tablets in warm water to remove stubborn lime deposits.

AFTER washing

● Air-dry dishes for best results. Place them vertically in a dish drainer. Make sure the handles of stainless steel and silverware all point down.
● Dry dishes while they are still warm to prevent watermarks and bring out the shine. Use tea towels made from an absorbent material, such as cotton or linen, and wash them several times before use.
● Protect porcelain by placing paper towels between each plate before storing them in the cabinet.
● Allow Thermos flasks to dry thoroughly and store them with the top off so they don't smell musty.
● Hang cups if possible to save space.
● Turn the tops of tureens, sugar bowls and teapots upside down to protect any protruding parts.

If you are using a dishwasher, check that the dishes you are putting in are designated 'dishwasher safe' by the manufacturer.

Doors

Internal and external doors are in constant use, so it's not surprising that signs of wear and tear appear. With a little attention, however, it's easy to keep them looking good and in perfect working order – which is important for security and privacy, and to contain noise and smells.

Ideally, every door in your home should open and close efficiently and silently. Rid yours of squeaky hinges, tricky locks and other door problems in a jiffy.

AROUND the door

● Look for a build-up of debris in the track of a sliding door if it sticks. If that is the problem, clean the track with a wire brush and lubricate it with a little petroleum jelly so it glides smoothly.
● If too much paint is causing the door to stick, sand the edges of the door and frame lightly, then reapply a thin coat of paint.
● Silence squeaky doors by lubricating the hinges with candle wax or petroleum jelly, or by applying a little oil.
● If a door sticks, try rubbing the edge with candle wax or another lubricant and it should move more easily.
● Fix loose doors by applying draught excluder inside the frame. Make sure that the surface is clean and free from grease before applying.
● Add draught proofing around the top and sides and a brush draught excluder at the bottom if your exterior door is letting in cold air.

LOCKS & fittings

● Use powdered graphite around a tight latch but never use oil as dust adheres to it and can clog the mechanism.
● Muffle the sound of a noisy door latch by gluing felt strips into the door frame at the level of the lock mechanism.
● Install a mechanical door closer so that the door doesn't slam when it shuts.
● If the lock mechanism sticks, either spray some liquid lubricant into the cylinder or use a pencil to

Spray a little sewing-machine oil or other liquid lubricant into the squeaky hinge on a heavy door.

apply some powdered graphite to the key and use it to lock and unlock the door repeatedly.
● Rub candle wax onto a stubborn key and it will slide into the lock more easily.
● Take off the door and apply a little lubricant, such as petroleum jelly, soap or wax, between the two parts of a squeaky hinge. Re-hang the door and wipe off any excess lubricant.

Fabric dyeing

New colour can refresh well-worn clothing and rejuvenate faded fabrics. You can buy synthetic products, but it's worth remembering that effective natural dyes found in your home or garden have been used for centuries.

Before dyeing, check that your washing machine is suitable for the job or use a plastic washing-up bowl. The type of fabric to be dyed also makes a difference.

GROUND rules for dyeing

Only fabrics made from natural fibres such as cotton, rayon, linen, half-linen or a mix of natural and man-made fibres can be dyed; avoid synthetic fibres such as polyester and acrylic that don't absorb colour.
● Wear rubber gloves and an apron when dyeing. Protect work surfaces and floors with newspaper.
● Garment colour will affect the final result. For instance, blue dye + yellow garment = green result.
● Light-coloured clothes are easiest to dye, and always choose a shade that's darker when dyeing.
● Don't dye high-performance fabrics such as Gore-Tex, microfibres or down-filled clothing.
● If you are using synthetic dyes, colour fabrics in the washing machine not by hand.
● You will need to use salt as an additional fixative.
● Don't dye anything valuable.
● Always follow the manufacturer's instructions to the letter.

PREPARATION for dyeing

Machine dyeing is ideal for bulky items.
1 Pre-wash garments to be dyed then place in the washing machine and start the wash cycle at 40°C. If you put around 1kg of fabric in a large machine, you'll get a medium shade. If there is less fabric in the machine, the colour will be darker.
2 Add the dye to the machine's dispenser, following the manufacturer's instructions. With most dyes you will also need to add a fixative.
3 After dyeing, run it through a wash cycle with detergent at the hottest temperature, then again with the machine empty.
● If dyeing by hand in a plastic washing-up bowl, weigh the fabric first and then prewash it. You will need 2 litres of prepared dye for every 100g of fabric.
● For hot water dyes, wash item and leave it damp. Submerge in the dye mixture, bring slowly to the boil

Only natural fibres will react to the dye.

and simmer. Cool in the water, then rinse in warm water.
● For cold-water dyes, dissolve the powder in very hot tap water. Add the fixative and salt. Soak for about an hour before rinsing. (For wool, substitute vinegar for salt as the fixative.) This method can also be used for tie-dyeing.

DYES from nature

These can all be used with the basic dye formula (see panel). The more berries or flowers you use and the longer the boiling time, the darker the final colour.
● Oak leaves: dark beige to olive
● Blueberries: purple
● Goldenrod flowers: yellow to golden-yellow
● Beetroot: carmine red
● Onion skins: orange
● Walnut shells: medium brown

Plant dyes

ONE Boil 200g plant leaves or flowers for an hour in 2 litres of water.

two Dissolve 10g alum in 2 litres of water for every 50g of material to be dyed. Put the fabric into the solution, heat it up to 70°C and then leave it to cool.

three Wash fabric well, then place it in the plant brew and let it simmer on the hob for about an hour. Place the fabric in vinegar and water to set the colour.

Floors

How you care for your floors depends on the material they are made from. Some, like shiny, sealed floors, take less effort to keep clean than unsealed ones, while delicate materials, such as wood or laminate, need special attention.

Floors are made to walk on, so naturally they need frequent care to keep them clean and in good condition. Tile, wood and stone floors require more specialised treatment to bring out their characteristic highlights and avoid damaging them.

TILE floors

● Sweep or vacuum the floor before mopping it. Mop in a wavy line, without lifting the mop from the floor, starting at one end and moving towards the door. If the floor is very dirty, you may need to change the cleaning water.
● Choose a sponge mop for cleaning tile floors as they do a much better job of cleaning seams and small irregularities in the tiles. For extra stubborn dirt, use a scrubbing brush and wash the tiles by hand – sometimes there is no substitute for elbow grease.
● Clean stone floors by adding a small amount of ammonia to the water. This combination also makes dull tile floors shine like new.
● Seal porous terracotta and unglazed natural stone tiles with linseed oil as soon as they have been laid (before grouting), and avoid mopping for two weeks. You can use this type of waterproofing for areas that are subject to heavy use, such as doorways and the kitchen.

● Remove liquid stains (such as tea, coffee, cola, red wine, fruit juices and ink) from porous tiles by dabbing them with a little regular stain remover, available at any chemist or DIY store.

WOOD floors

● Ensure as little moisture as possible is left behind by a wet mop as this will cause wood floors to swell and warp. Beware of using extremely hot water as it may cause the wood to crack and split.
● Tackle serious stains and streaks on sealed wood floors by adding a shot of ammonia to the water.
● Sweep up sand and small stones at once as they are abrasive and could cause damage.

Removing scratches from wood

one Rub out scratches with a mixture of equal parts lemon juice and olive oil, applied on a soft, lint-free cloth.

two Rub in some furniture wax, or mix a little medium-brown shoe polish with basic floor polish.

three Rub the mixture into the wood until the colour matches the floor.

- Remove scratches in a wood floor with a little homemade shoe polish (see box on page 125).
- Mop sealed wood floors with black tea to add a matt sheen and an attractive colour.
- Avoid waxing floors too often. The trick is to add 60ml furniture polish and 250ml white vinegar to your mop water occasionally.
- Scrub oiled wood floors with a warm soda solution (3 tablespoons of bicarbonate of soda per litre warm water), then mop them with tap water. Repeat until the solution is mopped up and floors are clear. You should occasionally recoat the floor with a thin layer of linseed oil.
- Carefully scrape off ground-in dirt with a knife, working in the direction of the wood grain. Then lightly rub the area with a dab of white spirit before washing and polishing with a soft cloth.

OTHER materials

- Sweep and damp-mop laminated floors. Too much moisture will make the material swell.
- Damp-mop then rub dry sealed cork flooring. Apply wax sparingly twice a year and occasionally polish until shiny. You don't have to dry cork flooring that is vinyl coated.
- You can best protect a stone floor by applying a cement sealer and wax. Clean slate and stone flooring with water and a sparing amount of household cleaner, but be careful – if you use too much cleaner it can remove the colour. After mopping and drying, apply some lemon oil to ensure the floor shines like new. Remove any excess oil with a dry rag.
- Wash polished limestone flooring with a low-pH all-purpose cleaner or it will become dull. Use a cleaner with as few detergents as possible (10-20 per cent) and no more than 4 per cent phosphate as both are non-biodegradable.
- Do not use vinegar for cleaning or washing natural stone floors, such as marble or limestone. They can dissolve when exposed to acidic cleaners.
- Clean linoleum floors with the water strained from boiled potatoes.
- Remove scuff marks and dirt from skirting boards covered with polyurethane or oil-based paint with a sponge and a grease-cutting, all-purpose washing-up

liquid, then clean with a cloth dampened with tap water. You can also use a household spray cleaner – but remember to spray the cleaner onto a clean cloth, not the skirting board, to prevent streaking and avoid getting it on the floor.
- If you encounter really tough stains on the skirting board, test an inconspicuous corner with scouring powder placed on an all-purpose, plastic scrubbing pad. If the test does not cause any damage, then apply the method to the entire skirting board.

Before mopping a parquet floor make sure it is free from dust and debris.

Footwear

Allow feet to breathe and keep athlete's foot at bay by wearing shoes made from leather or another breathable material. Feet need air. After all, they contain about 250,000 sweat glands and can produce up to 250ml of perspiration a day.

New shoes can be a little painful, but there are a few handy tricks that should soon resolve the problem. Then follow these traditional tips for cleaning and maintaining them, and they should last for years.

NEW shoes

● Buy shoes in the afternoon. Feet swell during the day and you don't want to purchase a pair in the morning only to find that by midday they are uncomfortably tight.
● Before wearing them for the first time, spray or rub new shoes with a thin coating of colourless water repellent to prevent staining.
● If shoes are tight, stuff them as tightly as possible with crumpled wet newspaper or peeled potatoes. Leave them to dry out slowly, then wear.
● Avoid wearing shoes two days in a row to give them time to air out.
● Make cork heels more durable by coating them with clear nail polish.

PROTECTIVE storage

● Don't jumble shoes together. If you do, they will lose their shape and become mouldy more easily due to poor ventilation. The best option is a wardrobe with compartments specifically for shoes.
● Never put shoes into a plastic bag as they won't be able to breathe. Instead, opt for a cloth bag or store your shoes inside old pillowcases.
● Use shoe trees to keep shoes in shape. A cheaper but equally effective alternative is to stuff them with old tights that have been filled with scrunched-up newspaper.
● Similarly, instead of using boot trees, you can prevent sagging in the leg of a boot by stuffing it with a couple of layers of rolled-up newspaper.

CLEANING & maintaining shoes

To make your own shoe-shine kit, you'll need a fairly stiff brush for cleaning the shoes, some soft rags for applying polishes (one for each colour), and a soft brush or cloth for polishing. Then follow these cleaning and maintenance tips.
● Repel water from leather soles by coating them with castor oil occasionally.
● Rub stiff calfskin with a mixture of water and milk to make it soft again, then polish.

Always polish your leather shoes in the evening so that the polish can work overnight, then brush and shine the shoes in the morning.

Homemade shoe polish

It is easy to make colourless shoe polish for smooth leather using the following ingredients:

150g petroleum jelly
10g beeswax granules
2 teaspoons boiling water
5g soap flakes

Melt the petroleum jelly and wax in a bain-marie, then add the boiling water and soap flakes. Stir with a cooking spoon or whisk until blended. Store in a glass jar or can. The mixture will keep for about six months. Wash utensils thoroughly after use.

Change footwear daily and store shoes and boots on shelves or in a well-ventilated shoe rack to allow them to dry out thoroughly after wearing.

- Remove scuff marks from patent leather shoes by rubbing them with a cottonwool ball dabbed in acetone-free nail polish remover.
- Produce a bright shine on patent leather shoes by rubbing them with either petroleum jelly or baby oil. Old patent leather also recovers its shine when rubbed with half an onion. Remember to buff them well afterwards.
- Warm up patent leather shoes before wearing them in winter to prevent the finish from cracking.
- Keep patent leather flexible by applying a little glycerin.

- Clean suede regularly with a suede brush.
- To remove oily stains from suede, try covering the stain with a little talcum powder and let it stand overnight before brushing off.
- Use a rubber to remove stains from suede and leather shoes, but make sure that you don't rub too hard.

COMMON problems

- Dry wet shoes by stuffing scrunched-up balls of newspaper inside, which will help them to maintain their shape. But avoid placing the shoes near any direct heat source as the leather will stiffen and crack.
- Once dry, remove any dirt with a brush – including dirt on the sole – and then rub in some shoe polish and buff.
- Remove water and other sorts of stains from smooth leather shoes by wiping them with half an onion or a little lemon juice. Let it work briefly and then buff it off.
- Hand wash white fabric or canvas shoes, or put them in the washing machine – but never use the dryer.
- Dip the ends of frayed shoelaces in clear nail polish or wrap them with clear tape to make them easy to thread through the eyelets again.
- Sprinkle talcum powder into sweaty shoes, let it work overnight then shake it out.
- You can also clean insoles with talcum powder. Remove them from the shoes, apply the powder the same way, let it sit and then brush it off.

Furniture

As household items are made from different materials, no single cleaning agent will work on all of them. But there are a variety of excellent, gentle everyday ingredients you can use to keep furniture bright and clean.

First, clear out all harsh chemical cleaning products stored under the kitchen sink and replace them with the natural, safe cleaning agents now available.

ANTIQUE furnishings

● Never expose antique wood to direct sun or treat with conventional polishes. Because of its age, it is delicate. Rather than using oils that can degrade the finish, polish antique furnishings with beeswax granules (available from a hardware store) made to treat antique wood for long-lasting protection.

WOODEN furniture

How you maintain wood furniture depends on the type of wood and how it was manufactured. Veneered furniture, too, should be treated according to the type of veneer.

● Shine teak or rosewood furniture with a mix of 300ml beer, 1 tablespoon melted beeswax granules and 2 teaspoons sugar. Apply thinly with a brush, leave to dry and buff with a wool cloth.

● Bring out the shine of a fine wood veneer:

1 Pour 200ml olive oil and 4 teaspoons filtered fresh lemon juice into a small glass bottle.

2 Seal the bottle with a cork and then shake vigorously for a minute.

3 Open and apply the mixure to a wad of cottonwool, cover it with a linen cloth or tea towel and polish the veneer using a circular motion.

4 Dry it with a clean cloth.

Furniture polish

For light wood:
2 teaspoons beeswax granules
100ml soya bean oil

For dark wood:
½ teaspoon beeswax granules
1 teaspoon lanolin
4 teaspoons soya bean oil
1 teaspoon white spirit

Melt the beeswax in a heatproof bowl then beat in the other ingredients. Allow it to cool and then pour it into a metal or glass container and seal. Apply sparingly with a soft cloth and buff with a clean one. It can be kept for up to six months.

● Maintain oak by applying warm beer, then go over it with a wool rag. Rub remaining stains with white spirit and apply another coat of protective varnish.

● Bring out the grain and colour of walnut by rubbing it occasionally with milk. Rubbing scratches with a walnut cut in half can work wonders.

● Grease from your hands, or oil, can leave stains on untreated or dulled wood as the wood will absorb it. Press a cheesecloth on it, then apply a warm iron to help loosen and absorb the stain.

● Rub water stains with white toothpaste mixed 50:50 with bicarbonate of soda. For light-coloured wood, try rubbing a Brazil nut over the spot. For dark wood, dab a mixture of ash and vegetable oil onto the stains using a cork.

● Remove rings left by glasses by mixing mayonnaise with a little ash, then rub it on the affected area.

● Remove scratches from light-coloured wood with petroleum jelly or a mixture of 1 teaspoon oil and 1 teaspoon white vinegar. For dark wood, use red wine instead of the vinegar.

Shine teak or rosewood with a mixture of beer, beeswax granules and sugar.

RATTAN & wicker furniture

● Clean untreated cane, rattan or bamboo by vacuuming the furniture, then cleaning with a cloth dipped in a mild soap solution. (For extra cleaning power, add a dash of ammonia.) For varnished wicker, cane or chairs with raffia or straw seats, soap and water should do the trick. A small, stiff brush is often needed to get into cracks.
● Increase the durability of rattan or wicker by brushing it down once a year with salt water.
● Lighten the colour of wickerwork by rubbing it with half a lemon or a mixture of salt and vinegar. Rinse thoroughly after treatment and dry naturally.
● Get sagging wicker seats back into shape by dampening them with hot water and then leaving them to dry outdoors for at least 24 hours. As the fibres dry, they should tighten and shrink back to their original shape. However, make sure they dry in the shade as direct sunlight could bleach them.
● Treat the underside of furniture with lemon oil to keep it from drying out.

SMOOTH surfaces

● Use a soap and water solution with a dash of vinegar to clean furniture with plastic tops.
● If the glass panes of display cases or bookcases are very dirty, moisten a cottonwool ball with surgical spirit and rub the glass in circles to avoid streaking.
● Marble is extremely porous and must never be cleaned aggressively. Wipe it with a soft cloth moistened with 1 tablespoon of liquid soap in 1 litre of water. If it is not of great value, try treating stains with a little lemon juice or lemon rind and salt. Let the juice work for just a few seconds or the acids will attack the stone. Apply a mixture of 3 tablespoons of bicarbonate of soda and 1 litre of warm water to dull marble. Leave it for 15 minutes, rinse and rub dry.

Removing wax from wood

1 Use a hair dryer switched to the lowest fan speed and highest heat setting to melt the wax residue on the wood.

2 Rub the wax off with a piece of kitchen roll and then wipe with a solution of equal parts white vinegar and water.

Glass and mirrors

They attract dust, dirt and scratches, and every touch leaves a fingerprint.
But when mirrors and glass surfaces gleam, the whole house sparkles.
Here are some simple methods to keep them looking good.

A few standard household items and a little elbow grease should keep glass surfaces and mirrors clean.

GLASS surfaces

● Wipe tabletops with lemon juice, drying them with kitchen roll before polishing with newspaper to keep them shining like new.
● Polish away small scratches by smearing on a little toothpaste, then buffing it off with a soft cloth.
● Remove dried-on dirt specks by wiping them with ammonia and water.
● Allow wax to harden before carefully lifting it off with a razor blade. Use surgical spirit to remove the remainder.
● Smooth a sharp chip in a glass surface by covering the edge with a thick layer of clear nail polish.
● Clean flat-screen monitors and televisions with a soft, dry microfibre cloth (the kind you use to clean spectacles), wiping it gently. Don't scrub as pushing on the screen can burn out the pixels. Instead, barely dampen your cloth with distilled water.

MIRRORS

● Keep mirrors gleaming with a mixture of warm water and vinegar. Combine 250ml white vinegar with 1 litre of warm water and rub onto the mirror with a soft cloth. Use crumpled newspaper to wipe the mixture away in slow circles. Hey presto. No streaks.
● Clean a mirror by rubbing it with a potato cut in half, then rinsing with water and polishing. It will keep bathroom fog at bay, too, which is an added bonus.
● Keep mirrors clean without leaving streaks by wiping them with cool, strong black tea and drying with a chamois leather.
● Use a little surgical spirit to leave a nice lustre after polishing. This will also remove the sticky film left by hairspray.
● De-fog mirrors with a hair dryer after a shower. A quick blast should clear up the problem.
● Prevent them from fogging up by coating them with shaving cream or toothpaste and wiping them dry with a clean towel before showering.
● Avoid placing mirrors in direct sunlight as that can dull their surface.

TO ELIMINATE STREAKS USE 250ml VINEGAR in 1 litre WARM WATER

Glassware

People have collected luminous stemware and beautiful crystal drinking vessels for centuries, handing them down from one generation to the next. Being fragile, they need a little special attention and should be handled with care.

By observing a few basic rules, you can banish smudges and water rings on decanters, vases and glasses – without breaking them.

- Hand wash high-quality glasses rather than putting them in a dishwasher, which may mark them.
- If you must use a dishwasher, slide stemmed glasses in sideways to ensure they don't break.
- Place a tea towel in the sink when you're washing delicate glassware to prevent chipping.
- Remove lipstick marks with table salt before washing glasses.
- Clear a cloudy glass by dipping it in an ammonia solution then washing it out thoroughly.
- Add a dash of vinegar or lemon peel to the washing-up water to make glasses gleam.
- To keep cold glasses from cracking, wash them in warm water only. Scalding hot water will leach the shine from crystal glasses and dull the gold or silver rims on glassware.
- Wash glasses with painted detail as quickly as possible. If you keep them in the water for too long the decorations may dissolve.
- Rinse out beer, wine and champagne glasses with warm, clear water only. Detergent residue can change the taste of a drink as well as taking the fizz out of champagne and the head off your beer.
- Dry glasses with a linen tea towel to avoid leaving lint flecks.
- Polish glasses by rubbing on a thin paste of bicarbonate of soda and water. Rinse thoroughly and buff with a soft cloth.
- Store wine glasses upright. There is a risk that the rims could be damaged if the glasses stick to the cabinet shelf.

SPARKLING clean vases & carafes

- Help crystal maintain its shine by washing it in warm water only, then rinsing and drying it. Or wash it in hot, soapy water then rinse with hot water and dry.

Cloudy glass vase remedy

You can combat serious dirt or clouding on glassware with a salt and vinegar paste made from:

80g salt
2 tablespoons vinegar

Mix the salt and vinegar to form a paste. Spread the paste on the inside of the vase using a finger or a soft brush. Let it sit for 15-20 minutes, then empty and rinse with clear water.

- Clean narrow-necked glass containers with a bottle brush moistened with vinegar and water.
- Remove hard-to-reach dirt in a decanter or container with crumbled eggshells and lemon juice. Leave it to stand for two days, shake it back and forth a few times, then rinse.
- To combat algae in a vase, add a handful of black tea leaves and douse them with vinegar. Tip the vase back and forth until the green coating disappears.
- Buff leaded crystal vases with a chamois leather after they have been washed.

Store wine glasses upright with the bowls open to the air, or they may develop a musty smell.

HOUSECLEANING

The key word when it comes to housework is 'organisation'. Follow a regular schedule that includes daily, weekly and occasional tasks, as well as spring cleaning, and it won't seem so daunting. It will also save you time and energy in the long run.

FROM TOP to BOTTOM, from BACK to FRONT

Daily

Keep your household orderly so that it is easier to keep clean. These daily household tasks can be accomplished in no time at all.

- Pick up scattered clothing, shoes, toys and newspapers – you're well on your way to a tidy home.
- Shake out bedding and air your bedroom thoroughly.
- After brushing your teeth in the morning, rinse out the bathroom sink with water to keep lime from building up.
- Wash all dishes, cutlery, pots and pans used to prepare and eat meals during the day.
- Wipe down the work surfaces in your kitchen to prevent dirt and germs from taking hold.

Weekly

Make time for these weekly housecleaning chores, then any major spring clean will be so much easier than if you had let the work build up.

- Dust, vacuum and mop the floors throughout the house.
- Clean the bathroom thoroughly: scrub the sinks, shower and/or bath and toilet; wipe off the tiles in wet areas; and wash the floor.
- Clean the mirrors.
- Wipe out your refrigerator and thoroughly clean the cooker and sink.
- Take out rubbish regularly and put the bin out every time there's a refuse collection.
- Put out recyclables on collection day.
- Change and wash household linens.
- Sweep the porch and/or patio.

Dust, vacuum and mop ...

Occasionally

Devise a cleaning schedule to give you an overview of tasks that require less frequent attention. Follow the schedule to ensure that nothing gets overlooked. Be sure to include the following:

- Wash windows
- Wash curtains and blinds
- Clean doors and door frames
- Clean upholstered furniture thoroughly
- Maintain wood furniture
- Wipe down kitchen cabinets
- Ensure kitchen appliances are serviced as recommended
- Clean lamps
- Clean rugs and carpets
- Dust walls
- Look after all cleaning equipment

Just one manageable task each day

A system for cleaning

Before you start, make sure you have to hand all the cleaning materials you require. You won't need heavy-duty chemicals, even for your annual spring clean. A mild, all-purpose cleaner, vinegar, lemon and furniture polish should do the trick.

Set aside adequate time for housecleaning and wear old, comfortable clothing that you don't mind getting sweaty and dirty. You might also want to wear nonslip shoes and gloves, along with eye protection, depending on the type of work.

If you want a time-honoured system for cleaning, follow these three rules:

- from top to bottom;
- from back to front;
- clean the same items at the same time (for example, clean all the glass, wash all the surfaces and so on).

Start by clearing the cobwebs from the ceiling, then polish the wood and glass, dust, clean upholstered surfaces, finishing with the floors. Always work from the back corner of a room towards the door. Right-handed people usually work more effectively from right to left, left-handed people from left to right.

Spring cleaning

Take on just one manageable area every day. For example, on the first day the bedroom, the next the bathroom, and so on.

- Move the furniture and clean neglected corners.
- Clear out your wardrobe and weed out clothes you haven't worn in a long time. A good rule of thumb: if you haven't worn it for a year, get rid of it.
- Check the freezer for out-of-date foods and discard them, then defrost (if necessary) and clean.
- Check storecupboard foods for freshness and throw out any that are past their best.
- Clear out the loft, garage or garden sheds, earmarking any reusable items to give to charity or put in a car-boot sale.

Household aids and accessories

The daily battle against dust and dirt can be won only if cleaning equipment is in proper working order. Maintain it regularly to ensure your house looks its best at all times.

SOAK DISHCLOTHS in LEMON or VINEGAR so they SMELL FRESH

Follow these suggestions to ensure that brushes and brooms are fit for the job, and cloths and sponges don't become a breeding ground for bacteria.

BROOMS & brushes

- Use brushes and brooms with natural bristles.
- Clean broom bristles in detergent and warm water. Rinse thoroughly in cold water and lean the broom head-up, handle-down against an outdoor wall. The sun's rays will help to kill any lingering bacteria.
- Soak new natural bristles briefly in salt water before the first use to make them last longer.
- Wash older brooms and hand brushes with natural bristles occasionally in warm, soapy water.
- Soak natural bristles in vinegar and water if they have become too soft.
- Soften stiff natural bristles by dipping them briefly in a solution of 30g powdered alum and 1 litre water. Alternatively, mix water and milk in a 1:1 ratio.
- Hang brooms on the wall with the handle pointing down to preserve the elasticity of the bristles.
- Apply steam to bent bristles to straighten them.

RAGS, sponges & cloths

- Machine-wash all cloths used for cleaning and dusting at 60°C in mild detergent and allow to dry completely. It'll help to increase their life span. However, don't wash them with clothes – it's unhygienic and the colour of classic yellow dusters runs for years.
- Fibres in microfibre cloths become matted from frequent washing, so pop them into the dryer after every third wash and they'll become fluffy again. Don't hesitate to replace them frequently.
- Leave wet dishcloths to dry in the sun. The ultraviolet light disinfects and kills bacteria.
- Boil natural sponges in vinegar water to combat unwanted germs. Alternatively, place a wet sponge in the microwave for a minute. Be careful – it will be piping hot. Never put a dry sponge in the microwave as it could catch fire.
- Pick up dust and lint from dust rags by dipping them in glycerin after they've been washed. Let them dry and store them in a plastic bag.

DUSTERS

- Wash feather dusters in a sink filled with warm water and a squirt of baby shampoo. Rinse, then carefully squeeze out and dry with a hair dryer.
- Shake out microfibre dusters before washing them in soapy water. Rinse and leave to dry.

Cleaning brooms

1 Wash the bristles of brooms and brushes in warm, soapy water.

2 Comb the washed, rinsed bristles with a sturdy old comb. Allow to dry thoroughly.

Ironing

With today's modern steam irons, ironing boards and other aids, ironing can be done in a jiffy. But that doesn't mean you can't learn from some of grandma's old tricks, which make even delicate textiles and tricky seams or pleats easy to handle.

Whether you iron while watching football on Sunday afternoon or rush to press a shirt or blouse before heading to the office, the same basic rules apply.

EQUIPMENT

● Help save your back from injury by investing in an adjustable-height ironing board.
● Place aluminium foil under the ironing board cover to reflect heat. It will save time and energy.
● Fill the iron with boiled, cooled water to help to avoid limescale. Let it steam off two or three times to expel lime deposits (or clean as the manufacturer suggests). Repeat this each time you iron.

BEFORE ironing

● If you have a tumble dryer, put cotton fabrics, including shirts and blouses, into it for 15 minutes at a moderate temperature and then hang them up to dry. This makes ironing easier than if they are dried completely.
● Spray dry fabrics with warm water before ironing – this spreads through fabric more quickly than cold.
● Sprinkle clean laundry with water, then roll it up and place it in a plastic bag to keep it damp until you are ready to iron it.

IRONING basics

● Iron just to the edges of seams and stitches to ensure they don't push through.
● Iron button facings on a soft underlay from the back side to avoid marking the front of the fabric. Don't iron the buttons.
● Protect delicate fibres by placing a cloth or tissue paper between the iron and the fabric.
● Press heavy fabrics such as wool and flannel with a damp cloth between the iron and the fabric. Use a pressing not a sliding motion.
● Iron the outside of collars first, then the inside from the tip to the middle to avoid wrinkles on the seam. Iron the cuffs the same way, followed by the arms and then the front and back.
● Allow laundry to air before storing it in a drawer or mould spots may form.

De-scaler for irons

It is easy to make an economical de-scaler from:

100ml vinegar
100ml water

Mix the vinegar and water, then fill the reservoir of the iron. Plug it in and wait for it to begin steaming, then shut it off. Let it work for an hour, then empty the reservoir and rinse thoroughly with warm water.

Jewellery

We often value our jewellery for sentimental reasons as much as for its beauty. Use these few simple tricks to ensure your treasured possessions retain their sparkle.

A small, soft brush – such as a child's toothbrush – is helpful when it comes to getting into the crevices while cleaning jewellery. But take care not to scrub too hard, as it's important not to disturb any settings.

GOLD

Before cleaning gold, you need to know what is real gold and what is gold-plated.

● Clean gold-plated jewellery with water only. If you are not sure if an item is solid gold or not, treat it as plated or you may ruin it.

● Clean pure gold with a little bicarbonate of soda on a moist cottonwool ball before rinsing it.

● Rub matt-gold jewellery with onion juice to bring out its shine. Rinse it to get rid of the smell, and then buff it.

● Clean dirty solid-gold jewellery with a solution of 2 tablespoons ammonia and 1 litre water, then rinse and dry. For serious dirt, wash with a solution of 1 tablespoon surgical spirit and 1 litre of soapy water, then rinse with water and dry.

● Add three drops of washing-up liquid to a medium-sized bowl of warm tap water. Place gold jewellery in the bowl and leave it to soak for 10-15 minutes before rinsing and drying. If more power is

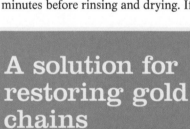

USE a soft TOOTHBRUSH to REMOVE TARNISH

required, use carbonated water instead. Gold plate can be cleaned in this way, as well.

● Store gold jewellery in soft paper envelopes to prevent it from getting scratched.

SILVER

If you are considering cleaning silver remember that it oxidises, especially if you haven't worn a piece of jewellery for quite a while.

● It is easy to remove tarnish from silver. Line a deep dish with aluminium foil, place the jewellery in it, sprinkle with salt and pour boiling water over it. Wait an hour before removing the jewellery, then buff it until it shines brightly.

● Protect silver jewellery from tarnishing by wrapping it tightly in aluminium foil.

● To remove a little tarnish, dab some toothpaste on your finger and rub the tarnished part, then rinse and dry with a clean cloth. For larger jobs, make a spreadable paste of bicarbonate of soda and water, and rub it on with a damp sponge.

● Leave silver in beer overnight, then rinse and polish to recover its sparkle. For an alluring shine, pour potato water over silver jewellery, then rinse it with warm water and rub dry with a soft cloth.

A solution for restoring gold chains

one Dissolve a little washing-up liquid in warm water.

two Add 1-2 teaspoons of powdered chalk and stir.

three Pour through a funnel into a small bottle or covered container.

four Place the gold chain in the solution, close it and shake well for a minute.

five Remove the chain and rinse with water.

Direct heat or sunlight can cause aquamarine to fade.

PEARLS

Wear pearls frequently – contact with skin helps them to retain their colour and lustre.
● If any hairspray, skin cream or perfume settles on your pearls, wipe it off immediately.
● Pearls dissolve on contact with acids.
● Clean pearls with a soft brush and a mild detergent. Wash off and repeat if necessary. Dry them naturally on a barely damp cloth – when the cloth is dry the pearls will be, too (and the string).
● A small velvet envelope or soft cloth will protect against scratches. Never hang them up as the string will stretch.

AMBER

● Remove dirt with warm water, then dry the stone immediately.
● Add 2 drops of olive oil to a clean flannel cloth and rub the amber until it shines brightly. If any oil remains, polish with a clean cloth.

COSTUME jewellery

● If costume jewellery is very dirty, sprinkle bicarbonate of soda onto it and scrub it carefully with a toothbrush, then wash and dry.
● Buff jewellery with a baby-size toothbrush and pick out residue from encrusted stains with a wooden, non-scratching toothpick.
● Avoid dark discoloration on your skin and clothing by applying clear nail polish to the back.

CORAL

● To clean coral, dab it with soapy water using a linen rag. Rinse with water and buff with a chamois leather.
● Put 2 tablespoons of household ammonia and 500ml of softened or boiled water in a pan, then place the jewellery in the solution. Simmer for about 10 minutes, then drain, rinse with cold water and polish with a chamois leather.

GEMSTONES

● Clean hard gemstones (diamonds, sapphires, rubies) with a solution of 1 litre of water and 2 tablespoons of ammonia, or dip them briefly in surgical spirit and then rinse.
● Never immerse emeralds in water. They are a fairly soft stone and sometimes contain cracks, which will absorb the liquid.
● Delicate opals don't respond well to major temperature fluctuations, so don't wear an opal ring when you are doing the dishes, for example. It is best to polish opals with a chamois leather.
● Many gemstones, such as emeralds, opals, lapis lazuli, turquoise and pearls, should not be exposed to direct sunlight. Remove them before sunbathing.

Keep jewellery in a box lined with velvet and divided into multiple compartments to prevent damage.

Kitchen care

The kitchen is the heart of the home, the room where the family gathers and everyone relaxes with a cup of tea, a mug of hot chocolate or a glass of wine. Guests often congregate in the kitchen, too, so it's a part of the home that needs regular attention.

Generally, you don't need special equipment or chemical cleaners to keep the kitchen looking good if you stick to a routine and deal with spills and other mishaps as soon as they occur.

SURFACE care

Work tops and other well-used surfaces in your kitchen require daily attention to keep them clean and free of mould and bacteria.
● Plastic or granite surfaces can be washed off with a sponge and a soap solution, or even with one part vinegar and one part water. Wipe dry immediately to avoid streaks.
● Wipe down large surfaces with a cloth in each hand: one for cleaning, the other for drying.
● To eliminate unwanted germs from work surfaces, scrub unsealed wood surfaces regularly with salt or a mixture of 4 tablespoons of bicarbonate of soda and 2 tablespoons of lemon juice.
● Rub wood surfaces with a little olive oil or linseed oil after cleaning to make them dirt resistant.
● When cleaning cabinets, inside or out, add a little vinegar to the soapy water to cut through grease.
Note: for advice on removing watermarks from wooden furniture, see Furniture, page 126.

SINKS

● Use a bar of soap to make stainless-steel sinks spotless. Rubbing them with potato peelings, lemon juice or bicarbonate of soda are also proven methods.
● Use a couple of dashes of lemon juice on a dishwashing sponge to rub down a discoloured sink.
● Rub out heat marks in the sink by sprinkling on a little bicarbonate of soda and then rinsing it off.
● Limescale spots disappear when you treat them with a mixture of vinegar and salt. Place a paper towel over the spot, sprinkle with the solution and leave it to set. Remove the paper towel and rinse.

Cleaning the hob

Ensure that your hob gleams by cleaning it with this mixture. It can be made up in larger quantities and stored as it will keep for up to four months.

Boil 250ml water then add 1 teaspoon soft soap.

Mix ½ teaspoon glycerin and 10 teaspoons vinegar in a bowl and pour the boiling water into it.

Stir in 6 tablespoons of whiting (dry, ground calcium carbonate) then transfer to a bottle and seal tightly.

COOKER top

● Wipe up splashes and spills on your cooker top immediately – it will save you a lot of extra work.
● If food gets burnt onto an electric heating element, dampen a cloth with soapy water, place on the cold element for 2 hours and then wipe it clean. Deal with spills in the grooves of the heating element by slightly heating the element and sprinkling it with a little bicarbonate of soda, then rubbing it in with a sponge. Wipe it off with a damp sponge or cloth.
● Ceramic glass hobs are especially easy to clean: simply wipe them with a damp sponge. If food is burnt on, sprinkle a little lemon juice on it, let it set for a few minutes, wipe and, if necessary, remove any residue with a glass scraper. To preserve an attractive shine, polish your hob with a little vinegar. And to avoid scratches, lift pots and pans from one burner to the next, rather than sliding them.
● Lightly rub dried-on deposits on a gas cooker's non-removable parts with a moistened dishwasher tablet, then wipe dry. Wear protective gloves.

OVEN & grill

● While the oven is still hot, put a heatproof container of hot water inside; the moisture will make it easier to wipe clean.
● Put aluminium foil or a baking sheet under a baking or roasting pan – it will save you some elbow grease if it boils over.
● If your oven is not self-cleaning, while still warm, remove burnt-on foods with salt and wipe the surface dry with a piece of newspaper or a paper towel. Use a

damp cloth to soften any particles that remain so they can be scrubbed away easily.

● Loosen baked-on deposits by filling a glass bowl with half a cup of full strength ammonia, putting it into a cold oven and leaving it overnight.

● Rinse cake tins with washing-up liquid and water. Use salt and vegetable oil for tough stains.

● Scrub burnt-on sugar with newspaper and salt, then wash with soap and water.

● Place food on foil before cooking it to help to keep grill pans clean. Put the foil matt side up to avoid 'sparking'.

REFRIGERATOR & freezer

● Look after rubber seals by rubbing them with talcum powder so they don't become brittle.

● Clean the inside of your refrigerator regularly with vinegar and water, or wipe it down with a solution of bicarbonate of soda and water.

● If you don't have a frost-free freezer, mini-icebergs may form. If so, it is time to defrost. Empty the contents into a cool box or wrap it in old blankets, then place a pot of boiling water inside and close the door until the ice melts. Wipe with washing-up liquid, vinegar and water.

● To prevent rapid ice build-up, after you defrost, wipe down the inner freezer walls with cooking oil or glycerin. When you defrost next time, the ice will come away from the walls easily.

Keep your kitchen sparkling with the right cleaning techniques.

Kitchen gadgets and utensils

A coffeemaker, toaster, blender or microwave make life easier, while a good range of utensils makes the preparation and cooking of food easier, faster and more enjoyable. Regular maintenance and a bit of traditional care will keep your kitchen in fine working order.

Most kitchen appliances run on electricity, therefore special care should be taken when cleaning them. The first step is always to pull the plug out, then detach all removable parts and wash them by hand or put them into the dishwasher.

CLEANING, de-scaling & maintenance

● De-scale the espresso machine or coffeemaker regularly using a mixture of equal parts vinegar and water. Add the solution to the reservoir and run it through just as if you were making a pot of coffee. Then repeat the process twice using plain water.

● Fill your kettle with the same solution (equal parts vinegar and water) to remove limescale. For significant calcium deposits, bring the mixture to a boil, leave it for 30 minutes, empty it out and rinse thoroughly.

● Before cleaning your microwave, add a slice of lemon to a bowl of water and heat until steam forms,

then simply wipe out the appliance with a cloth. Vinegar and water work just as well.

● Rub a little cooking oil into the rubber seals of kitchen appliances occasionally to ensure they close tightly.

● Make the hand mixer's beaters easier to insert and remove by putting a tiny drop of olive oil into the installation sockets.

● Empty the crumb tray of the toaster and shake out the crumbs over the bin.

● Use a toothpick to remove food particles trapped in the spray arm of your dishwasher. Once a month, clean the filters and run the machine on empty.

● Waste disposal units are self-cleaning, but they can get smelly. To keep them running smoothly, operate with a full stream of running cold water that will flush the ground-up debris away. At the first sign of an unpleasant odour, chop up some orange or lemon peel and run it through the system.

De-scale when water takes a long time to run through or the coffee tastes different.

● It is important to disinfect wooden cutting boards thoroughly after cutting up poultry, meat or fish to avoid cross-contamination of bacteria. Rub boards with a bleach and water solution, then wash as usual and dry.

● Wash wooden handles promptly and allow them to dry naturally. Occasionally rub in some olive oil and wipe off any excess with a clean cloth or paper towel.

● Prevent grease from sticking to cooking spoons by holding them under cold water just before using them.

● Remove dough stuck to a rolling pin by sprinkling a little salt on it, rubbing it off, then washing and drying it.

KNIVES

● Invest in stainless-steel kitchen knives and sharpen them regularly. If you don't own a knife sharpener, use the unglazed base of a porcelain teacup.

● Knives cut better if you warm the blades up before you use them.

● Use only top-quality knives for cutting foods.

● Use either a plastic or a wooden cutting board – glass, metal or stone aren't suitable.

● To protect your knife blades and avoid injury, keep knives in a knife block, on a magnetic knife strip or in a roll.

● Protect yourself from sharp, pointed implements by capping tips with corks.

● Remove stuck-on residue with a dishwashing brush or a cork dipped in salt.

METAL utensils

● Use a toothbrush to clean difficult-to-reach areas in graters and garlic presses.

● Clean metal flour sieves immediately after use in cold water – warm water will make the flour stick like glue.

● If metal gets rusty, sprinkle it with salt and rub with bacon rind.

● Stick metal shish kebab skewers into a cork; they'll stay together, making them less dangerous.

WOODEN implements

● Don't put wooden spoons in the dishwasher, as heat and cleaners can damage wood. All you need is water and washing-up liquid. Soak heavily encrusted spoons in a solution of one part bicarbonate of soda to ten parts water.

Ladles, whisks and salad spoons are easily reached when you stand them handles down in a container.

Laundry

Nowadays, the twist of a knob is all it takes to get clean clothes. But back in the days when doing the laundry was an eight-hour shift, there were some helpful tricks to hand – some of which are still useful for saving energy and protecting the environment.

WASH WORK CLOTHES SEPARATELY

We all like to keep our whites clean and bright, but laundry products such as bleach contain harsh ingredients that are tough on lungs and the planet. Traditional methods allow you to keep your utility room bleach-free and provide the special care required for coloureds, wool, lace, velvet and silk.

WHITEN clothes without using bleach

● Let your whites soak in a basin of hot water and lemon slices (tied in muslin) for 1 to 2 hours, then wash as usual. If clothes are particularly dingy, boil the water, turn off the heat, add your clothing and lemon slices and soak overnight.
● White vinegar works well, too. Pour 125-250ml into the wash with the detergent. It will whiten, help to wash away detergent or soap residue, and soften the fabric.
● One more idea: the old-fashioned power of sunlight helps to brighten whites and gives them that wonderful fresh outdoor scent.

PROTECT the environment & save energy

Get your washing beautifully clean in an environmentally friendly way with an energy-efficient washing machine and the following ground rules.
● Do not add too much detergent and never use more than the manufacturer recommends as this will result in a soap overflow.
● Take water hardness into account – the softer the water, the less detergent you need.
● Ensure that the water level is set to match the laundry load. This protects the machine, especially during the spin cycle.
● Avoid the prewash cycle, which is unnecessary for most laundry.

GETTING the laundry ready

● Soak yellowed or greying laundry in a natural bleach (see above).
● Put delicate articles into a cloth bag or an old pillowcase before washing.

● Before washing, close zips and rub them with graphite or a little grease so they don't jam later on. Also, undo buttons and turn pockets inside out.
● The collars and cuffs of shirts and blouses don't always come out of the wash clean. Before washing them, use a nailbrush to apply a little shampoo formulated for greasy hair or some liquid detergent. Leave for 15 minutes before washing in the washing machine at the hottest possible setting. For a quick treatment, wash the collar only, by hand, rinse well, rub as dry as possible in a thick towel, then iron dry.

Cleaning velvet

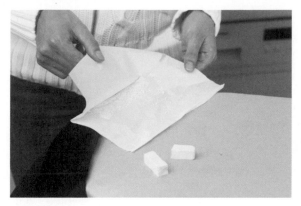

1 Grate pieces of chalk and sprinkle onto an absorbent paper towel. Fold the chalk into the paper.

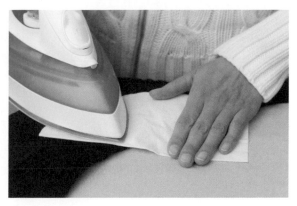

2 Place the paper towel onto the velvet and go over it with the iron to draw out the grease.

● If possible, remove stains before doing the laundry.
● Soak very dirty laundry overnight in soapy water before washing. Make use of oxygen bleach powder and modern stain removers.

FABRIC conditioner

● These can be pricey, so consider a natural softener. Add 1 tablespoon salt to the last rinse, or pour 250ml vinegar into the last rinse cycle.

WHITES

● Add 130g bicarbonate of soda to the main wash cycle. It will make your white washing whiter and give it a fresh smell.
● Soak discoloured white laundry in fresh milk until it turns sour, then rinse and wash as usual.
● Use dishwasher detergent in your laundry loads instead of washing powder. However, this can be hard on clothes so should be used only on heavy-duty clothing, sheets and towels.

COLOUREDS

Always check the colour fastness of new fabrics. To do this, wet the fabric and then rub it with white paper. If any dye comes off on the paper, hand wash the item with soft soap. Then, follow these tips to keep coloureds looking their best.
● Prevent new coloured fabrics from running by adding a dash of vinegar to the wash water.
● Add sugar to the rinse water to ensure coloured laundry comes out bright.
● Never hang coloured laundry to dry in bright sunlight as the colours will fade.

● Protect jeans from wear before washing them by soaking them in a salt solution (1 tablespoon salt in 1 litre of water).
● Do not starch coloured laundry when it is still hot, or the colours may run.

LACE, velvet & silk

● Hand wash blouses and pillowcases made from natural silk with a special wool or silk detergent, then rinse thoroughly and hang up dripping wet to dry.
● Allow black silk to retain its shine by washing it with black tea and a little mild detergent.
● Washable velvet will shine after being laundered if you brush it down with a little salt.
● Wash raw silk by swishing it around in warm, soapy water before rolling the garment in a towel to extract as much of the water as possible. Lay it flat to dry and iron while still damp on the reverse side.

WOOL

● To banish sweat smells, moisten two cloths with a weak water and ammonia solution, place the article of clothing between them and steam it with the iron.
● Woollens stay soft if you add a few drops of vinegar to the last but one rinse and a similar quantity of glycerin to the final rinse.
● Shampoo is excellent for cleaning wool and preventing it from becoming matted.
● To soften a scratchy sweater, dissolve 1 tablespoon of hair conditioner in a sink filled with water and swish the item around for 2 minutes. Rinse and dry.
● Wool items lose their shape in the spin dryer. Roll them gently in a dry towel, squeeze out excess water then hang them out flat on a clothes rack to dry.

Leather

Leather is strong, durable and will last a long time if it is treated gently and you avoid using harsh cleaners. There's no need to use expensive, chemical-based products to keep it looking its best – just try these traditional solutions.

Many of us regularly wear or carry leather products, including jackets, gloves, belts and bags. Follow these basic rules to ensure that they look and feel great.

PROPER storage

● Leather needs to breathe, so never store it in a plastic bag.
● Hang belts in your wardrobe or roll them in a cloth bag.
● Store leather garments in fabric clothing bags.
● Fill soft handbags with newspaper to help them to keep their shape.

GLOVES

● Rub black leather gloves with the inside of a banana skin to make them shine like new.
● Wash suede gloves in soap and water with a few drops of ammonia. Squeeze the water through the gloves carefully by hand, and then squeeze it out again before laying them flat to dry.
● Rub a little castor oil into old, stiff leather gloves to bring back their flexibility.
● Place tight gloves between damp cloths for a few hours, then put them on while they are still damp to stretch them. There's no need to keep them on until they're dry, just wear them for a few minutes.

LEATHER garments

● Never leave leather items in the sun for long because intense sunlight can leave a mottled effect on the surface.
● Allow wet leather to dry at room temperature but not in the sun or near a heat source as this may cause the leather to become brittle or, in some cases, even release toxic compounds.
● Brush suede well after drying to lift the nap.
● Use a rubber or suede eraser to rub down dirty collars and cuffs.
● Clean white leather with a little milk.

BELTS & bags

● Clean coloured leather belts and bags occasionally with a damp cloth, then allow them to dry at room temperature. But keep them out of direct sunlight or they may lose their colour.
● Clean dusty, dirty leather suitcases with a solution of equal parts milk and white spirit.
● To preserve leather suitcases, apply castor oil generously. Let it soak in for a few hours, then buff with a soft cloth.
● Clean a leather briefcase with 1 teaspoon surgical spirit mixed with 250ml water.
● For grease spots on coated leathers, apply a little unscented talcum powder, leave it to absorb the grease, then rub off.

Washing leather gloves

1 Check the leather is washable then fill the sink with warm water and washing-up liquid. Add a dollop of glycerin to make the leather softer and stretchier.

2 Put on your leather gloves and wash them in the soapy water.

3 Rinse with tap water and lay on a flat towel at room temperature to dry. When they are half dry, knead them to make them soft again and lay them flat to complete the drying process.

Lighting

Lamps add to the ambience and decor of a room as well as providing light, so make sure they are not shrouded in a layer of dust. Keep light bulbs and lampshades clean to increase brightness and save energy.

Safety is always important when electricity is involved. Before cleaning, always pull the plug on lamps or switch off the power at the mains. If you use a ladder to access light bulbs and lampshades, make sure it is sturdy and not standing on a rug. If possible, get someone to steady the ladder.

SHADES, bulbs & cables

● Dust fibre and rattan shades regularly or gently vacuum the shades with a soft upholstery nozzle.
● Paper shades shouldn't be cleaned with a wet cloth, but discoloured vellum shades respond well to soapy water and surgical spirit. Use a rubber to remove surface spots.
● Lampshades made from parchment don't get dirty as quickly and are easier to clean when they are sealed with clear varnish. Simply wipe them down with a damp cloth.
● Vacuum fabric shades or run a lint brush over them. Alternatively, have them cleaned professionally so that they don't shrink.
● Wipe silk lampshades with equal parts of vinegar and warm water. Wring the cloth out well before you start so that it is just slightly damp.
● Don't use cleaning agents on Tiffany glass lamps. Instead, wipe them down with a soft, damp cloth and clean the joints with a soft toothbrush.

Metal chandeliers shine beautifully when you add a little vinegar to the wash water.

● Dust glass, plastic and metal lampshades before washing them thoroughly inside and out with washing-up liquid and water. Polish them with a dry, lint-free cloth. Clean crystal chandeliers with surgical spirit applied directly to a lint-free cleaning cloth.
● To reach those in high places, invest in a duster with a telescopic handle.
● Remove ceiling and wall lights with half shades several times a year to shake out the insects and dust and clean them thoroughly.
● Wash glass shades in warm water with washing-up liquid.
● Dust-covered light bulbs lose up to half their brightness. Clean them regularly to ensure you have the best possible lighting.

Use a feather duster or soft brush on hanging lamps or cable systems.

Odours

Few homes can escape the less desirable smells that occasionally emanate from kitchens, bathrooms, bedrooms or hall cupboards. But armed with a bit of traditional wisdom, it's easy to control them and ensure that friends and family receive a fragrant greeting.

The kitchen and bathroom tend to produce most of a home's least desirable smells. But tobacco and other fumes may also be brought in from outside. Proper ventilation is the primary weapon against kitchen odours, but there are also a few simple tricks you can use to help during the cooking process.

COOKING

● Mask odours by boiling water and a little cinnamon in a small pot while cooking.
● Stretch a microfibre cloth moistened with vinegar over the pot in which you're cooking fish and other strong-smelling foods.
● Cooking cabbage? Drop a bread crust into the cooking pot to absorb odours. For cauliflower, try half a lemon.
● If milk boils over sprinkle it with salt to prevent it from boiling over again.
● After cooking, combat food odours by briefly moistening the cooker's heating elements with a little vinegar, or boil some vinegar in water. You can also try wrapping a few cloves and a cinnamon stick in aluminium foil and placing the package on the hot heating elements.

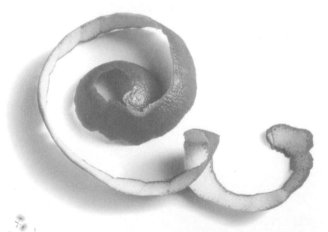

Burn a piece of dried orange peel so a pleasant fragrance permeates the living room or kitchen.

KITCHEN odours

It's not just the food itself that creates strong odours. Where we store and cook food also contributes to the problem.

● Combat musty smells in kitchen cabinets by washing them with vinegar and water, or bag up coffee grounds from the coffeemaker and place them inside to neutralise any remaining odour.
● Place a bowl of white vinegar on the rack to eliminate odours in the oven, or put some orange peel in a warm oven. If you want to eliminate nasty smells in a hurry, rub down the inside of the oven with half a lemon.
● Similarly, to neutralise microwave smells put half an orange or lemon in water in a microwave-proof dish and heat on high for 2 minutes.
● Avoid stomach-churning odours in the refrigerator by always wrapping foods separately and making sure that containers are tightly covered. If a bad smell does seep through, try placing a bowl of vinegar, half an apple or a little bicarbonate of soda in an open bag or on a plate or bowl in the fridge.
● Ensure the dishwasher smells fresh and clean by including a little lemon juice with each wash. Lemon is a natural deodoriser and degreaser.
● If unpleasant odours do develop, sprinkle a little bicarbonate of soda in the bottom of the dishwasher next time you run it. Or, just as effective, add a few drops of concentrated orange oil during the next wash cycle.
● The smell of onions or garlic can cling to a wooden chopping board. To freshen it up, rub it with half a lemon and rinse. Lemon will work wonders for your hands as well.
● Eliminate musty smells that can develop in clay and ceramic pots by rinsing them out thoroughly with hot vinegar.
● Refresh stale-smelling Thermos flasks by rinsing them with tap water and then filling them with hot water and a little bicarbonate of soda. Let the solution sit in the Thermos for 5 minutes before rinsing it with tap water.
● Empty the refuse bin frequently and, if possible, compost all fruit and vegetable peelings (but not meat or cooked foods, as these can attract vermin).

If bad odours do develop in your waste bin, add some orange or lemon peel or a little bicarbonate of soda.

BATHROOM smells

● Burn matches or light a candle to help curb stale smells in the bathroom. A potpourri of fragrant herbs and spices – such as cinnamon, rosemary, thyme, cloves or lavender – also does an excellent job.
● Add a dash of toilet cleaner or vinegar and water to the bottom of the brush holder to keep the toilet brush from getting smelly.
● Hang towels and flannels to dry and launder them frequently so they don't smell musty.

TOBACCO odours

● Place saucers of bicarbonate of soda in a room. Ideally, close the door and leave them to work overnight. Sprinkling fabric with bicarb also works; shake it off outside or vacuum in the morning.
● Rinse ashtrays with a vinegar solution and sprinkle coffee grounds in them to help absorb the smell.

CLOTHING

● A piece of soap or a small sachet of lavender, cedar chips or even coffee grounds will add fragrance to a wardrobe. Also, try to hang coats and jackets outside to air before returning them to the wardrobe.
● Unsurprisingly, well-used sports shoes tend to become smelly. One simple remedy is to sprinkle a little bicarbonate of soda into shoes, leave it to stand overnight then vacuum it out in the morning. If you have particularly sweaty feet, put talcum powder inside each time you wear them and change your shoes regularly. When you're not using trainers, stuff them with a pair of socks filled with cat litter to absorb the odour and any moisture. A final option is to place shoes in a zip-lock bag in the freezer overnight to kill odour-causing bacteria.

OTHER odours

Bad smells can occur anywhere in the house. These helpful hints will help to keep other parts of your home fresh and fragrant.
● Control musty odours by burning a bay leaf, placing a piece of citrus fruit peel on a hot heating element, hanging a vanilla bean in the room or using fragrant candles.
● If exhaust from the vacuum cleaner smells unpleasant, change the bag and vacuum up some lavender flowers or peppermint leaves. Or dab a cottonwool ball in a little perfume or aromatic oil and vacuum that up.
● Add a piece of charcoal to flower-vase water to keep it from smelling foul.
● You can eliminate smells from plastic containers if you wash and dry them, fill them with crumpled newspaper or coffee grounds, then freeze them overnight.
● Wipe musty-smelling leather suitcases with a vinegar-soaked cloth and leave them open in fresh air for a few days.
● Eliminate the smell of oil-based paint by setting bowls of peeled, sliced onions around the room.
● Lavender bags impart a delightful smell. Place one in the linen cupboard and enjoy its fragrance. An alternative is to fill several bags with a fragrant potpourri mixture.

Absorb the smell of animal litter by adding a little bicarbonate of soda to the litter tray.

Pests

Humans have been sharing their homes with creepy-crawlies for centuries. As a result, a number of simple remedies have been devised to help rid our homes of uninvited visitors. If these don't work, you may need to call in the professionals.

They may be after your food, your blood, the wood that provides the framework to your home or even your clothing. Here are some natural ways to deter or banish those biting and bacteria-spreading creatures from your property.

ANTS

● Ants follow a scent trail, marching in a straight line one by one. Disrupt the trail by sprinkling dried mint leaves, crushed cloves or chilli powder at the point where they enter the house. Once they lose the scent trail they can no longer find their way in.
● Draw a line with a piece of chalk or baby powder through the ants' route. The tiny gatecrashers won't be able to cross it.
● Ants will gobble down bicarbonate of soda sprinkled on the floor and feed it to their young. This causes their stomachs to rupture and so reduces that pest population. They can't resist a solution of sugar, yeast and water, either, which has the same effect.
● Effective baits against ants also include honey, water and syrup. Set them out in a shallow dish; ants get trapped in the sticky solution and die.
● Prevent ants coming indoors – put tomato leaves and stalks over their nests.

FLIES

● The best solution is to install screen windows and doors and try to keep food covered.
● Set out bowls of vinegar (replace them daily).

Try driving mice away with chamomile flowers or peppermint.

AN ORANGE STUCK with cloves on the windowsill KEEPS FLIES AWAY

● Use blue tablecloths – flies avoid this colour.
● The smell of basil, peppermint, lavender or tomato plants also wards off flies.
● Coat meat with lemon balm or basil before grilling it and the flies will stay away.

FLEAS

● If fleas have taken up lodging in the sofa, sprinkle it with borax, leave overnight, and then vacuum.
● A dish of lemon slices placed in a cupboard will keep fleas away.
● If a rug is infested, sprinkle it with salt, let it work for a few hours, then vacuum.

WOODLICE

● Put a bottle of sweet liqueur or leftover wine outside the door. They will climb in and become intoxicated. Discard them and repeat as needed.
● Leave hollowed-out potatoes or turnips as a lure. Crush them along with the woodlice that have crawled inside.

MICE

● Peppermint oil, turpentine and camphor are scents mice find unappetising. Dab a little on cottonwool balls and place them where mice might enter.
● If you do decide to set traps, lure mice with peanut butter or chocolate.

MITES

● Mites dislike both fresh air and light, so shake out bedding and blankets regularly.
● Vacuum rugs or beat them outside, and wash them frequently.
● Replace carpets with wooden floorboards or tiles.

MOTHS

- Help drive away moths with dried citrus peel or cedar chips in small bags – they dislike the smell.
- Moths won't take up residence in fabrics that are clean and used frequently, so shake out and launder linens and clothing regularly.
- Store wool and cashmere clothes in sealed plastic bags immediately after wearing.

MOSQUITOES

- When abroad in hot countries, make sure there is a screen to keep mosquitoes out. A mosquito net will help to prevent them from biting you while you sleep.
- Keep the pests away from patios and balconies by hanging up a cloth sprayed with a few drops of clove or laurel oil. Alternatively, pour the oil into small bowls or an oil lamp.

CARPET beetles

- Vacuum regularly to stop hair and lint collecting and providing food for carpet beetle larvae.
- Seal cracks in parquet flooring and spray neem oil (available from health-product suppliers) along floor mouldings. This makes the larvae stop eating and prevents them from growing and reproducing. Be warned, it will take time.

COCKROACHES & weevils

- For a minor infestation, use a cloth moistened with wine or beer as bait. When the insects have gathered on the cloth, pour boiling water over it.
- Combat cockroaches with a lethal 'roach dinner'. Mix equal parts of powdered boric acid, white flour and white sugar and place in bowls under the fridge, in the backs of drawers and behind the cooker.

- Remove food supplies immediately if they become infested with weevils and wash out kitchen cabinets with vinegar and water.

SILVERFISH

- Sprinkle a little borax on damp cloths and place them in the bathroom or kitchen at night. In the morning, shake them and their load outside.
- Grate a potato on a piece of newspaper to attract silverfish, then fold the paper up and throw it away.

WASPS

- Wasps make themselves scarce when they detect the smell of heated vinegar. Lemon slices studded with cloves will also keep them away.
- Make a wasp trap by filling a narrow-necked bottle with diluted fruit juice and a little detergent and vinegar. They will fly in but can't get out.

BEDBUGS

- Try spraying surgical spirit where bedbugs thrive. It will kill some bugs on contact.
- Placing clothing, shoes and boots, toys, stuffed animals, backpacks and other non-washables in the dryer for 20 minutes or more on a high temperature may kill bedbugs.
- Wash and dry nightwear and bedclothes at as high a temperature as possible.

Keep pests away from patios and balconies by lighting a citronella candle

Pots and pans

Today, there is a wide selection of pots and pans manufactured from a variety of materials. A good set can last a lifetime, so take into account the advantages and disadvantages of each type when deciding what to buy.

It's important to remember that special care is needed to clean and maintain different types of cookware. Aluminium is lightweight, conducts heat well and is fairly inexpensive while the most popular form of cookware is probably stainless steel. This is inexpensive, long-lasting and resists wear and tear. Some traditional care methods will go a long way to protecting any set of pots and pans.

GENERAL care

● Avoid using metal implements when you can, or use them carefully as they can cause surface damage, particularly to nonstick pans.
● Wash dirty cookware with washing-up liquid and water, using a sponge or brush. For enamel and stainless steel, use a plastic scourer for heavy-duty cleaning or follow the manufacturer's instructions.
● Soften dried-on and burnt-on food remnants overnight with water and a little salt. Boil the mixture in the pan, let it cool, wipe it out and rinse.
● Take special care when cleaning cookware with wooden handles. Wood may swell and split if it is soaked.
● Don't use steel wool on cookware unless you are sure it can take it – some materials are more delicate than others.

Get better cooking results and save money in the long run by buying good-quality cookware.

STAINLESS-steel cookware

● Rinse out immediately after cooking as salt can attack the surface.
● Stainless steel shines like new after rubbing with a moist cloth sprinkled with bicarbonate of soda.
● Rub stains and small scratches off cookware with a paste made from 1 tablespoon of washing-up liquid and 1 tablespoon of bicarbonate of soda. Rinse it off with cold water.
● To clean off burnt food, fill the pan with water and leave for an hour. Cut up an onion and place it in the pan with enough cold water to cover the burnt area. Add 1 tablespoon of salt and boil for 10 minutes. The burnt residue will dissolve and can be removed with a dishcloth. Boil for longer if the stain remains.

CAST-IRON pots & pans

● Before you use cast-iron cookware for the first time, clean pots and dishes, rub in some vegetable oil and heat them in the oven on the top rack. For

GOOD TO KNOW ✓

Avoid burning foods

Prevent food from catching on the bottom of pots and pans as it cooks by lightly coating them with oil and salt after each washing and wiping them with a paper towel. Foods won't stick so much and subsequent washing will be quicker and easier.

cast-iron saucepans, add 1-2 tablespoons of oil and heat them on the element until the oil smokes. Wipe them out with paper towels when cold.

● After use, wash pots and pans in warm dishwater, dry and rub with cooking oil.

● Store cookware in a dry place so it doesn't rust.

ALUMINIUM pots

● Never wash aluminium pots in the dishwasher or with abrasive cleaners as you risk discolouring or scratching the finish. Scrub lightly with a little soapy water and a scourer.

● To restore an aluminium pan's shine, boil rhubarb in it for a few minutes. But don't eat the fruit – aluminium forms toxic compounds with fruit acids, so you should not use aluminium pans for cooking fruit or vegetables unless they have been treated with a non-stick or similar coating.

● Don't soak aluminium pots for long and avoid storing food in them as aluminium has a tendency to discolour.

ENAMEL pots

● Before you use enamel pots for the first time, they should be boiled for an hour with a vinegar-salt solution (50g salt, 2 tablespoons vinegar, 1 litre water) to increase their durability. Faulty enamel coating allows harmful heavy metals to leach into food, so throw them away if they are chipped or damaged, or recycle them as plant pots.

● Pouring cold water into hot enamel cookware will cause the enamel to crack.

● Coat any stains on enamel pots with a paste of bicarbonate of soda and water. Let it sit for an hour, then boil for 20 minutes (leaving the paste in the pot) and wash with hot, soapy water. Boiling orange and/or lemon peel for 20 minutes in a pot three-quarters full of water also removes stains. Finish by rinsing thoroughly.

COPPER cookware

● Immerse new pots and pans in boiling water and allow to cool.

● Rub any tarnished spots until shiny with half a lemon and a little salt.

● Try a paste of vinegar and salt to make copper shiny. Let it work for 30 minutes, wash off with cold water and then dry with a chamois leather.

Removing rust from cast iron

If, despite your careful treatment, rust spots form on cast-iron cookware, use a lemon solution to remove them. Mix together:

1 tablespoon lemon juice or citric acid
500ml water

Mix together the lemon juice/citric acid and water, then dip a brush in the solution and coat the rust spots. When the rust has been removed, maintain the cookware as usual.

COATED pots & pans

● Wash with warm water and rub sparingly with vegetable oil.

● Remove dried-on remains from pots and pans by boiling 3 tablespoons of bicarbonate of soda in 150ml water in the pan. Pour out the liquid and wipe out the bicarbonate of soda residue along with the leftover food.

● Place paper towels between pans when you store them to avoid scratching them as you take them out or put them back.

TRY A PASTE OF VINEGAR & SALT to make COPPER SHINY

Sewing and mending

Do you have any missing shirt buttons, trousers that need a patch or socks that could do with being darned? Long considered a craft, sewing is a skill that everyone should acquire. It's easy to learn and will help to extend the life of your clothing.

KEEP HANDY sewing and darning NEEDLES IN A PINCUSHION

THE sewing basket

You don't need a large, well-stocked basket for simple clothes mending. Basic sewing items will fit into a shoe box or biscuit tin. This is what you will need.

● Keep sewing and darning needles of various sizes in a pincushion or piece of soap. It coats needles, helping them to pass easily through thick fabrics.
● Try using a magnet as a pincushion; you can also use it to gather up lost needles and spilt pins.
● Store pins and safety pins in small matchboxes.
● Keep a selection of silk and cotton sewing thread, plus darning thread in common colours, as well as some yarn for thick fabrics.
● Add a thimble, darning mushroom, stitch unpicker, tape measure, scissors and perhaps a needle threader to complete the sewing kit.
● String buttons together by colour and size to save time and prevent them getting muddled at the bottom of the sewing basket. Store extra ones for specific articles of clothing in transparent, labelled bags.

SEWING on patches

Elbows, knees and the crotch area are usually the first areas to require a repair. Here are a few things to consider when sewing on patches.
● Hand stitch iron-on patches along the edges when patching stretchy fabrics.
● Sew together holes in knitted items before applying a patch to ensure they don't get larger.
● Sew on a patch backing made from previously discarded denim when patching a favourite pair of well-worn jeans.
● Use a sewing machine to attach patches to tough fabrics such as denim.

BUTTONS & buttonholes

● Before you lose a loose button and have to find a matching replacement, sew the original on properly.
● Avoid damaging a favourite blouse when cutting off buttons by inserting a comb between the fabric and the button before snipping the threads.

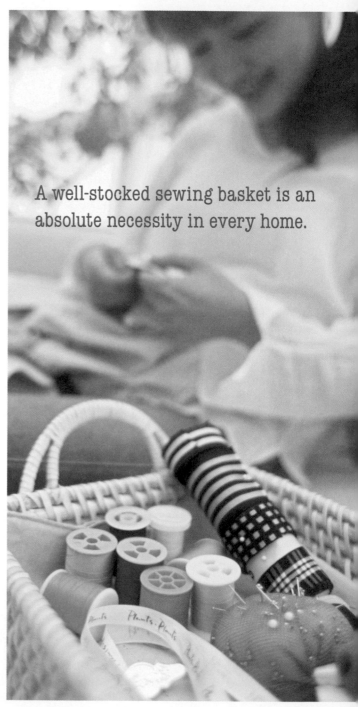

A well-stocked sewing basket is an absolute necessity in every home.

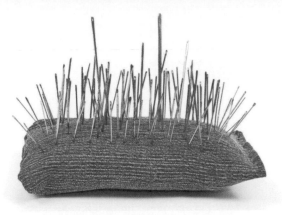

- Stiffen thread by rubbing it with candle wax, which makes it easier to thread onto a needle.
- Sew on a button with four holes in a crisscross shape so if one thread breaks, the second keeps the button in place.
- Use a matchstick as a spacer between button and fabric. Once you've fastened the button, remove the matchstick. Then secure the button with multiple strands to reinforce it and finish with a couple of extra stitches on the back of your fabric.
- Avoid tearing out fabric on the button facing, especially on coats, by sewing an additional smaller flat button onto the inside when sewing on a button.
- Sew a frayed or oversized buttonhole with a couple of horizontal stitches at the upper and lower edges of the hole so that the head of the button just fits. Sew up the ends of the thread on the wrong side.

REPAIRING a seam

This is how to repair a small gap in a seam:
1 Press the folded seam and secure it with pins.
2 Sew the folded inner edge of the seam in a cross-stitch pattern, from the back side to the top of the face fabric.
3 Keep stitches small and neat to ensure that they are not visible on the outside of the fabric.

4 Avoid pulling them too tight. This will keep the seam flexible and protect it from tearing out again.
5 Use an iron and damp cloth to steam the seam.

PROTECT hems from wear & tear

Iron-on selvage is quite stiff and may prevent trousers or a skirt from hanging properly. In order to protect hems, sew on a selvage band as follows:
1 Wearing your regular shoes as you try on the skirt or trousers you want to adjust, mark the new hem with chalk or straight pins, then lightly iron in place.
2 Use a sewing machine to sew on the selvage band along the ironed edge so that about 1mm protrudes beyond the hem.
3 Hand-sew the hem like any other seam.

QUICK help

These extra sewing tips could come in handy.
- Spray the end of thread with hairspray so it passes through the eye of the needle easily.
- Stop runs in knitted fabric by adding a little clear nail polish or glue. But bear in mind that you won't be able to sew them later, so use this technique only when you have no other option.
- Rub zips that stick with beeswax.
- Replace elastic in waistbands or elsewhere by fastening the new elastic to the old with a safety pin. As you pull the old one out, the new one is pulled in.
- Rub thread with beeswax to make it more durable.
- If you don't have the right-coloured thread choose a shade darker than the original, never lighter.
- Use new shoelaces to make loops to hang up heavy jackets. Cut them to the appropriate length and then attach them with strong thread inside the collar.

How to darn

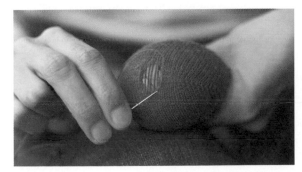

1 Stitch over then under the hole in the first undamaged row to form parallel threads lengthwise.

2 Weave parallel threads running crosswise through the lengthwise threads until the hole is filled.

Silver, brass and pewter

Silverware is treasured over generations, but with age comes tarnish. Pewter and brass items, too, lose their lustre. Thankfully, a little traditional know-how and a bit of tender loving care can help silver to regain its shine and breathe new life into other metal objects, utensils, fixtures and fittings.

SILVER storage

Storing and cleaning silverware correctly is key to maintaining treasured belongings.

- Always dry silver thoroughly before storing it. For silver place settings and other silverware, a drawer lined with velvet that closes tightly is a good choice.
- Buy rolls with felt pockets for storing silverware. Felt also prevents silver from tarnishing quickly.
- Keep silverware in tightly sealed plastic bags or wipe on a very thin layer of glycerin to prevent tarnishing. Store it in a place that isn't subject to major temperature fluctuations.

- Tuck a sachet of activated charcoal into the plastic bags to act as an air filter – look for it at a local pet store. Replace the sachet once a year.
- Silver is a soft metal easily scratched by harder steel, so store it separately.

FIRST aid for tarnished silver

- When silver shows the first traces of tarnish, coat it with toothpaste, work it up into a lather under warm water then rinse it off.
- Treat minor spots of tarnish with a paste of cornflour or bicarbonate of soda and water. Let the paste work for a few minutes, then rinse it off and wipe dry with a soft cloth.

SILVER place settings & silverware

Silver tarnishes upon contact with egg, broccoli and fish, so wash it immediately after use.

- Egg spots on silver vanish quickly when you rub them with damp salt or a sprinkling of bicarbonate of soda. Then rinse thoroughly with tap water.
- Remove spots on silver coffeepots with a little vinegar on a soft cloth.
- To clean a silver teapot, fill it with boiling water and add a little bicarbonate of soda. Leave it to work overnight, then rinse.

DECORATIVE items of silver

- Clean embossed or detailed silver pieces with lukewarm soapy water and a soft brush. But don't rub hallmarks too hard or you could obliterate them.
- Wipe silver pieces with intentionally darkened decorations with a cloth for polishing silver – that way the darkened indentations remain dark.
- Use a jeweller's rouge cloth for buffing up silver.

BRASS

Brass, which is an alloy of copper and zinc, is characterised by its yellow-gold colour and shine. Keep it looking gleaming with a variety of these traditional cleaning methods.

Fine ornamentation and edges are best cleaned with toothpaste and a soft cloth.

- To remove smears, wipe brass items with a damp chamois leather. To clean them thoroughly, use a vinegar paste. Mix together equal parts of vinegar, salt and white flour, and spread it on the brass. Leave for a while, wash it off and then buff.
- Where items have corroded, wipe this away with a mixture of 500ml buttermilk and 1 tablespoon salt. Rub it repeatedly and forcefully over the corroded spots until they disappear, then rinse and polish with a soft cloth.
- Create a beautiful shine by rubbing brass with the cut surface of a potato, then buff.
- To remove green corrosion or verdigris, soak pieces for several hours in a strong solution of washing soda, rubbing occasionally with an old cloth and using an old toothbrush to get into any crevices. Alternatively, rub with a cut lemon dipped in salt. Rinse well and dry.

PEWTER

Pewter is a malleable metal alloy with a shiny silver-grey surface that scratches easily. Here are some traditional tips for keeping your pewter pieces in perfect condition.
- Rub pewter pieces with the outer leaves taken from a head of cabbage and then rinse with clear water. Finally, treat the surface with a piece of cut leek and rinse again.
- Remove ingrained dirt by immersing a pewter object in a glass or bowl of warm beer and then scrubbing it with a soft nailbrush.

Make your own silver dip

one Half fill a sink with warm water.

two Put a piece of aluminium foil in the sink along with 2 tablespoons of salt and 3 tablespoons of bicarbonate of soda.

three Dip your silver in the solution. Most tarnish will simply slide off, but you may have to soak some stains for 5 minutes or so.

four If stains remain, remove silver from dip, clean with soap and a cloth, and insert again.

five Remove the silver and wipe with a soft cloth until shiny.

Remove tarnish and brighten brass fixtures by rubbing them with a cloth moistened in olive oil after each polishing.

- Common horsetail (also called scouring rush) is an effective cleaner for pewter. It can be found in fields and by the sides of roads. Mix 1 handful of the crumbled 'needles' and stem with 1 teaspoon of vinegar. Rub onto the pewter and leave overnight. Then rinse and polish with a soft cloth.
- Clean wax stuck to pewter candlesticks by placing them in the freezer and then carefully flaking off the wax. If any wax remains, melt it with a hair dryer set to hot and carefully wipe away any excess.
- Rubbing pewter with a paste made from cold white fire ashes or barbecue ashes, mixed with a little water, will give it a sheen.

Stain removal

We've all been irritated by a mark on a new tablecloth or shirt, but don't despair – milk, fruit, coffee and other typical stains may seem disastrous but with a bit of know-how they can often be remedied quickly and easily.

BASIC rules

Not all stains should be treated the same way. As a general rule, the fresher and damper the spot, the easier it will be to remove.

● Every stain remover – whether shop bought or homemade – should be tested on an inconspicuous part of the fabric first. If it has no adverse effects on the fabric or colour, use it directly on the stain. Dab it on undiluted and wash the garment afterwards.

● Always try the gentlest treatment first before resorting to stronger remedies. Take special care with delicate fabrics – some of these remedies are strong (especially those that use ammonia).

● When removing a stain, work from the outside towards the middle. Avoid using hot water as you risk setting the stain (especially if you don't know what it is). Blow-dry the wet area to avoid leaving an outline.

● Water-soluble stains are the most common. All you'll need to remove them is tap water, at least when they are fresh. Treat protein-based stains (blood, mayonnaise or egg) with cold water.

● Treat older stains with a mixture of 2 tablespoons of water and 3 tablespoons of vinegar. Leave the mixture to dry and then rinse.

● If possible, scrape a dried stain with a spoon or soften it with glycerin before treating it.

THE ABCs of stains

Just about any substance from the refrigerator, and other foodstuffs, can stain. The garden and garage are also full of potential hazards.

● Remove beer stains with a dilute soap solution containing a little ammonia, then rinse with water.

● Wash bloodstains on clothing immediately in cold water – hot water will cause the protein in the blood to congeal and attach firmly to the fibres. For stubborn stains, moisten the clothing in cold salt water. Dry bloodstains should be soaked in cold water, then treated with salt water or a solution of bicarbonate of soda. When cleaning delicate fabrics, use a paste of water and potato flour or cornflour.

Test a hidden section of fabric first. If the colour is unaffected, spread it on the stain, leave it to work for a few minutes, rub it off and rinse thoroughly.

● Gently rinse burn marks on washable fabrics with cold water, sprinkle them with salt and then dry in the sun. Treat burn marks on delicate fabrics carefully with diluted vinegar. There is no guarantee that these remedies will work, and you need to be especially careful with these marks as burns will weaken the fabric.

- Makeup, butter, mayonnaise, cooking oil and engine oil are among the substances that can leave grease stains. Promptly sprinkle them with cornflour to absorb the grease, then brush away the saturated starch. You can also try rubbing off the stains using hot water mixed with a little washing-up liquid to dissolve the grease. For delicate fabrics, place a paper towel on both sides of the stain and iron it. Stains are best removed from wool by rubbing them off with a little mineral water and a towelling cloth.
- Remove tough lipstick stains by dabbing them with eucalyptus oil, leaving it to soak in before washing. Boil white table napkins, handkerchiefs or facecloths marked by lipstick stains.
- Rub tar stains with lard before washing the item. For an extra boost, add 2 tablespoons of bicarbonate of soda to the laundry detergent. Oil, tar and grass stains can also be treated with a few drops of eucalyptus oil.
- Treat fresh grass stains with ammonia, but first test the sensitivity of the fabric on an inconspicuous spot. Alternatively, apply a halved potato to the grass stain to allow the starch to dissolve the stain, then wash as usual. Soak older grass stains on white fabrics with a mixture of one part egg white and one part glycerin before washing.

For dry, caked-on mud, shake the garment outdoors and rub off as much dirt as possible. Then soak in cold water before washing with detergent and warm water.

- Soak coffee and cocoa stains in cold salt water while fresh, then wash with a detergent containing enzymes. Dab older stains with glycerin and wash them out or, for upholstery and rugs, pat them dry.
- Remove chewing gum by putting the affected clothing into a plastic bag in the freezer. Once the gum is frozen, it is easier to chip off.
- For clear glue spills, try cologne or oil-free nail polish remover; in other cases, white spirit, surgical spirit or lighter fluid may do the trick.
- Rinse milk spots with cold water before washing. Dab non-washable fabrics with cold water then ammonia and finally with warm water.
- Treat fruit stains while fresh by holding the soiled item over a bowl and pouring a little very hot water onto it. Alternatively, soak it in buttermilk and wash

Universal stain remover

250ml surgical spirit
100ml ammonia
2 teaspoons lighter fluid

Mix the ingredients together in a sealable bottle, then test on an inconspicuous area of fabric before dabbing it onto the stain undiluted. Be especially careful with delicate fabrics. Wash the garment afterwards.

Keep out of the reach of children.

as usual. For dried fruit stains, sprinkle with lemon juice and rinse after 30 minutes. If the stain still doesn't come out, try treating it with an ammonia solution (2 tablespoons ammonia in 1 litre water) or glycerin solution (equal parts glycerin and water).
- Apply lemon juice, vinegar or carbonated water to red wine stains, or pre-soak the stained item in a biological detergent for 30 minutes before washing.
- For red dye stains from ice lollies or maraschino cherries, mix equal parts hydrogen peroxide and cool water in a spray bottle; spray onto the stain and leave for 30 minutes. Rinse with equal parts vinegar and water. Peroxide is a bleach, so test a spot first. Repeat the treatment if it doesn't work first time.
- Treat sweat stains with a mix of 2 tablespoons vinegar to 3 tablespoons water, or use ammonia solution.
- Scratch off wax spots then place a paper towel under and over the spot and iron until all excess wax is absorbed. If necessary, replace the paper towel. Remove any remaining stain from coloured wax by dabbing it with surgical spirit, always rubbing from the outside in.

TRADITIONAL CLEANERS

There's a vast array of synthetic cleaning products available in supermarkets today, some with hefty price tags. Many were originally modelled on existing natural cleaning products, all of which can still be used to help keep your home clean and germ-free.

Most of the cleaning materials mentioned in this book can be bought at the local chemist or supermarket, and you can easily mix up the solutions yourself. But wear gloves and a face mask when working with ammonia, talcum powder or turpentine products. Avoid swallowing, breathing or absorbing them through your skin, and store your homemade cleaning materials in well-labelled containers out of reach of children (just as you would shop-bought products). It is always a good idea to check their shelf life regularly.

Vinegar

The word vinegar comes from the Old French *vin aigre*, or 'sour wine'. The substance – made by fermenting ethanol – has been used for seasoning food since earliest times. In fact, traces of it have been found in 3,000-year-old Egyptian urns. But vinegar doesn't just liven up your salad, it is an excellent natural cleaning product, disinfectant and deodoriser. To make an all-purpose cleaner for tables, work tops, baths and tiles simply mix equal parts vinegar and water in a spray bottle, or pour the substance directly into the toilet bowl to remove discoloration. Just be sure to test it on an inconspicuous area before use. Vinegar should never be used on marble and, when improperly diluted, it may eat away at tile grout. But for the most part, vinegar is effective, accessible and cheap.

Glycerin

This colourless, odourless, rather thick non-toxic alcohol is also known as glycerol. Glycerin is commonly used in pharmaceutical and personal care products like cosmetics, soaps and toothpaste, as well as in certain food products. Glycerin, in its pure form, can be used to treat a number of minor medical conditions, including calluses, bedsores, rashes and cuts. But it is usually used at home to soften up tough, dried-on fabric stains from substances like coffee, berries and lipstick. It is also used as an antifreeze on windowpanes; rub windows with glycerin before the temperature drops and they won't freeze over.

Plain white toothpaste makes an easy and

Use traditional products to keep tennis shoes in match condition.

Ammonia

Ammonia, or ammonium chloride, is the ammonium salt of hydrochloric acid and a crystalline solid. It has been used in everything from fertiliser to rocket fuel. That may make it sound like a dangerous chemical but, in fact, ammonia has also been used for a long time as a household cleaner. Commercially, it's most commonly found as the watery solution ammonia (ammonium hydroxide). Use liquid ammonia at home for stain removal or when cleaning stainless steel, glass and porcelain – it leaves a streak-free shine. It is also used to combat mould and pests.

Turpentine

Turpentine is a mixture of resin and essential oil from various species of pine. Turpentine oil, produced by distilling turpentine, is effective at dissolving grease. The colourless to yellowish fluid has numerous applications in the home, for example as a floor polish, shoe polish or as a solvent for stain removal. Don't dispose of turpentine products by pouring them down the drain, which is damaging to pipes and the environment. Ask your local authority about proper disposal methods.

Warning: exposure to turpentine through splashes, inhalation of fumes or swallowing can lead to serious health problems. Wear protective clothing, including a face mask, and ensure a room is well ventilated if you are using it. For many applications, it is cheaper and just as effective to use white spirit, which is derived from petroleum, to do the same jobs.

Talc

Also known as magnesium silicate hydrate, talc is the softest mineral. It feels soapy, which accounts for its alternative name: soapstone. It is ground up to form talcum powder or a base for make-up. In the home, it can be used as a gentle scrubbing agent and to treat rubber seals or to silence squeaky wooden floorboards and stairs. But be careful not to breathe it in as the substance can cause serious inflammation in your breathing passages.

economical silver polish.

Upholstery

An upholstered seat in the living room is likely to get slept on, jumped on and have things spilt on it. Given the amount of abuse it takes, the maintenance of its upholstery and fabric is important to ensure it looks good and lasts.

It is not surprising that upholstered furniture requires regular care, considering how many hours we spend on the sofa gazing at the television, reading a book or socialising.

USE a CREVICE NOZZLE to CLEAN BETWEEN the seat and arm of the SOFA

MAINTAINING upholstery

- Clean upholstered furniture regularly with the vacuum cleaner, but reduce the suction to avoid damaging the underpadding.
- If possible beat out the dust outside to prevent it from spreading through the room and settling on other furniture.
- Before beating large, upholstered furniture, cover it with a damp cloth to catch the dust being released. If you moisten the cloth with vinegar and water, the colours will look fresher afterwards.
- If chairs have sagged or settees have visible indents where people have been sitting, moisten these pressure points with a little hot water, cover with white paper and iron dry. But be careful not to burn these spots.
- Clean dirty sofa cushions every year with a vinegar solution made from one part vinegar and one part water. Apply it with a cloth and then wipe it off with tap water.
- Machine wash or dry clean removable covers when necessary. Observe the care directions: brocade, silk, chintz, velvet and wool tweed, for instance, must be dry cleaned.
- After washing cushion covers, iron them from the inside and put them back on the cushions while still damp; they will stretch better and dry without wrinkling.
- Rub soiled spots on wool or linen covers with a soft rubber.
- Clean dark velvet upholstery covers with a brush moistened in cold coffee. Then moisten a cloth with tap water and pat the velvet to pick up any excess.
- Clean synthetic covers by dipping a cloth dampened with water and a little bicarbonate of soda and gently rubbing the cushion with it. Go over it again with a water and soap solution. Test this on the reverse side first (or a corner) to make sure that it doesn't leave a mark.
- After using water on fabric upholstery, cover the area with paper towels, weight them down and leave overnight to dry.
- Remove lint and pet hair with a damp sponge or a piece of Sellotape wrapped around your hand with the sticky side out.
- Try to train pets from a young age to keep away from upholstery. Leather is least vulnerable to pet damage. Keep cats from scratching upholstery by providing a scratching post. If that doesn't work, cover vulnerable areas with plastic or attach pieces of old carpet. They particularly hate bubble wrap.

REMOVING stains from upholstery

- Remove residue and vacuum up spills at once. Remove stains by working from the outside towards the middle to avoid leaving an outline.
- Sprinkle fresh grease and oil stains with talcum powder or cornflour. Leave it to set and absorb the grease, then brush it off.
- Dab older grease stains with an ammonia solution, surgical spirit or cologne, then carefully rub with water.

● Treat milk spots immediately with cold water or a lather of moisturising soap and luke-warm water. To finish, pat dry.

MAINTAINING leather upholstery

● Clean washable leather with a soap solution (1 teaspoon liquid soap in 1 litre of water). Wring out the cloth thoroughly before wiping the leather. Allow the furniture to dry, then buff.
● Maintain dark leather by rubbing castor oil into it once or twice a year.
● Treat light-coloured leather with petroleum jelly. Leave it to work for about an hour before removing the excess with a soft cloth.
● Restore the shine to scuffed leather upholstery by treating it with a mixture of equal parts beaten egg white and linseed oil.
● For a natural leather polish boil some linseed oil, let it cool and mix with an equal amount of vinegar. Apply with a soft cloth and buff.
● Special nourishing creams for older leather can be purchased at furniture stores or online. Allow them to work their magic for about 24 hours after application, then buff. But make sure you wipe them well so nothing rubs off on clothing.

REMOVING stains from smooth leather

● Remove water-soluble stains with a damp cloth and moisturising soap foam; wipe with warm water.
● To treat older grease stains on colourfast leather dip a cloth into hot water, wring it out, sprinkle on a small amount of bicarbonate of soda, then carefully

Revitalising upholstery

25g soap flakes
500ml water
100ml glycerin
100ml surgical spirit

Heat the water and dissolve the soap flakes. Let it cool, then stir in the glycerin and surgical spirit. Store the mixture in a container. When needed, dissolve 1 tablespoon of the mixture in water, stir with a whisk and apply the foam with a sponge.

rub the grease from the edge towards the middle. Go over it with a warm, moist cloth.
● For stubborn dirt or stains, work up a foam with saddle soap and rub it in using a sponge in circular motions. Go over the spot again with tap water and apply leather conditioner after it is dry.
● Pour a little milk on a clean cloth and dab it on leather to help remove marks from a ballpoint pen.

REMOVING stains from full-grain leather

● Sprinkle fresh grease stains with talcum powder. The powder will absorb the grease, then you can simply brush it off. Repeat as needed.
● Use a rubber to remove grease stains. But don't rub for too long or you risk damaging the leather.
● Remove water stains by allowing them to dry, then roughening them with a brush.

Dab a little water on an inconspicuous area first before using it to clean leather. If the water soaks in, then don't use it to clean your furniture.

Walls

Regular maintenance keeps walls and ceilings clean and makes rooms appear light and airy. Whether walls are painted, panelled, brick, rendered or wallpapered, keep them in good condition to avoid unnecessary redecoration and to ensure every room in your home looks inviting.

USE A SOFT paintbrush to CLEAN CORNERS and CURVES.

Keeping walls and ceilings dust and cobweb-free should be part of the regular housekeeping routine. However, certain types of surface will require more detailed care. Protect the floor before you embark on any drastic wall cleaning.

MAINTENANCE routine

- Dust walls and ceilings regularly. If you don't have a feather duster, simply tie a duster to a broom.
- Suck up cobwebs with the crevice nozzle on the vacuum cleaner.
- Test all cleaning products on an inconspicuous spot before removing stains. If the product stains the wall, at least it will be hidden from view.

Spray a feather duster with water so cobwebs stick to it.

- If you plan to put up wallpaper, go for the washable, scuff-resistant variety. This will make cleaning much easier and will cut down on time spent on maintenance.
- Wash walls using a damp, soft sponge and an all-purpose wall cleaner, working in gentle, circular strokes.
- If you have to wet-wash the walls, be extra cautious and switch off the electricity.

BRICK

- Brush brick walls occasionally with a scrubbing brush and sweep up loosened grout.
- Do not wet-wash brick walls – the moisture will soak into the porous masonry and could cause mildew or other damage.

PLASTER & render

- Dust plaster mouldings and decorations regularly, preferably with a feather duster.
- Wet-wash render only when necessary. First check that the finish is solid enough, then spray it with a little soap and water so you are able to reach even the cracks. Finally, spray with a small amount of water and dab up all liquid with a dry cloth.

PAINTED surfaces

- Remove dirt or stains on walls painted with oil-based paint with soapy water.
- Wash walls painted with acrylic paint with soapy water containing a couple of dashes of ammonia. Wipe with water and pat dry. A soft scrubbing sponge is helpful around light switches (but remember to turn the power off first).
- Never wet-wash whitewashed walls as it will take the colour off.
- Use a rubber or a fresh piece of white bread to remove new stains from painted walls.

WALLPAPER

● In heavy traffic areas such as the kitchen or bathroom, choose a washable wallpaper.

● If a section of wallpaper is very dirty, it might be easier simply to remove it and replace it with a new piece. Make sure the remaining wallpaper and the patches abut each other, rather than overlapping at the edges.

● A trick for repairing wallpaper is to tear rather than cut the edge of the patch so that it blends in seamlessly with the 'background'.

● Dust textured vinyl wallpaper before cleaning, then wipe it with a damp cloth, sponge or soft brush and warm detergent solution. Rinse it with water and dry thoroughly, but don't let it get too moist.

● Dust grease stains on textured wallpaper with talcum powder. Allow it to work and then brush it off. Wipe off any excess powder with a damp cloth or suck it up with a quick vacuum.

● Never soak fabric or cork wallpaper as they will swell. Instead, carefully wipe them off with a damp cloth. If too much water is applied, pat dry with a clean cloth.

● Treat cork wall tiles like washable wallpaper if they are sealed with a matt varnish. This is one way to avoid the potential problem of soaking the cork. If the tiles are not sealed, it is worth taking the time to varnish them.

● Carefully vacuum hessian wallpaper on low power and avoid wet-washing it.

PANELLING

● Don't ever let waxed surfaces become too wet when you are cleaning them. Treat them as you would waxed floors or furniture.

● Wipe sealed, painted or varnished panelling with soapy water. Avoid using an abrasive powder that could leave scratches.

● Treat wood coated with clear lacquer once a year with furniture polish.

● If minor mould spots appear on wood panels, dry them with a hair dryer and scrub with a soft brush. Finally, polish them with a soft cloth and furniture polish. Note that mould can be dangerous if it grows out of control. Make sure that the room is well ventilated and be sure to wear a protective face mask. If you are ever in doubt about the severity of the mould, call in a professional to deal with what could be a serious problem.

● Remove layers of old polish with a coarse sponge and white spirit. Rub with the grain of the wood.

Removing grease stains from wallpaper

one Place an absorbent paper towel on the grease stain or wax crayon scribble.

two Go over it with a not-too-hot iron for as long as it takes for the spot to disappear. Move the paper around occasionally and replace it if necessary.

three Try dabbing any remaining stain carefully with some bicarbonate of soda on a clean cloth.

Windows

You don't need special cleaning products to keep your windows clean and streak-free, or your sills and frames looking as if they have just been installed. With the proper technique and correct accessories this housekeeping chore can be simple and rewarding.

ADD half a raw ONION to WASH WATER for GLEAMING WINDOWS.

Before window cleaning begins, a little preparation is needed to help make the job easier. First, remove the curtains from the windows and any knick-knacks from the windowsills, then cover sensitive surfaces to protect them from drips. Do yourself a favour: buy a real chamois leather. It absorbs water more quickly and is easily squeezed out. With proper care it will last for years.

GOOD TO KNOW ☑

Adverse weather conditions

Avoid washing windows in the sunshine as the water dries too quickly on the panes and causes streaking. It also makes it hard to see if the glass is clean. Cold days aren't ideal, either, because the glass is likely to be brittle. If you must wash windows in cold weather, add a dash or two of surgical spirit to the wash water.

LONG-LASTING chamois

● Keep your chamois leather soft and smooth by using it only with water or solutions made with water and vinegar or alcohol. Detergents remove the leather's oils and leave it stiff.
● Rinse out a chamois with warm salt water after every use to keep it soft.
● Never wring out a chamois leather. Instead, squeeze it gently, open it up, shake it out and let it dry slowly in the air.

WINDOWSILLS

● Apply a soft soap solution on windowsills to battle everyday dirt.
● Remove water stains with a soft cloth moistened with a solution of equal parts surgical spirit and tap water.

WINDOW frames

● Vacuum the window frame joint with the appropriate nozzle on the vacuum cleaner before starting to clean with liquid.
● Wipe wood window frames treated with clear glaze or varnish with a damp cloth. Replace the water frequently and to get them extra clean rub them with a solution of equal parts surgical spirit and tap water.
● Clean painted window frames with a solution of 2 tablespoons of ammonia in about 500ml of water.
● Clean wooden window frames with a barely wet cloth then dry with a clean, soft cloth.
● Rub off specks on wood frames using a rough cloth moistened with water. Another option is to clean them with a mixture of reduced-fat milk and cold water in equal proportions.
● Stick with hot, soapy water for washing aluminium or plastic frames. Scouring powder will scratch them.

WINDOWPANES

● Keep them free of condensation, which attracts mould and can make wooden frames rot easily. Wipe them every morning if necessary.
● Clean dirty panes regularly with a vinegar solution of 1 litre warm water and 250ml white vinegar. The vinegar dissolves the dirt and helps keep flies away.

● Clean extremely dirty windowpanes first with warm dishwater, then wipe them down with tap water, making sure you don't leave streaks.

● Add a squirt of glycerin to the water to help windows resist dust and to ensure they don't fog up in winter.

● Mix a few drops of ammonia with the water to keep the frost off windows.

● Expel streaks and leftover drips with a lint-free cloth, newspaper or chamois leather.

● When squeegeeing windows, moisten the edge of the squeegee to keep it from squeaking and also to improve contact with the glass.

● Polish washed windowpanes with newspaper or an old pair of tights to bring out the shine.

● Clean small windowpanes (skylights, louvre windows, transom windows) with a chamois leather only. Thoroughly wet the chamois in the wash water, squeeze it out gently and work from the edge towards the centre of the glass. Immediately wipe dry to prevent streaking.

● Wash skylights when it's raining hard for maximum effect. Carefully tilt the wet window and lather it with wash water, then shut it again and let the rain rinse it off.

● Rub dulled windows or mirrors with olive or linseed oil to get their shine back. Leave the oil on for an hour, wipe dry with tissue paper then clean as normal.

● Eliminate grease stains with half an onion. The sulfides in the onion are powerful cleaners.

● Remove specks from glass panes with a clean cloth moistened in warm black tea. A couple of squirts of surgical spirit will also dissolve them, making it quick and easy to wipe them off.

● If glazing compound has been left on the glass from when it was installed, rub it off with a little ammonia or white spirit on a cottonwool ball.

● Scrape stickers off new panes by soaking them with warm water and then removing them using a glass scraper. If any old, brittle stickers remain, rub them with olive oil until the pane is clear.

● Remove fresh paint splashes with white spirit or nail polish remover. For dried paint, use a glass scraper, such as those for cleaning ceramic hobs.

SPECIAL cases

● Modern stained glass is so strong that it can be cleaned as you would normal glass panes. Take greater care with old stained glass: simply wipe it carefully with a damp cloth. If the old glass is actually painted, don't wash it at all or you risk removing the colour. Instead, dust the panes with a soft brush.

● Clean frosted glass with hot vinegar water to give it a dull sheen, then carefully wipe dry.

● Dust etched glass with a soft brush, and clean the textured side of the glass with a chamois leather. You can simply squeegee the smooth side.

When windowpanes are especially dirty, lather them generously with a water and detergent solution and squeegee them free of streaks.

Home cooking

Traditional ways of preparing foods – such as drying or pickling, making jellies or liqueurs – have come back into fashion as people discover how economical they are and how delicious and nutritious the results can be. Discover how they work in the modern kitchen.

Baking, biscuits and cakes

If you are wary of attempting your grandma's favourite recipes because of fear of failure, just follow these tried-and-tested tips. Before you begin, read the recipe and preheat the oven in good time. Measure ingredients exactly and follow directions carefully. With a little effort, you can turn out delicious baked goodies every time.

Once you master the art of making batter and pastry, and basic biscuit and cake-making techniques, there is no limit to what can be prepared in the kitchen.

Making shortbread

Butter is an essential ingredient of shortbread, a Scottish speciality, while rice flour or semolina gives it a delicious crunchy texture.

100g plain flour
50g rice flour or semolina
50g caster sugar, plus extra for dusting
100g salted butter, refrigerated

1 Sift the flours into a bowl, add the sugar and grate in the butter using a coarse grater. Rub in the butter until it resembles breadcrumbs. Press into an 18cm straight-sided sandwich tin and level the top. Prick all over with a fork.
2 Chill in the fridge for 1 hour. Heat the oven to 150°C/gas mark 2 and bake for 1 hour or until straw coloured. Cool in the tin for 10 minutes then turn out onto a wire rack. While still warm, mark into 8 wedges. Dredge with caster sugar and separate before serving.

Don't overwork or overcook the dough or it will become tough and brittle. When cool, store in an airtight tin.

TIPS on batter & dough

● Partly replace butter with a vegetable oil such as corn oil, or add 1 tablespoon vinegar to cake batter to make cakes especially light. To obtain the same result, you can substitute mineral water for half of the full-fat milk.

● Before baking, prick shortcrust and puff pastry several times with a fork to prevent bubbles forming.
● Shortcrust pastry should be smooth after kneading; if it crumbles, it is too dry.
● Don't open the oven door during the first 20 minutes of cooking or the cake may sink.
● To help a cake to rise, place a heat-resistant container of water in the oven with it.
● Insert a thin skewer or cocktail stick to test if a cake is done. If no batter clings to it, the cake is ready.
● To prevent cakes from collapsing, leave them in their tins for a few minutes to cool then turn out onto a wire rack.

FATLESS sponge cake

When making a fatless sponge, always ensure the sugar is thoroughly dissolved. Nowadays, an electric mixer makes light work of this task.
1 Separate four eggs. Beat together 125g caster sugar with the egg yolks until light, then add 4 tablespoons water and 1 teaspoon grated lemon zest.
2 Sift together 125g plain flour and 1 tablespoon baking powder. Add to the batter.
3 Beat the egg whites until stiff peaks form, then fold gently into the batter.
4 Put the mix into a greased cake tin and fill it no more than two-thirds full. Bake in an oven preheated to 160°C/gas mark 3 for 30 minutes.

CLASSIC baking-powder biscuits

These simple biscuits take just minutes to make and don't require extra time for leavening. Eat them warm from the oven.
1 Sift 250g plain flour then add 2 teaspoons baking powder and ½ teaspoon salt and sift again.
2 Cut in 140g butter or margarine with a pastry cutter (or two knives) until you have pea-sized crumbs.
3 Gradually add 175ml milk, stirring until a loose dough forms.
4 Turn out onto a lightly floured board and knead lightly for about 30 seconds.

5 Roll out your dough to about 2.5cm thick and cut out shapes with a 5cm cookie cutter.

6 Bake on an ungreased baking sheet in an oven preheated to 200°C/gas mark 6 for 12-15 minutes.

7 Leave them on the baking sheet for a couple of minutes, then use a palette knife to transfer them to a wire rack to cool.

● For biscuits containing little fat you should grease the baking sheet first to prevent sticking. Alternatively use a nonstick sheet or a lining of baking parchment to prevent them from sticking

● Biscuits scorch easily, so you should check them 5 minutes before the recommended baking time is up. And you don't want to overbake them as they will crisp up after they are removed from the oven, .

KEEPING cakes, breads & biscuits

● Most cakes, scones, muffins and teabreads can be frozen without any deterioration in quality, but they should be wrapped well to prevent moisture loss. However, many filling and toppings – including glacé icing – don't freeze well, so it is better to freeze a cake plain and then fill or decorate it after thawing.

● Freezing cooked biscuits is not advisable as they lose their crispness. But most biscuit doughs can be frozen raw, rolled and wrapped in cling film or baking parchment and overwrapped in foil. When you want to bake the biscuits, remove the dough from the freezer and thaw until it can be sliced. It will cook perfectly when only just defrosted.

Roll out your dough to about 2.5cm thick and cut out shapes with a cookie cutter.

TOSS chocolate chips IN FLOUR to prevent them from SINKING

Bread

For thousands of years, humans have prepared and cooked bread. There's nothing quite like a warm, fragrant loaf fresh from the oven. And it doesn't take many ingredients, or too much effort, to create delicious breads.

To make bread dough, in addition to flour or whole grains you need water, salt and a leavening agent such as yeast, sourdough or baking powder. Generally the more thoroughly you knead the dough, the looser and finer the crumb. Make sure you give dough enough time to rise undisturbed at an even temperature of about 22°C. The volume of the dough determines when the bread is ready to bake: it should roughly double in size.

BASIC bread

Bread dough isn't difficult to make. You can select from a wide range of flours and add seeds, raisins, nuts, sundried tomatoes, herbs or other flavourings. This easy-to-follow recipe makes two loaves.

1 Dissolve 15g fresh yeast, or 2 teaspoons dried yeast, plus 1 teaspoon sugar in 500ml lukewarm water and mash well with a fork. Leave to stand for 10 minutes or until frothy, then stir well. (Or use a 7g sachet of easy-blend/fast-action yeast which needs no pre-treatment; just follow the instructions).

2 Sift 750g white bread flour and 1 tablespoon salt into a large mixing bowl. Add the prepared yeast mixture, or the easy-blend yeast direct from the packet with the amount of water specified.

3 Mix with one hand, drawing in the flour to make a stiff paste. Add a little more flour if necessary.

4 Turn onto a floured board and knead for 10 minutes, flouring your hands as necessary, until smooth and elastic and with a slightly blistery surface. Give the dough a quarter turn with each hand movement as you knead, working rhythmically.

5 Lightly grease a bowl with vegetable oil, add the dough and cover with lightly oiled cling film. Leave in a warm place for 1½-2 hours or until it has doubled in volume.

6 Punch down and knead well again. Grease two 450g loaf tins. Divide the dough into two and roll each portion into a loaf shape to fit the tins. Put into oiled, sealed plastic bags and leave to rise (prove) for about 40 minutes at room temperature.

7 Heat the oven to 230°C/gas mark 8. When the bread has risen, brush with egg glaze if wished. Bake for 35 minutes.

8 When it is baked, the loaves will have shrunk from the sides of the tins and sound hollow when tapped on the base. Remove from the tins and place on a wire rack to cool completely.

Note Yeast dough cannot withstand draughts or cold, so keep doors and windows closed while it rises.

● The dough will keep in the refrigerator for a day. In the freezer, it will keep for up to five months.

Knead the dough carefully and thoroughly with the heel of your hand for a few minutes.

• For rolls: divide the dough into 16-20 pieces, shape and prove for 15-20 minutes. Glaze, then bake for 12-18 minutes at 220°C/gas mark 7.

TIME-SAVING tips

• Use a mixer with a dough hook (3 minutes on low speed) or mix the ingredients in a food processor – the ball produced will not require as much kneading.
• Use wholemeal flour – because it has less gluten it requires less kneading.
• Speed rising by using the microwave. After kneading, put the dough into an ungreased bowl and cover with lightly oiled cling film. Heat on high for 10 seconds then leave the bowl to stand for 20 minutes in the microwave or a warm place. If not doubled in size, heat on high for another 10 seconds, then leave for another 10 minutes.

SUCCESS with dough

• Avoid mixing salt and yeast on their own as the salt can stunt the activity of the yeast.
• You can add any number of flavourings: try caraway seeds, some chopped olives or sundried tomatoes, or mix in 2 tablespoons of sunflower seeds for a tasty change.
• Press your fingertip firmly on the dough. If it leaves an impression, it is risen and ready to bake.
• Prevent air bubbles from forming while the dough rises by punching it down vigorously before shaping it, then kneading it again carefully and thoroughly with the heel of your hand for a few minutes more.

• Use bicarbonate of soda or baking powder for leavening quick breads, which don't need to be left to rise. However, they must be baked in a loaf pan or the dough will spread all over the baking sheet.
• For sourdough bread, you will need to buy a pre-made sourdough starter or make one yourself to use as a leavening agent. To do this, measure out equal quantities of plain flour and warm water, seal the mixture in a jar and leave it to ferment for about a week.
• Bake dough within 3 hours or there is a risk that it will collapse.
• Form rolls by shaping the dough into a fairly thick rope, cutting it into pieces and rolling each one into a smooth ball with your cupped hand.
• Place a bowl of water in the oven during baking to ensure bread and rolls come out crisp (but not hard)
• Brush loaves or rolls with a little water or egg yolk for an attractive shine.
• Keep yeast breads for one to three days, as they taste best fresh.

Yeast dough

1 Add 500g sieved white bread flour to a bowl, and form a crater in the middle.

2 Mix 40g yeast and 250ml milk. Stir in 35g flour and a pinch of sugar. Pour mixture into the crater.

3 Cover with a tea towel and leave it in a warm place to rise for 30 minutes.

Cold storage

A cellar or pantry was once the only place to store fresh produce. We now have many more options at our disposal, but it's still best to buy fresh, seasonal fruit and vegetables – and some of yesterday's cold storage methods still work best.

Even if you don't grow your own produce, it's still a good idea to store fresh, local fruit and vegetables, and to take advantage of seasonal bargains offered on more exotic fruits. As well as saving space in the refrigerator, keeping fresh food in cold storage helps to ensure a healthy diet and acts as a back-up in case anything prevents you from buying fresh food at the supermarket.

ASK STORES for old cardboard FRUIT TRAYS – IDEAL FOR STORAGE.

THE best conditions

For storing fresh produce, all you need is a cool, dark, dry, well-ventilated storage room or a frost-free garage. The optimal ambient temperature for storage is 4-5°C.
● Before storing produce, wash your shelves with bicarbonate of soda and water or vinegar to stop mould from spreading.
● Store only undamaged fruit and vegetables. If you want to keep any type of cabbage in cold storage, be sure to remove all rotting leaves and any that appear unhealthy in order to avoid deterioration.

● Check fruit and vegetables regularly during storage. Remove damaged or decaying items immediately and clean the storage space if necessary. Remember that decay can spread quickly.

STORING fruit

With the exception of apples and pears, most fruit is not suitable for long winter storage as it will suffer cold damage if the temperature drops below 10°C.
● Place apples individually with stems up on wooden shelves, wrap them in tissue paper or store in wooden boxes. They also store well in large cardboard fruit trays. Make sure the fruits don't touch and that there are no rotten ones – one bad apple really will spoil the whole batch. Apples should last for about six months after the harvest depending on the variety.
● Treat pears in the same way. Store them while they are still firm. As soon as they yield to slight thumb pressure on the stem end or start to change colour, move them to a warmer place. They should reach full ripeness in a couple of days.

SQUIRREL away vegetables

● Store potatoes in a wooden box or plastic dustbin that is well-ventilated top and bottom, or in a wire basket. Keep them in the dark, as light will produce green spots on the potatoes that are harmful to health if eaten. Throw out any that are affected or have rotted. Potatoes can withstand temperatures above 2°C. If it gets colder, cover them with straw or hay. Potatoes can lose their vitamin C content if they are stored incorrectly.
● Cover carrots with damp sand in a box to keep them fresh and edible for longer.
● Hang squashes or marrows in a pair of old (clean) tights to lessen the strain of their own weight. Or place in big plastic trays, keeping them well apart.
● Tie cabbages together by the stems and hang them on wires or beams with the heads hanging downward if possible. They can also be stored on wooden shelves with good ventilation.
● Store parsnips and beetroot in a bed of sand on the floor.

Store carrots and other root vegetables in boxes filled with damp sand. Remove all leaves first.

● Tomatoes stay fresh for longer if placed separately, without touching one another, on boards on the floor. In the autumn you can even pull up green tomatoes along with the stalks, tie the roots together and hang them in a cool, well-ventilated area. In time, you'll have vine-ripened tomatoes.

STORAGE in the garden

If you are short of indoor storage space and you don't have a garage, store produce in the garden.
1 Carefully place small amounts of vegetables or fruit in wooden boxes and bury them in a pit in the garden.
2 Line the box with fine-mesh wire to keep mice and other creatures out.
3 The moisture in the ground keeps the produce fresh. Apples, carrots and celery can be stored for many months in this way.
Note If you grow carrots and parsnips, leave them in the ground over the winter, covering them with fleece in frosty weather – then simply dig them out as required.

EXTRA-dry conditions

Produce such as garlic, onions and citrus fruits can't tolerate moisture, so require extra careful treatment when being stored.
● Dry onions on a wire rack in a patch of sun for a few days, then hang them individually, head down, in old tights with a knot separating each one. Store in a cool place and, when you need one, just snip it off.

● Alternatively, tie onions onto long strings, French style, or hang them up in string bags. Check them regularly and remove any that have softened or rotted.
● Dry and store garlic in the same way as onions. Never store garlic in the fridge or a sealed plastic container. It will quickly soften and could go mouldy.
● When wrapped in tissue paper and stored in cardboard boxes in a cool and dark environment, oranges and clementines also stay fresh for a long time. Check the fruits periodically and remove any that begin to rot.
● Choose oranges and lemons that are heavy for their size. These are full of juice and will not dry out as much during storage.

GOOD TO KNOW ✓

They don't all go together
Be careful with certain combinations in the storage area: apples, pears and tomatoes should be stored away from other fruit and vegetables. They give off ethylene, which hastens the ripening of other produce. Cabbage, for example, can develop leaf yellowing and rot, and carrots may acquire a bitter taste.

Cooking tips

Do you steer clear of cheaper cuts of meat because you are worried they will be tough and dry? As food gets more expensive it is worth going back to old and more creative ways of turning less popular joints into delicious dishes.

Whether a meal is fried or roasted, grilled or braised, these helpful hints can make your kitchen experience easier, less time consuming and not as taxing on your wallet.

BRAISING, stewing or frying

Braising is a good way to turn less expensive cuts of meat into an appetising meal.

1 First, brown the meat on all sides. Whether braising or stewing, searing meat before cooking helps to lock in moisture and flavour.

2 Then add a little simmering liquid such as wine or stock. Use about 1 tablespoon of liquid at a time and scrape up the tasty bits on the bottom of the pan. Keeping the amount of fluid to a minimum helps to retain the meat's flavour. Cover tightly.

Steaming is a quick and easy way to prepare seafood and vegetables.

● If you want a thick stew, dredge the meat in flour before browning it.

● Before frying or grilling rump steak, marinate it for a few hours to add flavour and tenderise the meat, then cut across the grain when serving.

● Bring out the tenderness in ribs by parboiling the meat before grilling.

THRIFTY fish chowder

Fish can be expensive, but a creamy fish chowder with plenty of hearty potato chunks is a good way to make it go further.

1 Cut two rashers of streaky bacon into 5mm cubes and sauté in a large saucepan until the fat is liquefied (3-4 minutes).

2 Add half a large onion, chopped, and cook over a medium heat until it is translucent. Then add 2½ tablespoons of plain flour and stir vigorously until the fat is evenly browned but not burned.

3 Slowly whisk in 250ml of stock, making sure there are no lumps. Keep stirring until the mixture comes to a boil, then add 3 tablespoons of white wine.

4 Add 250g potatoes, scrubbed, peeled and diced, and simmer for 15 minutes, or until you can pierce them easily with a knife.

5 Stir in 175ml single cream that's slightly warm and bring the chowder back to a simmer before stirring in 250g white fish, such as less expensive whiting or pollack. Heat gently for 5 minutes until the fish is cooked (don't let it boil).

6 Season with salt and pepper and serve with fresh, crusty bread.

FRYING meat, poultry & fish

● Take meat from the refrigerator at least 30 minutes before cooking, whether roasting or frying.

● Sautéing is basically quick cooking in a small amount of fat. Choose oil with a high smoking point, such as rapeseed or groundnut oil.

● To prevent the soft flesh of fish breaking up during frying slash the skin at an angle in several places.

● When deep-frying, keep an eye on the temperature. Excessively hot oil gives rise to dangerous toxins but if the oil is too cool more is absorbed by the food. After cooking, allow deep-fried food to drain briefly on a paper towel.

ROASTING & grilling

Tender cuts of meat are better suited to roasting or grilling, which allow them to maintain the maximum amount of flavour.

● For the tastiest roast always choose meat that is marbled and has a rim of fat. Remove any connective tissue because it becomes tough at high temperatures.

● When turning a roast, avoid poking it too much and losing the juice.

● To seal juices in, brown a roast then add some root vegetables, pour in a little stock and cook in the oven.

● Use the meat stock and juices from the roast as the basis for gravy.

● Insert a meat thermometer to determine if the meat is done.

● You can also use a metal skewer. If it feels warm when you pull it out, the meat is still rare; if it's hot, the meat is cooked through.

● Roast on a rack when possible to allow even heat circulation and browning.

● Roast beef or lamb with the fat side up to allow natural basting.

● Large roasts continue cooking for up to 10 minutes once removed from the oven. Let the roast sit for at least 10 minutes before carving for the muscle to relax and become tender.

● Add seasoning before grilling but salt afterwards to avoid toughness.

● Always baste a chicken with its own juices. A mixture of olive oil, lemon, pepper and garlic gives it a Mediterranean twist.

GOOD TO KNOW ✓

Healthy cooking

One of the gentlest ways to cook food is by steaming it in special slow cookers or in a heavy-based casserole in the oven. Cheaper cuts of meat not only remain juicy but conserve their nutrients and flavour. The bonus: you can pop a meal in the oven for a few hours or, better still, leave it in the slow cooker all day and come home to a cooked meal. This is particularly convenient after a busy day at work or whenever you know you'll be home too late to prepare dinner.

● The drier the duck or chicken's skin, the crispier it becomes when cooked, so pat skin dry and leave the chicken to sit loosely covered in a cool place for several hours before cooking, if possible.

● You can insert seasoned butter between the skin and the breast meat of the chicken. This makes basting unnecessary and the skin still turns out crispy and the meat tender and juicy.

Before grilling, marinate lean meat, fish and vegetables or brush them with oil.

Drying food

This natural form of preservation concentrates and enhances the flavour of fruit, vegetables and herbs. With a move back to local and home-grown produce, drying food ensures it can be enjoyed well beyond its season.

Warm air is used to remove the water and harmful bacteria that cause spoilage when drying foods. Removing the water also concentrates the flavour and aroma of the fruit and vegetables.

Check dried fruit and vegetables regularly for signs of mould, as the skin still contains a fair amount of water – unless they were dried at high temperature, which unfortunately sacrifices taste. After drying, small plums are usually little more than skin and stone that are hard to separate. You are better off choosing larger, late plum varieties and halving and stoning the fruit so it dries faster.

WHAT can be dried?

● Dry only high-quality, fresh, ripe produce, preferably organic.
● Harvest produce for drying only on dry, sunny days. Wet produce decays quickly and takes extra time and effort to dry.
● Pick herbs in the late morning or early afternoon, when their water content is lowest.
● Windfall fruit isn't suitable for drying.
● Fruits with stones or seeds are easy to dry, although you may wish to peel seeded fruits such as pears and apples.
● Avoid the tedious job of picking one berry at a time by drying the entire cluster; the berries will then fall off by themselves.
● Whatever the variety you choose, make sure the berries have a full, ripe fragrance or the dried fruits won't have any taste.

GOOD TO KNOW ☑

Well preserved

Dried products take up less space than fresh ones, but they need a cool, dry storage area with low humidity. For that reason, store only the things you will need in the near future in the kitchen. To postpone the deterioration of dried goods, keep them in a cool, dark place in opaque containers. Check them regularly for mould.

● The aroma of herbs changes when they are air dried. Many, like dill, chervil, tarragon, basil and coriander, lose their scent altogether.

PREPARING goods for drying

● Items should be of comparable size and thickness so they dry at an even rate.
● Place fruits such as apples and pears into lemon water (1 teaspoon lemon juice and 500ml water) immediately after cutting to preserve their colour.
● Alternatively, dissolve 225g caster sugar in the same amount of water to make a sugar bath. Bring the mixture to the boil and let it cool before immersing the fruit briefly.
● Always lay out produce for drying in a single layer with the cut side facing up.

Thread rings of apple onto a cane and dry in the oven.

TIE HERBS in bundles and hang to dry IN A well-ventilated PLACE.

- Pit apricots and peaches before drying.
- Blanch peas and beans in advance in order to retain their colour and aroma, and onion rings for 30 seconds. You can also blanch other vegetables before drying.
- Thread apple slices, mushrooms and chilli peppers onto cotton threads and hang to dry in a warm airing cupboard.
- Dry the plucked leaves of herbs or hang the stems in bunches and pick the leaves off after drying.

AIR drying

- Dry in the fresh air or in an attic or garage during hot, dry months. Arrange the produce to be dried on trays or, even better, on gauze stretched over a wood frame and covered with muslin.
- Allow air to circulate freely around the produce during the drying process.
- Drying takes two to three days outdoors in hot sun or up to two weeks indoors.
- If no juice seeps from the fruit when you break it into pieces, it is dry and ready to store in containers.

ACCELERATED drying in the oven

- Place produce directly onto the oven rack when oven-drying (baking sheets interfere with air circulation). Put small items onto a piece of muslin stretched over a cake rack. Prop the oven door slightly ajar with a wooden spoon to allow the moisture to escape.
- String apple rings onto bamboo canes cut to fit the width of the oven.
- Dry fruits and vegetables at a maximum of 120°C/ gas mark ½. Because of their essential oils, herbs should be dried at the lowest oven setting. The lower

the temperature at which produce is dried, the fewer vitamins are lost.
- Turn the produce once during the drying process. Return the rack in a different direction if possible to allow for even heat distribution.
- Mushrooms are dry when they look wrinkled and feel leathery.

Dried mushroom risotto

Serves 4
25g dried mushrooms, such as porcini
3 tablespoons olive oil
25g butter
1 small onion, chopped
300g arborio rice
50ml white wine
1 litre hot vegetable stock
50g Parmesan, freshly grated
Bunch of parsley, chopped

Soak mushrooms in hot water for 30 minutes. Strain off the liquid and set it aside. Heat the butter and oil in a large pan and fry the onion until soft. Add the rice and stir until it begins to change colour. Add the mushrooms, soaking liquid and wine. When the liquid has been absorbed, add the stock a ladleful at a time, stirring, until the rice is al dente. Add half the Parmesan and the parsley, check and adjust the seasoning if necessary. Serve with the remaining cheese sprinkled on top.

Entertaining

Careful preparation is key when entertaining, whether you are planning a cocktail party, a formal dinner or a get-together with good friends. The better prepared you are as a host, the more successful the party will be – and the more time you'll have to relax and enjoy it.

Once you've decided to host a party, work out how many people it is practical to invite, what you will need and a timetable for action.

BEFORE the party

● For a small circle of friends, an invitation by phone or email is fine. For larger or more formal events, a written invitation is customary. Send them out in plenty of time and include a final date for replying.
● Make a list of what you will need well in advance, such as buffet tables or additional glasses or silverware. Arrange to hire glasses or folding chairs and tables, which can be enhanced with tablecloths and chair coverings. Only buy them if you are likely to need them again, can pick them up at a reasonable price and have somewhere to store them afterwards.
● Once you have a shopping list, pick up drinks and non-perishable foods a week or two before the party and perishables one to three days before.
● Make place cards and table decorations before the party, as long as you are not using fresh flowers.

MEAL planning

● Take into account the season and the number and types of courses you are serving when preparing a menu. The more courses you plan, the smaller the portions and the lighter they should be.
● Make small, easy-to-grab appetisers for a cocktail party as people usually have only one hand free.
● For parties of six people or more, avoid pan-frying or other dishes that require last-minute preparation as you will be chained to the kitchen.
● Avoid making dishes for the first time when entertaining. This is especially true for people with less cooking experience.
● Lighten your workload on the day of the party by planning to serve food that can be prepared a day or two in advance. Soups, many desserts and salad dressings can be kept in the refrigerator until guests arrive.

THE day of the party

● Pace yourself – a special checklist will help.
● Set and decorate the table early – even the night before. You never know what hurdles you may encounter on the day.
● Have bowls and plates at the ready in the kitchen, and keep serving utensils near the table.
● Make sure the drinks, glasses and cold snacks are on the tables before guests arrive for a cocktail party.
● Ensure white wine is well chilled and open red wines at least 2 hours before the meal so that their flavours can develop.
● Take cheese for a cheese board out of the refrigerator 1-2 hours before serving it to enhance the flavours and textures. And place a few sugar

GOOD TO KNOW ✓

Decanting red wine

Young red wines will be smoother if you pour them into a carafe beforehand to allow their aroma to develop through oxidation. Use a wide-bottomed decanter. Old wine benefits from decanting as a way of removing deposits or dregs that form in the bottom of the bottle. Pour wine into a carafe slowly and carefully.

Even an informal get-together requires some planning so that everyone can relax and enjoy it.

cubes on the board – they can prevent cheese from sweating and absorb surplus moisture – but don't forget to remove them before serving.
● Light your candles shortly before guests arrive and check one last time that everything is ready.

GARNISHING foods

Delight guests and stimulate their appetites with creative garnishes made from a variety of fruits and vegetables, eggs and cheese.
● Make tomato baskets: cut two long slices across the top of a tomato so that a small 'handle' remains across the centre. Remove the core and cut the base flat to make a solid stand. Stuff the tomatoes with egg, vegetables, rice and/or cheese mixtures.
● Peel fine strips from raw, peeled carrots and put them in a bowl of ice water. Leave in the refrigerator for a couple of hours. Remove, drain the carrots curls and pat dry with a paper towel or clean cloth.
● Make four evenly spaced vertical cuts in the top of a cleaned and trimmed radish, then put it into cold water; it will open up like a rose.

WHEN the guests arrive

● Greet the first guests with a pre-dinner drink and a few canapés. Take the time to chat until everyone has arrived, which helps to break the ice, and introduce guests to each other.

● Before serving dinner, warm the plates in the oven for 5 minutes to prevent the food from cooling off too quickly once it is on the table. But be careful not to overheat the plates – guests should be able to hold them comfortably.
● Cover warmed-up bread with a large, clean table napkin to keep the heat in.
● Use an attractive wine bottle pouring spout to prevent drips that could stain your tablecloth.
● Leave the glass on the table and hold the bottle or decanter above the rim of the glass without touching it when pouring drinks.
● Fill glasses only a third to a half full so that the full aroma of a good wine can be appreciated.

Flavoured liqueurs

Herbal and fruit liqueurs are simple to make and an excellent way to use up home-grown produce or hedgerow fruits such as sloes. Homemade versions will be just as delicious as shop-bought brands, livening up a whole range of dishes and desserts – and they make great gifts, too.

When making liqueurs, use fresh, high-quality ingredients and keep utensils and the work area spotlessly clean. Store liqueurs and brandied fruits in bottles or jars. A German 'rumtopf' (rum pot) will age nicely in a ceramic container.

WHAT'S required?

● Container lids deserve particular attention: they must be clean and close tightly.
● For steeping fruit in liqueur, you also need large Kilner or preserving jars with wide necks and a capacity of at least 2 litres.
● Add sugar in the form of syrup, granulated white sugar or honey. They will all dissolve with time.
● Vodka is a good choice for steeping a liqueur or for preserving fruits. It has a fairly neutral taste, so helps to release the aromas and flavours of the ingredients. Depending on its strength, you may choose to dilute it with water.
● Other spirits such as fruit brandies, gin, tequila, rum or brandy can also be used.
● Fruits such as sour cherries, strawberries and sloes are particularly good for steeping in fruit liqueurs. The choice is yours: other berries, small yellow plums, papayas and lemon peel also make wonderful liqueurs.
● Fruits are often steeped with the peel intact, so be sure to wash them carefully. Dry well before use.
● If a desired fruit is not in season, frozen fruit works equally well.
● Filter liqueurs through tea or coffee filters, muslin, a fine metal strainer or through a normal strainer lined with one or two paper towels.

BASIC recipe for fruit liqueur

This is an all-purpose recipe suitable for a variety of different fruits.
1 Add the fruit to a large container along with about 280g sugar per 500g fruit.
2 Once the mixture draws out the juice, pour in the alcohol. Carefully stir the mixture then cover tightly.
3 Let the contents steep in a sealed jar for at least a month, then filter and bottle it.

VARIATIONS for liqueurs

● Give the liqueur a subtly different taste by using ingredients such as vanilla and cinnamon sticks, citrus peel, a piece of ginger, cloves, cardamom or a few drops of bitter almond oil.
● Use herbs or spices such as fennel, cloves, coriander, peppermint, anise, ginger, cinnamon sticks, vermouth herbs (available online), caraway or dandelion, if you like things spicier. Pour about 500ml vodka over 50g-100g herbs or spices, add sugar to taste and filter after two to three weeks.
● If the liqueur needs to ferment, you can also add a little high-proof brandy.
● If you want to adjust an excessively strong liqueur after it has steeped, add some fruit or a little sugar and let it steep again. It's best to add the sugar in the form of simple syrup (one part water to one part sugar, shaken in a bottle until the sugar dissolves), rather than less soluble sugar granules.

Always wash fruit, to remove any traces of pesticide.

BRANDIED fruit

Fruit steeped in alcohol is delicious by itself, as a garnish for meat and game dishes, and as a delectable complement to fine desserts. But if you are considering driving, remember that the fruit is saturated with alcohol.

● Add about 250g sugar per 500g fruit. If you want the fruit flavour to be more prominent, reduce the amount of sugar.

● Use brown or white sugar crystals to create a unique flavour.

● Layer the prepared fruit and sugar in jars and cover with alcohol.

● If you use a fruit brandy for steeping, make sure the fruit and brandy are a good flavour match.

● When making brandied fruits, avoid cheap brandies as they can spoil the taste.

● If you don't serve the fruit-infused alcohol with the fruits, you can use it for punch later.

STEEPING a rum pot

If you have plenty of fruit make a rumtopf (fruit preserved in rum), which is named after a traditional German pot, although you can use any large container with a well-fitting lid. Started in early summer, it will be perfect by Christmas.

● To start the rum pot steeping, use a sterilised, wide-mouthed jar. Add about 280g sugar per 500g fruit, and pour high-proof rum over the mixture. Seal the jar and store it in a dark, cool placc.

● Add other seasonal fruits in the summer and autumn (two parts fruit to one part sugar). Fill with enough rum to cover all the fruit each time.

● Add a few cinnamon sticks and star anise in October, then set the rum pot aside.

● By Christmas, use it to enhance ice creams, puddings or pancakes.

GOOD TO KNOW ✓

Store liqueurs, rum pots and brandied fruit safely and securely

Make sure that any alcohol and fruit mixes prepared at home cannot be reached by children. Your alcohol creations are best kept stored in a dry, dark place – under lock and key – to ensure that they are always enjoyed responsibly.

Food safety

Kitchen hygiene is just as important in the home as in a restaurant. With a little established culinary wisdom, you can prevent harmful bacteria multiplying on work surfaces and equipment, causing food to spoil and endangering health.

HAMBURGERS SHOULD be cooked THE SAME DAY THE MEAT IS BOUGHT.

Cleanliness in the kitchen begins with you. Wash your hands with soap and water before working or cooking, and between each stage of preparation – especially after handling raw meat or fish.

IMPORTANT rules

● Always wash kitchen utensils in hot, soapy water between each stage of preparation.
● Plastic cutting boards are more hygienic than wooden ones and can be put in the dishwasher.
● Use two different cutting boards: one for raw and one for cooked foods.
● Maintain the cold chain: pick up frozen and refrigerated foods last at the supermarket; put them in the fridge or freezer first as soon as you get home and before putting other groceries away.
● Wash or replace dishcloths, tea towels and sponges frequently because bacteria reproduce explosively in warm, moist environments.
● Don't leave raw foods or leftovers at room temperature for any length of time, or germs will multiply.
● Take out the rubbish regularly.

WHERE special care is required

● Be careful when preparing meat, fish, poultry and eggs to ensure harmful bacteria don't have a chance to breed.
● When thawing meat, fish or poultry, place it on a plate in the fridge, unwrapped but covered. Discard the liquid from thawing as it may contain harmful bacteria. Thoroughly wash the meat under running water, dry well and prepare immediately.
● Cook meat, hamburgers and sausages all the way through. The core temperature in the meat must be 70°C for 10 minutes to kill off salmonella.
● Check eggs are clean and not cracked when you buy them. Keep them in the fridge and use within the 'best before' date. Wash your hands, utensils and surfaces before and after contact with eggs.
● Shellfish should be cooked on the day of purchase. Wash thoroughly before cooking and follow the proper cooking times. Throw away any mussels that are already open or smell unpleasant.

Check eggs are clean and not cracked when you buy them.

Freezing

The freezer makes it possible to keep many food items to hand, ready to whip up a favourite dish. It also allows you to buy food cheaply in bulk or when it is in season, and to bake in batches and save separate portions.

Foods meant for the freezer should be fresh or, if cooked, packaged up and put into the freezer as soon as they have cooled down. Never refreeze foods that have already been frozen and thawed.

THE right way to freeze

● Use only sturdy bags, plastic containers, aluminium foil trays and special freezer containers for freezing.
● Never fill plastic containers to the top with sauce, soup or other liquids. They will expand and lift the lid or split the container.
● Only partially fill freezer bags so that the contents freeze faster and the bags are easier to stack.
● Prevent freezer burn by getting as much air as possible out of the bags before sealing them.
● It is best to freeze multiple small amounts and thaw packages as needed. This will decrease wastage, as once frozen food has been thawed it has to be consumed or thrown out, not frozen again.

WHICH foods to freeze & how to do it

● Individual meat portions should be no more than 10cm thick, while joints should weigh no more than 2.5kg. Depending on the fat content, meat can be frozen for 6 to 12 months. Lean meat lasts longest, as fat gradually becomes rancid at low temperatures.
● Place pieces of foil between slices of cooked meat to keep them from sticking together. They will keep for two months in the freezer.
● Freeze only freshly caught raw fish that has been scaled, cleaned and washed in advance. Use within six months.
● Shortcrust pastry can be frozen raw or baked. Allow yeast dough to rise once before freezing.
● Bread and rolls are easy to freeze and will keep for three months. If you freeze sliced bread, it can be defrosted as needed.
● Blanch vegetables (see box) to reduce the danger of freezer burn. Blanched vegetables will keep in the freezer for up to ten months.

Peas are one of the best vegetables for freezing. They taste nearly as good as when fresh.

● Mushrooms, red cabbage and legumes such as peas and beans become even easier to digest after being frozen. They lose the substances that often lead to flatulence.
● Freeze berries on a baking sheet, then drop them into bags. Fruit can be kept for up to 12 months in the freezer.

WHAT should not be put in the freezer?

● Whole raw or cooked eggs can't be frozen. Egg yolk and beaten egg white can, however, be frozen in appropriate plastic containers.
● Milk products don't belong in the freezer, the exceptions being cheese and (homemade) ice cream.
● Keep most exotic fruit away from the freezer. When exposed to extreme cold, bananas turn brown and citrus fruits get spots. However, peeled, cubed pineapple freezes exceptionally well.
● Don't freeze water-rich fruit and vegetables such as tomatoes, cabbage and onions, which take on an unappetising colour or become soft. Cooked spinach freezes well, however.

Blanching vegetables

one Bring 5-10 litres of water to the boil, then plunge prewashed vegetables into it. Bring back to the boil.

two Remove the vegetables from the boiling water after 2-3 minutes, drain well and plunge into ice-cold water.

three Transfer to a colander and let them drip-dry thoroughly before packing.

Grains

The seven main grain varieties – wheat, rye, barley, rice, corn (maize), millet and oats – are a source of nourishment for people all around the globe. In recent years we've also rediscovered traditional grains such as bulghur wheat, spelt and quinoa to enhance our diet.

Grains can be eaten either in their whole form (whole grain), crushed or ground. Only the hull or husk is removed from whole grains; with refined grains, the germ layer is removed as well. Whole grains are a healthier choice, because they have a high fibre content and are digested more slowly.

STORING grains

● Protect grains against moisture and insects by storing them in cool, dark places in clean, airtight containers.
● Store whole grains in airtight containers in a cupboard for up to a year.

PREPARING & cooking grains

● Rinse grains thoroughly before cooking to remove dirt and dust particles. Continue to rinse until the water is clear, and throw away any off-colour grains.

● Soaking grains such as barley and oats before cooking makes them quicker and easier to cook and digest. Add grains and water to a pot in a 1:2 ratio and soak until they are softer and swollen.
● Strain away any excess liquid that is not absorbed and save it for cooking with later as it contains many valuable nutrients.
● Grains need to be stirred constantly during cooking, as many types will stick to the bottom of the saucepan.
● After cooking, allow grains to sit briefly in the covered saucepan, then fluff them up with a fork as you would a pot of rice.
● To separate sticky grains, tip them into a colander and pour a kettle of boiling water over them.
● You can also cook grains ahead of time, then reheat them up in the microwave for just a minute or two. Add a sprinkle of water, cover the bowl with microwave-safe plastic wrap and fluff them again when the cooking time is up.

A small hand mill allows you to grind as much grain as you wish.

Muesli is an easy-to-make healthy breakfast and you can add fresh fruit according to taste.

OATCAKES

Historically, oatcakes were toasted on a peat fire. This hot griddle recipe makes about eight.

1 Mix 125g medium oatmeal, a pinch of salt and a pinch of bicarbonate of soda in a bowl then make a well in the centre.

2 Melt 2 teaspoons of butter in 4 tablespoons of boiling water and pour it into the well. Mix to form a stiff dough.

3 Turn out onto a floured surface and knead lightly. Roll out thinly and cut into eight triangles.

4 Cook on a hot griddle or in a dry, heavy frying pan until the edges curl up and the cakes are firm. Store in an airtight container and reheat before eating.

VARIOUS grain types

● In northern climates, wheat and its close relations, barley and oats, are most commonly grown.

● Spelt is a protein-rich grain that makes a particularly robust pasta, gives bread a nutty taste and is also good as a side dish and in stews. It is also good for wheat-intolerant people.

● Whole-wheat kernels should be presoaked and then cooked. They make a tasty rice substitute and can add crunch to a salad and a nutty flavour to chilli or stews.

● Rye is high in fibre and low in gluten. It can be made into flour for bread or served in flakes as a breakfast cereal (like oats).

● Because of its low gluten content, barley flour is unsuitable for baking. But you can make tasty soups, stews and pilafs with whole (pearl) barley.

● Millet is one of the oldest grain varieties. Use the kernels like rice or incorporate them into a salad. Bread made with millet flour is particularly crispy.

● From a nutritional point of view, oats are one of the most valuable grains in northern climates, mainly because they contain a protein and fat that are good for you and easy to digest. Oatmeal, oat bran and oat flour are made from oats.

● Bulghur (a form of processed wheat) can replace rice in most recipes, thicken soups and stews, and be used in salads, breads and even desserts.

● Once called 'the gold of the Incas', protein-rich quinoa has a fluffy, somewhat crunchy texture that makes it a wonderful rice substitute or wheat-free alternative to bulghur in tabbouleh and other salads.

BREAKFAST

Swiss-style muesli is a healthy breakfast food. It is easy to make and you can modify the ingredients according to taste.

1 In a large mixing bowl, combine 375g porridge oats, 50g wheatgerm, 30g wheat bran, 50g oat bran, 150g sultanas, 50g chopped walnuts, 4 tablespoons soft dark brown sugar and 30g unsalted sunflower seeds.

2 Mix well then store in an airtight container. It will keep for two months at room temperature.

GOOD TO KNOW ☑

Soaking & cooking times

VARIETY	COOKING TIME	SOAKING TIME
Millet	5-15 minutes	10-20 minutes
Bulghur	2-5 minutes	none
Spelt	30-45 minutes	30-45 minutes
Barley	30-45 minutes	30-60 minutes
Rye	30-45 minutes	30-60 minutes
Quinoa	10-15 minutes	none

Herbs

Herbs add a dash of colour and flavour to food and are a healthy alternative to salt as a seasoning. They can also help food to be digested more easily. Use strong-flavoured varieties sparingly and add tender green herbs to a dish just before serving.

Herbs lose their flavour quickly if handled incorrectly. Use them soon after they've been picked or purchased and always chop the tender leaves with care.

A herb chopper quickly cuts up delicate herbs and preserves their flavours.

HANDLING herbs

- Chop herbs on a moistened wood or plastic chopping board. The flavours can be dulled by the wood of a dry cutting board.
- Harvest herbs at midday, when their essential oils are most intense.
- Wrap herbs in a damp tea towel before putting them in the refrigerator's vegetable drawer, where they'll keep for one to three days.
- Dry or freeze herbs for longer storage.

COMMON kitchen herbs

- Parsley is the most popular herb and ubiquitous in most kitchens. It is used to season soups, stews, casseroles, salads, pasta and potato dishes. Just add scissored sprigs near the end of the cooking time, or tear up the leaves with your fingers.
- Fragrant basil has a pungent taste that works well in sweet and spicy foods such as stir-fries and spaghetti sauces. Simply tear it apart or cut it into strips to add to salads and marinades, or blend it in a food processor to make pesto.
- Coriander is one of the most widely used herbs in the world. Its strong flavour adds a boost to southwestern, Asian, Latin American and Middle Eastern cuisine. Sprinkle the fresh leaves on a finished dish.
- Rosemary has an astringent, clean scent and tastes great in Mediterranean cooking. Add the needle-like leaves to lamb, wild game and pork, or use the stripped branches as skewers to grill or barbecue meat, poultry or fish.
- Fresh chervil adds a delicious mild aniseed flavour to salads, soups and herb butter, as well as egg and cheese dishes.

● Dill complements fish, shellfish, cucumber salad, vinaigrette and pickles. Sprinkle on before serving.

● Tarragon is native to Siberia and North America. It adds a distinctive flavour to vinegar, mustard and béarnaise sauce, and goes well with poultry and shellfish. The herb's bittersweet and spicy aroma is lost when dried.

● Bay leaves are widely used in soups, stews and spaghetti sauce, but their aromatic flavour is also great with fish and in many East Indian dishes like biryani. Bay leaves stimulate the appetite and also act as a preservative. You usually add the leaves whole and remove them before serving.

● Marjoram aids digestion and goes well with meat dishes and summer vegetables such as aubergine, tomatoes and peppers. It can't be frozen but, when dried, it can be added during cooking.

● There are many different varieties of mint but spearmint and peppermint are the most frequently used. Mint highlights the fine taste of spring vegetables, peas, green beans and salads, and adds a

GOOD TO KNOW ✓

Special collection

A bouquet garni of fresh parsley, thyme, rosemary and bay, tied together with string, will add depth of flavour to a stock, soup or casserole. Simply remove it before serving.

fresh, elegant touch to desserts and fruit bowls. It is also a traditional accompaniment to lamb.

● Oregano belongs to the same family as marjoram and is regularly served up on pizza and in tomato dishes. Use the herb sparingly – it is easy to overdo it.

● Add peppery-tasting sage when cooking roast pork and poultry. It also goes well with tomatoes, beans and pasta.

● Chives taste like mild onions. Use the herb to dress up salads, vegetable soups, soured cream, potato and leafy salads and mashed potatoes, as well as scrambled eggs. The younger the shoots, the more tender the consistency and the more intense the taste.

● Use thyme in soups and stews, with vegetables and in casseroles and fish dishes. Fresh lemon thyme goes well with fish and poultry.

GROW HERBS in decorative ceramic POTS ON THE WINDOWSILL.

Freeze fresh herbs in an ice-cube tray ready to add to soups or sauces as required.

Jams and jellies

Many of us now opt for shop-bought jars but nothing beats a homemade jam or jelly – and making them is fairly simple. Buy fruit and vegetables when they are plentiful, saving money and restoring a traditional delight to your table.

Although all fruits contain natural pectin, the trick for getting a jam or jelly to gel (set) lies in striking the proper balance between acids and pectin – or you can buy natural pectin and add it as per the maker's instructions. Cleanliness is also crucial when processing fruit. Even the tiniest bit of decay in the fruit can make preserves go mouldy.

IMPORTANT tools

● In addition to a chopping board and knife, you need a set of weighing scales, a preserving pan or large saucepan, a wooden spoon, a skimmer and glass jars with lids.
● Copper pans are excellent heat conductors but they react with acids, so a stainless-steel pan is better.
● A wide jam funnel and ladle make it easier to pour jams and jellies into jars.

FRUITS gel differently

In principle, nearly all fruits are good for making jams and jellies. The higher the pectin content the quicker the set. For some fruits, you will need to add pectin in the form of lemon juice, another fruit or pectin concentrate.
● Apples, blackberries, redcurrants, gooseberries and citrus fruits are high in pectin so gel quickly. You can add plain granulated sugar when using them.
● Apricots, raspberries, blackcurrants, plums, nectarines and peaches have a moderate amount of pectin, so a little lemon juice helps them to gel.
● Pineapples, pears, strawberries, rhubarb, cherries, marrows and grapes have a low pectin content. Add the juice of a lemon to these fruits, combine them with other fruits that are high in pectin or use jam or preserving sugar containing pectin.
● A sugar thermometer is useful but not essential for judging the setting point of your jam. Jams and jellies will set at 103°C. Place the thermometer in a bowl of hot water before you use it to make it react quickly.

Sterilised jars with screw tops make ideal containers for jams and jellies.

MAKING jams

Making jam is easy: simply add one part jam sugar (a mix of sugar, pectin and citric acid) to two parts fruit. Sterilise jars and lids with boiling water then line them up in advance.

1 Wash, clean, chop or crush the fruit then weigh it and put it into a large, thick-based saucepan so the contents won't stick while cooking or boil over.

2 Simmer harder fruits such as pineapple, pears and apples in a shallow pan with a little water until soft before crushing.

3 Stir in a little lemon juice to preserve the bright colour of the fruit.

4 Add jam sugar to the fruit, stir well and wait a few minutes while the fruits draw out some juice.

5 Bring the contents to the boil, stirring constantly. Make sure nothing sticks to the bottom of the pan and burns. Adding a small knob of butter to the fruit mixture will help to ensure it doesn't boil over.

6 Simmer the contents for 5-10 minutes, stirring constantly, and use a skimmer to scoop off the foam.

7 Do the gel test (see box, below). Depending on the result either remove the pan from the burner or, if it is too runny, cook for a few more minutes and retest.

8 Pour the jam immediately into hot, clean jars, wipe any spilt jam from the rim and seal them securely.

● Instead of fruit, use marrows, pumpkins, carrots and tomatoes to make delicious jams and chutneys.

● Rich spices such as ginger, vanilla and cardamom can add a unique flavour to jams.

MAKING jellies

1 To obtain the fruit juice necessary for a jelly, add cleaned, chopped or crushed fruit to a saucepan with a little jam sugar to draw out the juice. There is no need to remove berry stalks or apple cores or skins.

USE a strainer to remove TINY SEEDS AFTER DRAWING out the JUICE.

The gel test

one Put a teaspoonful of the boiling fruit mixture onto a plate you have cooled in the freezer.

two Let the sample cool down by placing it briefly in the fridge.

three If it congeals and no water forms around it, then the jam or jelly is ready to put in jars. If you have a sugar thermometer, the mixture should reach 103°C.

2 Bring the mixture to the boil and simmer until all the fruit floats to the top.

3 Pour the fruit mixture and its liquid into a strainer lined with muslin and collect the juice in a bowl. Leave overnight. Don't squeeze out the juice or the jelly will be cloudy.

4 Boil the juice with an equal amount of jam sugar for 5-10 minutes (for 250ml juice use 250g sugar).

5 After the gel test (see box, left), pour into sterilised jars and seal.

● You can increase the amount of juice by returning the mixture to the pan after straining, covering it with water, simmering and straining it again.

● Fresh herbs such as mint or lemon balm can add a distinct flavour to jellies – especially apple jelly.

Juices

Citrus fruit is easy to squeeze by hand, but fruit presses and juicers are convenient for making larger quantities.

Homemade fruit and vegetable juices contain nutrients that can enhance your diet; turned into shakes and smoothies they will appeal to even the fussiest eater. Or add a little sugar and make a fruit syrup you can add to drinks or desserts.

The vitamins, minerals, carbohydrates and other nutrients contained in juices provide energy and help to strengthen your bones and immune system, and cleanse skin. But don't overdo it: fruit drinks contain a lot of natural sugar so are highly calorific.

FRESH is best

● Avoid prepackaged juices that are pasteurised to make them last longer. Many vitamins and minerals are lost in the process.
● Opt for fresh juices that contain no additives but include all the nutrients of the squeezed fruit as well as enzymes and fibre, which aid digestion.
● Drink the juice freshly squeezed or keep fresh juices in dark, sealed containers in the refrigerator for up to three days.
● Add a little ascorbic acid, vitamin C, from the chemist or health food shop to help the juice retain its colour. Lemon juice works just as well, and also helps to reduce sweetness.

MAKING juice

● Cold pressing fruit in an electric juicer is the best way to make fresh juice. Load the cleaned fruit or vegetables into the chute. A grater at the bottom shreds it, while centrifugal force spins the juice through strainers. The residue is collected in a basket.
● Alternatively, use the juicer attachment on a food processor to produce fast results.
● You don't need a machine to make juice. The traditional method involves stewing fruit in a pot with some water until it bursts or becomes soft. At that point, strain the mixture through muslin or a sieve, capturing the juice in a bowl. Boil the juice with sugar to sweeten, if necessary, then pour it into bottles while still hot and seal them.

TIPS for fruit & vegetables

● To avoid the risk of contamination from chemicals and pesticides in the skins and peels, try to use organic fruit and vegetables for juicing. If you're not using organic produce, it is advisable to peel all fruit and vegetables before you juice them.

● After cleaning and washing the fruits and vegetables, dry them to avoid diluting their juice.

● When using organic fruits, only peel those with a tough skin. Soft-skinned fruits can simply be cored and cut up. A fruit press is useful for firm fruit.

● If you find grapefruit juice tastes too sour, mix it with pear juice to take the edge off.

● Make vegetable juice more appealing to children by mixing carrot or tomato juice with fruit juice.

● Relieve heartburn, encourage hair and nail growth, improve your complexion and aid vision by drinking carrot juice.

● Stimulate your immune system with the juice of sweet peppers. They have more vitamin C than lemons and oranges.

● Beetroot juice is thought to have antibacterial properties. The betaine (a protein building block) it contains can also strengthen the liver and help the body to flush out toxins.

● Mix together apple and pear juice to provide a gentle remedy for constipation.

FRUIT syrup

Fruit syrups are easy to make and taste delicious added to sparkling mineral water, poured over pancakes or crêpes, and drizzled on ice cream or angel food cake.

● Use fully ripened fruit for producing syrup – it will yield more juice and has a much stronger flavour.

● Juice the fruit in the usual way, then mix it with an equal amount of sugar. Boil again and pour into sterilised bottles or jars.

● Fruit syrups make lovely gifts when presented in pretty bottles.

PEPPERMINT syrup

Diluted with water, peppermint syrup is often a favourite drink with children.

1 Wash one bunch of fresh peppermint leaves then place them in a pan and cover with 1.2 litres of water. Bring to the boil, simmer for 10 minutes then leave to stand for 30 minutes before straining the juice through a fine sieve.

2 Mix 500ml of the juice with 350g caster sugar. Pour this mixture into a pan and simmer for 15 minutes, stirring until the sugar has dissolved.

3 When cool, pour the syrup into bottles and seal.

● It's just as simple to make syrup from herbs such as lemon balm, mint, rosemary and lavender.

● Drink the syrup diluted with water or use it to add pizzazz to a variety of desserts.

Blueberry syrup

350g blueberries (or any berries of your choice)
375ml water
350g sugar
60ml golden syrup
1 tablespoon lemon juice

Sort and wash fruit, then run it through the blender or mash it and mix with the lemon juice. Add to a large pan along with the water.

Bring the contents to the boil, then lower the temperature and simmer for 15-20 minutes. Strain through a muslin-lined sieve. Return the berry juice to the saucepan and add the remaining ingredients, stirring to dissolve the sugar. Bring to a full rolling boil for 2 minutes, then pour directly into sterilised jars or bottles.

Fresh apple juice changes colour quickly, so use it quickly.

Kitchen safety

More accidents happen in the kitchen than anywhere else in the home. Cutting yourself with a knife, slipping on a wet floor or burning yourself with a hot pan are among the most common dangers. Minimise the risk of injury by taking a few precautionary measures.

A little care is key to kitchen safety. Learn how to handle sharp kitchen utensils safely and make sure you know what to do in case of emergency. When it comes to cooker safety, never leave a pot on the hob unsupervised, especially if you are cooking with gas. Turn off burners immediately you have finished using them.

PREVENTING burns

● Make sure that you position pots and pans carefully on the cooker so that you can't knock their handles as you move about the kitchen.
● When heating oil, watch the pan carefully as oil is so flammable. Deep frying is a common cause of kitchen fires.
● Step back and let the hot air escape when you open the oven door.
● Be especially careful not to spill hot liquid when removing pots and pans from the oven. Wipe up any spills on the floor at once in case you slip on them.

● Never place hot foods on the edge of the cooker, table or work surface where they may tip over.
● Keep tea towels well away from the burners that could set them alight.
● Wash or replace exhaust fan filters regularly. They can easily catch fire when saturated with grease.
● Keep your hands, face and body out of harm's way when using the steam release valve on a pressure cooker.
● For safety's sake, and for peace of mind, it's best to use a kettle that turns itself off automatically when the water boils.
● Never overheat oil in a deep fryer. Also, replace used oil as it is more likely to catch alight.
● Make sure the deep fryer's heating coils are totally immersed in oil or there is a strong risk that the oil will catch fire.
● Run cold water in the sink while draining potatoes or other vegetables so that the rising steam won't scald your hands.

TAKE care with knives & scissors

● Store knives and other sharp objects in a knife block or on a magnetic wall strip out of the reach of children.
● Don't drop knives in washing-up water; if they're obscured by suds, you may accidentally grab them by the blade.
● Wash knives with a brush straight after use. For dried-on food, let knives soak first in a container of warm water.
● Read the instruction manuals to find out how to safely remove the blades from food processors, shredders, slicers, meat grinders and other kitchen devices with blades.
● While cutting, guide the knife away from your body, with your fingertips held at an angle.
● Sweep up broken glass immediately and wrap it in newspaper before putting it into the refuse bin. Gather up tiny fragments of glass with a piece of sticky-back plastic wound around your hand.
● If a glass breaks in the washing-up water, drain the water through a sieve and rinse away any excess soap suds before removing the pieces.

Use oven gloves when removing hot food from the oven.

Make sure pans fit well and are stable on the hob.

● Treat bruises from a fall by applying a cold pack or a bag of ice wrapped in a towel. Never apply ice cubes directly to skin as it will burn.

Note For severe falls, seek professional advice.

● For a minor burn or scald, cool the skin for 10-30 minutes using cool or tepid water to prevent it getting worse. Do not use ice, creams or greasy substances.

● Cover the burn with a single layer of cling film; a clean, clear plastic bag is suitable for covering burns on the hand. Do not interfere with the burn or break any blisters.

Note For severe burns, go to hospital immediately.

● Stop bleeding by applying a clean tea towel or wad of kitchen paper and pressing hard on the spot. Clean it by holding the part of your body that is injured under cold running water. Then cover it with an adhesive bandage to prevent infection.

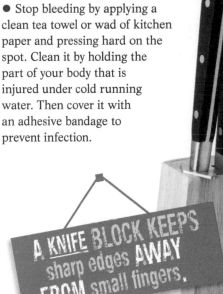

OTHER dangers

● Keep household chemicals out of the reach of children. Make sure that they are clearly labelled and don't store them near foods or in containers normally used for foods.

● Use a hand guard to protect fingers when using a grater or vegetable slicer.

● Wipe up splattered grease and spills on the floor immediately to ensure that no one slips.

WHAT to do in an emergency

● If something does go wrong, remain calm.

● Smother the flames of an oil fire with a thick, damp tea towel or a pot lid, then remove the pan from the heat source. Don't use water or a foam fire extinguisher, as this has the potential to splatter the oil outside the contained pan and spread the flames.

● Turn off the oven immediately in the event of an oven fire. Keep the oven door closed – the lack of oxygen will suffocate the fire.

A KNIFE BLOCK KEEPS sharp edges AWAY FROM small fingers.

Leftovers

In the UK alone, we throw away more than 8 million tonnes of food and drink a year, costing an average of £50 per household. Spare the environment and your pocket by following these simple ideas to banish waste and turn leftovers into tasty meals.

Casseroles and salads are the classic way to use up leftover potatoes, pasta and rice. But you can also transform them into delicious soups and puddings.

LEFTOVERS CAN OFTEN be used to CREATE TASTY OMELETTES.

USING up leftover potatoes, pasta & rice

- Try making potato patties: mash leftover potatoes, mix with an egg, salt, pepper, garlic, parsley and a diced onion, shape into rounds, coat in breadcrumbs and deep fry in vegetable oil.
- Purée leftover potatoes with stock and season to taste with cream or milk, herbs and/or grated cheese to make a delicious potato soup.
- Convert a whisked egg and a little cheese, leftover meat and potatoes into a delicious omelette.
- Add a can of black beans and some salsa to leftover chicken or hamburgers to create a tasty topping for tacos or a filling for burritos.

- Add leftover Parmesan cheese to breadcrumbs and use to top a fish pie for a great flavour. Because Parmesan can burn easily, cover the dish with foil for the first half of the cooking time.
- Add 2 tablespoons of capers or sliced olives, a can of tuna and some diced tomatoes to leftover pasta, then season with vinegar, olive oil, salt and pepper to make a light summer pasta salad.
- Potato, rice and pasta salads taste particularly good when you let the flavours blend before eating. The same is true of potato and rice salads.
- Make a hearty rice dish from leftover rice, eggs, pork or chicken, onions and frozen vegetables. For an Asian touch, use sesame oil for stir-frying and season with soy sauce and/or toss in a little curry paste.
- Transform day-old rice into rice pudding: mix 500ml milk per 250g cooked rice and simmer gently on the hob so the milk doesn't burn but the mixture is heated through. In a separate bowl, mix together 2 eggs, 55g sugar and ½ teaspoon vanilla. Whisk into the rice mixture and heat through before transferring to a casserole dish. Bake in a preheated oven at 180°C/gas mark 4 for 20 minutes.
- Whip together delicate rice pancakes by mixing leftover rice with flour, milk and eggs then cook like normal pancakes. Serve with cinnamon and sugar or marmalade.

VEGETABLE variety

- For a simple casserole, layer leftover vegetables with potatoes, pasta or rice in an ovenproof dish and pour cheese or béchamel sauce over it. Sprinkle with breadcrumbs and bake.
- Make a puréed soup from leftover vegetables: heat the cubed vegetables in a little stock, purée and stir in milk for a creamy consistency. Season to taste and serve with crunchy croutons made from day-old bread.

VARIATIONS for old bread

- If rolls or bread are not too stale, turn them into croutons for salads or soups. Just cut them into bite-sized cubes and toast in a pan with a little olive oil or butter. Heat them until they are crunchy and have browned slightly. Add herbs, crushed nuts or Parmesan cheese for extra flavour.

● Dip stale bread into egg whisked with milk then fry to make French toast.

● Soften stale rolls in vegetable or beef stock before using them in meatballs or meatloaf to make the dish especially tasty.

● In summer, when tomatoes are ripe and juicy, try this novel bread salad: chop stale slices of white bread into bite-sized chunks and mix with diced tomatoes, onions, olive oil and wine vinegar. Season with salt, pepper and herbs, and let it sit for at least 15 minutes.

OTHER leftovers

● Serve day-old roast chicken or turkey breast on a green salad with additions such as vegetables, chickpeas, walnuts, pine nuts, raisins, cranberries or cheese. Keep it simple – serve with a classic balsamic vinaigrette.

● Chicken pie makes the most of leftover chicken and vegetables. Make a thick sauce with milk, flour and butter, together with a dash of salt and pepper. Mix in diced chicken and leftover or sautéed vegetables and pour into a lightly greased baking dish. Top with ready-made pastry, sealing it to the sides. Cut a few vents in the pastry and bake until bubbly and golden brown.

● Cooked grains are a nutritious and filling addition to any soup.

Croutons made from leftover bread complement fresh salads or puréed soups.

Meat, fish and poultry

Cooking meat, fish and poultry successfully involves proper preparation. Use time-honoured methods such as marinating overnight and making tasty stuffings to ensure your dishes are tender, moist and appetising.

Once you have decided what to cook, calculate how much meat or fish you will need, taking into account that it may shrink during the cooking process. And always wash fish and poultry thoroughly to protect your own and your family's health.

HOW much to serve

- With red meat, generally allow about 200g per person, or 300g for a roast with the bone in.
- Depending on the size, one 1.5-2kg roasting chicken or large duck feeds three to four people. One goose is usually enough for six hearty eaters; allow about 350g per person. With turkey, allow about 500g per person, which will give you leftovers, too.
- With fish fillets, allow about 250g each, or 350g if using a whole fish.

A GOOD meat and fish marinade

Marinades serve a dual purpose: to tenderise meat and fish and add a mouthwatering flavour. Most marinades contain an acidic ingredient – such as lemon juice, vinegar or wine – that breaks down muscle fibres. The enzymes in ingredients such as onions, ginger and papaya also act as an excellent tenderiser. Marinades have yet another function: you can use them later for making a sauce or gravy.

- For a simple marinade, stir together 5 tablespoons each of oil and vinegar, and a pinch each of pepper and sugar. Place the meat or fish you are preparing in a deep container or sturdy freezer bag and cover it with the marinade.
- Use salt sparingly in a marinade to avoid drawing the moisture out of meat.
- Boost flavour by adding extra ingredients such as honey, mustard, wine, lemon, onion, spices, herbs or fruit such as apples and oranges.
- Marinate veal and offal for a maximum of 2 hours, other types of meat for 3-4 hours or overnight, and fish for 2-3 hours in the refrigerator. Game can steep in a marinade for one to three days.
- Don't use metal containers or bowls for marinating as they react with the acids in the marinade, which can be harmful and may leave the meat with an unpleasant flavour.

Garlic and herbs enhance the flavour and aroma of meat.

Filleting fish

1 Cut off the head just behind the gills. Hold the knife blade horizontal and cut across the body of the fish, using the backbone to guide the knife.

2 Release the upper fillet completely from the bones with a second cut, and set it aside.

3 With a final cut, guide the knife flat beneath the central bone and free it from the second fillet.

CUTTING, marinating & stuffing birds

When cutting up poultry, make sure you have a sharp knife and good poultry shears to hand.

1 Lay the bird on its back, cut through the skin between the drumsticks or wings and breast with the knife, then use poultry shears to snip off the bones at the joint.

2 Use the knife to separate the breast meat on both sides of the breastbone.

● Marinate white meat in white wine and herbs, dark meat in red wine.

● Give chicken a Mediterranean flavour by rubbing olive oil, lemon, garlic and herbs on and under the skin.

● Good poultry stuffings include savoury ingredients such as sausage meat and chestnuts, as well as traditional sage and onion. Or try parsley, onions, almonds and dried apricots.

● Season the abdominal cavity with salt and pepper before you start stuffing.

● When cooking large birds such as turkeys, only stuff the neck end, leaving the body cavity unstuffed to help air to circulate as it cooks.

PREPARING fish

● Scale a fish by holding it by the tail while you scrape from the rear towards the head, using a scaler or a knife with a serrated edge.

● Boning is more complicated. Open up the abdominal cavity and cut the perpendicular bones away from the flesh on both sides. Then, separate the central bone from the head and tail and carefully remove it along with the other bones.

● Cut the lateral (side) and dorsal fins off when serving a fish whole.

COOKING fish

Don't overcook fish if you want it to remain tender and juicy.

● To test to see if a fish is done, use a knife to part the flesh a little. It should flake easily.

● Alternatively, stick a cocktail stick into the fleshiest part of the fish. If it meets little resistance and comes out clean, the dish is done.

● Make it easier to cut fish into chunks for chowder or bouillabaisse by freezing it for 45 minutes first.

● Thaw frozen fish in the refrigerator or cook it direct from frozen; thawing at room temperature increases the likelihood of contamination.

● When roasting a whole fish, wash it thoroughly inside and out, salt, season with herbs and rub it with olive oil. Bake in a preheated oven at 220°C/gas mark 7. You'll know it's done when the eyes turn opaque and the flesh near the backbone flakes when prodded with a cocktail stick.

FRYING fish

● Dredge the fish in seasoned flour, beaten egg and finally breadcrumbs before frying.

● If you have no eggs available for a fish coating, use lemon juice instead. The coating will stick just as well and the fish will have a lovely lemony zest.

● Try chopped hazelnuts, flaked almonds or crushed cornflakes instead of breadcrumbs.

● Always bread fish shortly before cooking or the breadcrumbs will flake off.

● Absorb excess grease by placing the cooked fish briefly on a paper towel.

● If fish portions have skin, fry them skin side down first. Turn them over when nearly cooked through to finish. The skin will be crisp and delicious..

Non-alcoholic beverages

Water is the basic ingredient for all refreshing drinks, as well as being essential for good health. But water alone can be a little dull, so here are some traditional methods to satisfy your tastebuds while quenching your thirst.

Drinks should quench your thirst and refresh you without sending your blood sugar soaring. Commercially produced drinks often contain too much sugar, so borrow your grandmother's recipe book and make your own.

WATER, the elixir of life

Advertising by bottled water companies gives consumers the impression their product is safer and healthier than tap water. As a result, world consumption of bottled water has grown enormously in the past ten years. However, you may be better off drinking the elixir of life straight from the tap. Here are three reasons why.

1 Producing and delivering bottled water produces greenhouse gases, and most discarded plastic bottles end up in landfill sites.

2 Scientists found that many bottled brands contain higher levels of bacteria than tap water.

3 Drinking bottled water can cost up to 1,000 times more than tap water.

STORE COFFEE BEANS in a tightly closed container OR they will go STALE.

COFFEE beans

● The amount of coffee you need will depend on personal taste, water quality and the type of coffee maker you are using. Generally, when using a filter coffee maker, allow about 1 tablespoon of coffee per cup using 250ml water. With hard water that contains lime, use more coffee.

● Store coffee beans in a dry, cool, dark location and keep them in a tightly closed container or they will go stale.

● Store vacuum-packed ground coffee for several months if unopened. Once opened, keep in an airtight jar in the refrigerator.

● Ground coffee loses its flavour quickly and should be used within six months.

● Use a coffee grinder or spice mill to make freshly ground coffee.

TEA

After water, tea is the most widely consumed drink in the world. The term 'herbal tea' usually refers to an infusion of leaves, flowers, fruit, herbs or other plant material that doesn't include traditional tea leaves.

● Infuse black tea leaves with boiling water and then steep for 3-5 minutes. Strain before serving.

● Steep green tea in hot water at 80°C for only 2 minutes.

● When water has a high lime content, use strong tea blends or run the water through a filter.

● Store tea in a dry container with a tight-fitting lid. Special metal or porcelain containers and opaque bags are ideal for this.

Lemon and mint add a tangy flavour to tap water.

● Don't keep opened packets of tea in the refrigerator. The sensitive leaves can be damaged by fluctuations in humidity.

● For long-term storage, place unopened tea in the vegetable drawer in the refrigerator.

● Always store the same type of tea in a container as tea will quickly pick up other flavours, which can ruin its subtle nuances.

● Make delicious herbal teas with rose hips or mint. Depending on the season you can boost their flavour: in summer, use lemon or lime peel (mint); in winter, use cinnamon or cloves (rose hip).

FOR hot days

Iced tea is a wonderfully thirst-quenching drink for hot days and easy to make.

1 Brew 3-4 teaspoons of tea leaves in 1 litre of water and let it steep for 3 minutes.

2 Stir in the juice of 1½ lemons, sweeten to taste, then strain it over about 25 ice cubes. The tea cools quickly and doesn't become bitter.

3 Flavour it according to taste with apple juice, fruit nectar, citrus fruit or mint leaves.

HOMEMADE lemonade

There's nothing quite like the flavour of traditional homemade lemonade, which is refreshing and often healthier than shop-bought versions.

1 Boil the zest of 6 organic lemons with 280g sugar and about 500ml water. Simmer for a few minutes, cool, then add the juice from the lemons.

2 Strain the lemonade concentrate and pour it into a glass jug. Fill it up with ice cubes and cold water.

● Honey, ginger or rose water will give lemonade a highly original flavour.

● Limes and oranges can be also be used to make equally refreshing drinks.

HEALTHY fruit drinks

Get more vitamins and minerals into your diet by drinking freshly made fruit juices or whipping up thick and frothy smoothies from frozen fruit.

● Make a delicious fruit drink by cutting up your choice of fruits and whizzing them in a blender with the juice from half a lemon, together with 55-115g sugar. Adjust the amount of sugar according to the fruit used and the sugar content of the juice.

● For a berry punch that will serve about 15 people, put 1kg of raspberries and strawberries, the juice of a lemon and 1.5 litres of mixed fruit juices into a

punch bowl and cool for several hours, stirring from time to time. Shortly before serving, add two bottles of mineral water, followed by ice cubes.

● Keep your punch cold by freezing slices of citrus fruits and using them in place of ice cubes.

● Blend frozen fruit such as bananas, strawberries, milk and a touch of honey to make a frothy smoothie.

FREEZE CUBES of fruit to chill A DRINK without diluting it.

Oils and vinegars

Flavoured vinegars and oils have been used for generations to enhance the taste of everyday foods. Frying fish in olive oil with a hint of citrus adds a subtle tang, while herbal vinegars provide a fragrant finishing touch to salad dressings.

If you want to create flavoured oils, it is best to start with those that have little taste of their own, such as mild, refined (not virgin) olive, sunflower or corn oil. Good, clear wine or cider vinegars with at least five per cent acid are the best base for flavoured vinegars.

WHAT you need

● Clean bottles, tops, pots and other utensils plus fresh, washed herbs and other ingredients.
● Use tops made of cork, plastic or glass to seal the bottles or jars. Metal deteriorates, unless coated in plastic, due to the acid in the vinegar.

FLAVOURED oils

Here's a simple, basic method for home-flavoured oils.
1 Put lightly crushed herbs into a bottling or Kilner jar and pour oil over them. You will need about 150g herbs per litre of oil. If fresh herbs are not available, you can get similar results using dried herbs.
2 Let the oil steep for two weeks in the closed jar, stirring it once a day.
3 Strain the oil through muslin and pour it into bottles into which you have put sprigs of fresh herbs.

Chilli vinegar

1 litre vinegar
2 large mild chilli peppers
3 dried hot chilli peppers
1 teaspoon peppercorns
1 spring onion, sliced

Bring the vinegar to the boil, then pour it over the other ingredients and let it steep in a closed jar for two weeks. Shake daily. Strain and pour into glass bottles.

Add a couple of sprigs of fresh herbs to bottles of oil for decoration and to enhance the taste.

● Garlic, chilli peppers or citrus peel can be used, too.
● Make a basil oil by heating the oil to 40°C before pouring it over the fresh herb.
● Use sesame oil as a base for flavoured oils with Asian aromas or taste-free groundnut oil.
● Simply infuse oil with aromatic herbs such as rosemary or tarragon – there's no need to boil them.
● Oils can be kept for a year if unopened and stored in a cool, dark place. A smell and taste test will clearly show if they are still fresh.

FLAVOURING vinegar

Using a similar basic method, you can also create deliciously flavoured vinegars.
1 Heat the vinegar until it bubbles. Once it cools to 40°C, pour it over the crushed herbs in a bottling or Kilner jar (you need about 150g herbs per litre of vinegar). Make sure the jar is tightly closed.
2 Let it steep for two to three weeks, shaking from time to time, then strain and pour into bottles.
● In addition to herbs, you can also flavour vinegar with numerous spices. Thyme and tarragon go well with garlic, chilli flakes and peppercorns.
● Put the spices into a small muslin bag, add it to the boiling vinegar and let it simmer for 10 minutes.
● For a hint of sweetness, dissolve 2-4 tablespoons of sugar or honey in 1 litre of hot vinegar.
● Spiced vinegars can be kept for several years unopened. With time they will mature and become milder.

Pasta and rice

Noodles and rice have been staples on numerous international menus for centuries. Nourishing and extremely versatile, they come in many forms and can be combined with just about anything.

Nothing beats the taste of fresh, homemade pasta. While it requires some effort, it is not difficult and working with your hands to make food from scratch can be therapeutic.

PASTA dough

Here are some basic recipes for pasta dough.
For pasta without egg:
1 Mix 175g durum wheat semolina flour with 140g plain flour and a pinch of salt. Add 175ml water and knead into a smooth, elastic dough.
2 Add a little more water if the dough is too dry – it should be solid but not sticky.
3 Let it sit for 1 hour, then roll it out thinly, cut it into the desired shape and leave it in the air to dry.
● Pasta made from durum wheat is particularly good with thick sauces.

IT'S EASY TO roll out and cut pasta dough with a PASTA MACHINE.

For egg noodles:
1 Knead 260g pasta flour, 3 lightly beaten eggs, 1 teaspoon salt and 1 tablespoon olive oil into a smooth dough. (It is ready when the top is shiny and it loosens from the work surface.)
2 Let it rest, then roll it out, cut it up and let it dry.

TIPS for the pasta kitchen

● Brew coffee when you make pasta – it helps to have moist air in the kitchen.
● Turn dough over several times to make the process of rolling it out easier.
● Work on pasta dough in sections. Put the rest in aluminium foil to keep it from drying out.
● Egg noodles can be coloured and flavoured: beetroot juice turns them dark red; tomato purée light red; carrot or pumpkin purée orange; saffron yellow; spinach green; mushrooms brown and squid ink black. Mix these ingredients with eggs, oil and salt and knead them in with the flour. You may need additional flour to compensate for any extra moisture.
● To make tortellini, roll out the dough and cut small squares before adding a dab of filling in the middle. Carefully fold the dough over the filling. Moisten the edges in advance so that they stay together.
● Use a large, deep saucepan for cooking so the noodles can move freely while they boil.
● Pasta swells to about two-and-a-half times its original volume during cooking. Make sure you add it to water at a full rolling boil.
● Stir pasta around as soon as it is in the water so it keeps its shape and doesn't stick together.
● Don't add oil to the water. Although it prevents water from boiling over, it won't stop the pasta from sticking together and can prevent the noodles from holding the sauce well.
● Add a little stock to the water for a zestier flavour.

TYPES of rice & cooking methods

One of mankind's oldest cultivated plants, rice has shaped the culture, diet and economy of Asia. It is undoubtedly the world's most important food staple.
● Long-grain and basmati rice both have long, slender grains and a dry, glassy core. For best results,

wash them thoroughly before cooking. Then, bring one part rice and two parts salted liquid to the boil, lower the heat and cook, covered, for 15-20 minutes. **Note** There is no need to soak rice before cooking, but if you do so it will reduce the cooking time. Soaked rice should be rinsed to remove additional starch before cooking or the grains will not separate well.

● Before easy-cook or parboiled rice is husked and polished, hot steam is used to force about 80 per cent of the vitamins and minerals contained in the silver outer membrane into the rice grain so that it is high in nutrients. To cook it, bring one part rice and two parts salted liquid to the boil, then lower the heat. Cook, covered, for 20 minutes.

● Short-grain rice such as arborio is chalk white and soft and sticky at the core. It produces plenty of starch when cooked and can be used for risottos and sweet dishes. Soak it for 20 minutes prior to boiling. Use one part rice to one part water. Bring water to the boil, then reduce heat and cook, covered, for 10 minutes or until done.

● Brown rice is nutrient-dense and high in fibre. Soak it overnight or for 4 hours. Bring one part rice and two parts salted liquid to the boil, then lower the heat. Cook, covered, for 20-25 minutes.

TIPS for cooking rice

● Unless you're making risotto, don't stir rice while you're cooking it – the result will be a starchy mess.
● Most rice varieties triple in size when cooked. A side dish of cooked rice is about 125g.
● Rice is even tastier when prepared in stock or water and wine, as is usual when making a risotto.

Wild rice consists of the seeds of a type of grass that grows wild.

The water-to-rice ratio and cooking time vary for each type and brand of rice; always follow the packet instructions.

GOOD TO KNOW ☑

Valuable silver skin

The nutrients in a rice kernel are located mainly in the bran. For maximum nutritional value, choose natural or wholegrain rice that has not had the bran removed, or parboiled rice.

● Reuse cooked rice as fried or baked rice with the help of oil and spices, and additions such as eggs, leftover meat and vegetables. It's also good in soups.
● Add a pat of butter to the water to keep rice from boiling over.
● If rice is too moist after cooking, let it dry for 10 minutes in a baking tin in the oven.
● If the rice is sticky, drain it in a colander and pour over a kettle of boiling water to separate the grains.
● Store rice in a dry, well-ventilated place away from strong-smelling foods; it absorbs other flavours easily.

Potatoes

The ubiquitous spud, which originated in South America, has earned its place as the king of side dishes. The world's fourth largest crop is filling, economical and nutritious, packed with potassium, vitamin C and fibre. And with some kitchen know-how at your disposal, it is also extremely versatile.

Good potatoes smell earthy, not musty. In addition to carbohydrates and protein, they contain a long list of healthy minerals.

HINTS about spuds

● Texture varies considerably, from waxy to floury, so choose the right kind of potato for the cooking method you want to use. Waxy potatoes, such as Charlotte and Maris Peer, stay firm so are great in salads; drier, floury varieties, such as King Edward, Maris Piper and Desiree, are best for mash.

● Store potatoes in a dry, dark place. Do not keep them in the refrigerator.

● Store potatoes in a heavy brown paper bag or hessian sack. This will prevent them turning green, which occurs when they are getting too much light.

● As the skin is rich in nutrients, organic potatoes should be peeled sparingly or not at all.

● After boiling potatoes, drain and cover with a cloth to absorb the steam and keep them warm.

● Rather than mashing, crush potatoes lightly with a rolling pin. Dot on some butter and a sprinkling of herbs before serving.

● Try mashing potatoes with buttermilk or chicken stock instead of milk or cream. It contains less fat and fewer calories.

● Parsley and marjoram are particularly tasty herbs to use with potatoes. Pesto is also great in mash.

● Avoid wrapping baking potatoes in aluminium foil as this will hold in moisture, steaming the potato.

● To bake potatoes in the microwave, wash but don't dry them, pierce them with a fork, wrap them in kichen roll and cook them on a microwave rack. You can crisp them up in a hot oven for 20 minutes.

● Topping a baked potato with chilli, leftover stews, or vegetables au gratin turns it into a main meal.

● Potatoes are low-fat and low-calorie. To keep them that way, opt for low-fat toppers such as plain yoghurt, low-fat soured cream and chives or salsa.

GOOD TO KNOW ✓

Perfect mash

The best way to get perfect mash is by hand, using a masher, or by passing potato through a food mill. Always add milk and butter or oil while the potatoes are hot, as this will prevent lumps. After mashing, return potatoes to the pan and reheat to drive off any excess moisture. Serve at once.

● Leftover mashed potato is excellent for thickening soups, stews or sauces. Or, you can use it to top a shepherd's or cottage pie, if you also have some leftover meat and gravy.

POTATO cakes

Another good, traditional use for leftover mash.
1 Mix 250g mashed potato, 1 teaspoon baking powder, 75g plain flour and a pinch of salt until smooth.
2 Roll out to about 5mm thick and cut into 12 rounds with a pastry cutter.
3 Cook on a pre-heated griddle or in a heavy-based frying brushed with a little oil. Serve hot, with butter.

THERE ARE 5,500 varieties of POTATO worldwide; 80 are grown IN BRITAIN.

Preserves

Bottling is a relatively simple – and traditional – method of preserving fruit and vegetables that ensures you have a handy supply of tasty produce throughout the year.

In addition to spring-clip top or screw-top bottles or jars, you will need the following tools for preserving fruits and vegetables: kitchen scales, measuring jug, skimmer and funnel. A jar lifter makes it easier to remove jars from the hot water, and don't forget labels for the filled jars.

People who are sensitive to acids tolerate bottled fruit better than fresh fruit.

PREPARATION

- Choose purpose-made jars, such as Kilner jars, with rubber seals, glass lids and spring closures.
- Check all bottling jars for scratches and chips.
- Check the rubber seals used to keep the jars airtight for brittle spots.
- Before adding preserves, boil the jars and seals in vinegar and water and let them drip dry, bottom-up.
- You can also sterilise jars in the oven: place them on a baking sheet lined with kitchen roll for 10 minutes in an oven preheated to 160°C/gas mark 3.
- If you choose to boil them in a saucepan, make sure the preserving jars are all the same height and they don't touch the top edge of the pan.
- Place jars in warm water, then raise the heat; they can crack if placed directly into boiling water.
- When preserving in the oven, place the jars on a shelf or baking sheet. Don't preheat.

FRUIT compotes

In theory, all types of fruit can be preserved safely because they are naturally acidic and the added sugar will help to preserve them. Fruit compotes with ice cream or pancakes make a delicious winter treat.
1 Boil down about 2kg fruit with 1 litre of water and 300g sugar into a syrup.
2 Adjust the amount of sugar to the natural sugar content of the fruit so that the compote has a balanced taste.
3 Place your washed and chopped fruits in lemon water to keep them from oxidising and turning brown.
4 Layer the fruit no higher than 2.5cm under the rim of the jar.
5 Add sugar syrup, seal the jars and put them into a saucepan, or place them in the oven at 90°C for 30 minutes, or in a pressure cooker, following the manufacturer's instructions.

BOTTLING vegetables

You can get particularly good results from preserving cucumbers, carrots, peppers and tomatoes. Beans, peas, red cabbage and celery are also good choices. However, because they need heating to high temperatures, you must bottle non-acidic vegetables

Before preserving, remove the stems and stones from cherries for easy eating.

in a pressure cooker or you may end up with a case of food poisoning. Follow your cooker's instructions.

● Layer the vegetables tightly in the jars and cover with a boiling salt solution consisting of 1 tablespoon salt and 1 litre water.

● Leave a top space of about 1.5cm when filling. Too much space may result in discoloration of the top layer; too little headspace may cause the vegetables to swell and break the seal.

● Acidic vegetable mixtures containing tomatoes can be preserved at a temperature of 90°C for 90 minutes. A mixture containing beans needs 2 hours, but tomatoes with aubergines need only 30 minutes at 85°C or they become too soft.

● Shorten preserving time by blanching or half-cooking firm vegetables such as carrots in advance.

TROUBLESHOOTING

● If the bottled vegetables aren't crisp, the salt solution isn't strong enough.

● If they turn dark, use distilled water next time. Chemicals in tap water can produce a dark colouring.

AFTER preserving

● Remove the rings on bottling jars only after they have cooled completely (usually overnight).

● Check to ensure the jars are sealed by pressing down gently on the centre with your finger. If it pops up and down (it may even make a popping sound), it isn't sealed. If the lid has been sucked down, it has worked correctly.

● If there is no vacuum, put the jar into the refrigerator and be sure to eat the contents within a short period of time.

● You can keep preserved fruit or vegetables in a dark, cool location for up to a year.

● If the bottling jar resists opening, place it right side up in a pot of hot water to reduce the inner pressure. The jar will open more easily.

IN THEORY, ALL FRUIT CAN be preserved SAFELY.

Preserving in vinegar and oil

Fruits, vegetables and herbs are easy to preserve in vinegar or oil – as people have done for centuries. It puts home-grown produce to good use, produces delicious foods and gives bacteria little chance of thriving.

Select vinegars and oils carefully, as they will influence the taste of the finished product. And remember that cleanliness plays an important role in any method of preserving foods.

PRESERVING basics

● Choose mild, refined oils such as olive, sunflower, grapeseed and corn oil; they won't overpower the flavour of the ingredients.
● Add some herbs or spices for a stronger taste when pickling or preserving foods.

● Pick light, distilled varieties of vinegar with good flavour, such as cider, white wine or white malt.
● Practically anything with a reasonably firm peel can be pickled. Fruit such as pears and plums, or mixtures of vegetables and fruit taste wonderful in a sweet-sour mixture of ingredients.
● Never use dishes made of aluminium or copper when pickling. The acid in the vinegar will attack the surface and there is a risk of the pickles becoming contaminated.
● Before consuming the fruits of your labour, let the pickles or preserves steep for at least a month. After opening, keep jars in the fridge.

SOUR pickling

Use this basic recipe to sour pickle vegetables.
1 To keep vegetables crisp and prevent them from diluting the vinegar with their high water content, place them first in a salt brine for around 24 hours. Mix together 1 litre water and about 55g salt, and put the vegetables into a bowl and cover them with the brine. Rinse them thoroughly before use.
2 For the vinegar solution, boil up about 500ml vinegar and 250ml water, 1 teaspoon salt, plus any spices and herbs.
3 Briefly blanch about 1kg washed and sliced vegetables in the solution.
4 Remove them with a skimmer or slotted spoon and layer them in a jar to within 2.5cm of the rim.
5 Layer fresh spices between the vegetables. Commonly used spices include peppercorns, bay leaf, mustard seeds, whole garlic cloves and chilli flakes.
6 Rather than adding spices directly to the jars, you can also suspend them in the pickling liquid wrapped in a square of muslin tied up with string. Remove before filling the jars.
7 Boil the vinegar solution again and pour it into the jars. The liquid should cover the vegetables completely. Seal the jars and label them with the date and a description of the preserve.

Preserved delicacies in attractive jars
make unique and tasty gifts.

Sweet-sour pickled fruits must be boiled if they are to be kept for more than six months.

RECIPES & tips for sweet-sour pickling

● Add 2-3 teaspoons of sugar to the sour vinegar solution, according to taste, and then follow the same method as with sour pickling.
● Stone fruit such as plums can be layered in the jars raw, ideally halved and pitted. Fill up the jar with the hot vinegar solution and then seal.
● Good spices for sweet-sour pickling include cloves, ginger, star anise and allspice.
● Honey or fruit juice concentrate can replace some of the sugar.
● Sweet-sour fruits pickled in vinegar go well with meat dishes.

SEALING in oil

● Oil provides an airtight seal for the ingredients, thus improving their shelf life. Depending on the recipe (fruit and vegetables are the best ingredients to choose), additional preserving through salting, cooking or marinating in oil may be necessary.
● If you want to preserve 1kg vegetables, you need 500ml oil, 1 tablespoon salt, and spices and herbs according to taste.

GOOD TO KNOW ☑

Pricking pickled foods
Cucumbers and plums turn out spicier and crispier if you prick them all over with a cocktail stick before preserving.

● After washing and rinsing the produce, cut it into bite-sized pieces and steam it briefly. Cool before using.
● Dry tomatoes in advance in the oven. Lightly brown aubergine or sweet peppers under the grill. They don't need to be steamed.
● Layer the vegetables in the jars with spices and herbs and pour in enough oil – heated to about 75°C – to cover them.
● Press down firmly with a spoon so that all air bubbles rise and the jars can be sealed airtight.
● You can vary the flavourings: green peppercorns, thyme and lemon peel go well with mushrooms; tomatoes taste especially delicious when seasoned with basil, mint, rosemary and chilli; aubergines go well with garlic and lemon.
● As seasonings and herbs release their flavouring into the oil, it can be used for seasoning other dishes and for making marinades and salad dressings.

OIL PROVIDES an airtight seal FOR THE INGREDIENTS.

Rescues and solutions

We all have mishaps in the kitchen, no matter how experienced we are. But all need not be lost. Armed with some home remedies and a little improvisation, the burnt roast can be rescued, salty potatoes and watery soup salvaged – even if guests are at the door.

You can't prevent all misfortunes in the kitchen. Luckily, many mishaps can be remedied quickly and easily, and others are easy to avoid altogether.

PREVENTING cooking mishaps

● Milk won't burn so easily if the pan is rinsed out with cold water before heating it up.
● Add a little vegetable oil to the pan to prevent butter from browning.
● Beef stock will stay clear if you boil a clean egg shell with it.
● Boiled eggs won't split their shells if you make a small hole with a pin at the rounded end before you immerse them.
● Keep a roast from becoming tough by basting it regularly with its fat and juices.
● Prevent fried fish from sticking to the pan by dusting it with flour or putting a little salt into the frying oil.
● Fish holds together better during cooking if you pour a little lemon juice over it first and set the fish aside for a moment. The acid in the lemon juice cooks the flesh slightly, creating a thin seal.
● Prevent dishes burning in the oven by covering them with foil.

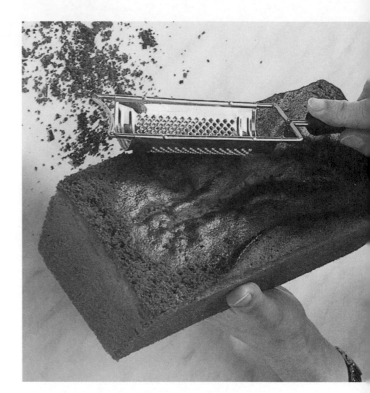

Scrape burnt spots off a cake when cool with a grater and cover with icing.

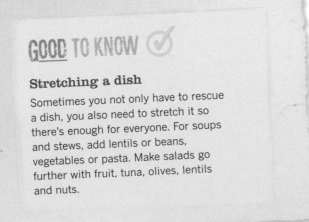

GOOD TO KNOW ☑

Stretching a dish

Sometimes you not only have to rescue a dish, you also need to stretch it so there's enough for everyone. For soups and stews, add lentils or beans, vegetables or pasta. Make salads go further with fruit, tuna, olives, lentils and nuts.

QUICK help

● Add a dash of vinegar or salt to the pan of water if an egg bursts while you are boiling it; the whites will thicken immediately.
● Wrap cracked eggs in aluminium foil before boiling.
● Tough stewing meat becomes tender when you add a dash of vinegar to the cooking liquid.
● To prevent homemade mayonnaise from congealing, beat an egg yolk and a little salt until stiff, then add the oil drop by drop.
● If mayonnaise begins to separate, add a tablespoon of warm water. Or start again with one egg yolk and gradually add the curdled mixture.
● Put excessively firm semolina dumplings into cold water for 10 minutes. They'll swell up more and become tender when boiled.

USE GRATED CHEESE
to rescue BURNED
CASSEROLES.

- Add a grated carrot or potato to overly spiced meat broth and bring to the boil.
- Add a sprig of parsley to a dish that is too garlicky for 10 minutes or so, to counter the taste.
- A little soured cream or plain yoghurt can often rescue bland soups.

TOO thick or too thin

- Thicken gravies or soups that are too thin. Make a paste with 1 teaspoon of cornflour or potato flour mixed with water into a smooth paste, then add it to the gravy or soup. Don't pour it all in at once; add small amounts and stir until you have the right consistency. Bring to the boil and season as needed.
- Gravies that are too thick can easily be thinned with water, wine, stock, milk or cream. Add seasoning if necessary.
- Pour lumpy gravy through a fine strainer to make it smooth again, or whisk it vigorously.

BURNT foods

- If potatoes boil dry and catch on the bottom of the pan, carefully remove all but the last unburnt layer, put them in a pot of fresh water to finish cooking, then add a dash of salt.
- If you burn the top of the roast, cut off the burnt bits and roast the meat in a clean pan with more fresh fat.
- If a casserole burns, remove the top layer, sprinkle it with cheese or breadcrumbs and dabs of butter and finish baking it.

- If noodles stick together, strain them into a colander and set it over a pan of boiling water. The steam will separate them.
- If gelatine clumps together, warm it up carefully while stirring constantly, then add it to a warm mixture.
- When you take a cake out of the oven, place the tin on a warm, damp cloth for 10 seconds to ensure that it comes out of the pan more easily when you invert it onto a plate or wire rack.
- To remove the fat quickly from an over-fatty gravy, sauce or soup, skim the top of the liquid with a couple of lettuce leaves or pieces of kitchen roll. The fat will stick to them.

TOO salty, spicy or bland

- In most cases you can neutralise saltiness with a mixture of cider or wine vinegar and sugar in equal proportions.
- Thin oversalted soups or gravies with water, wine, milk or cream.
- Grate a potato into soups or stews to reduce excessive saltiness. This will add thickening as well.
- Add 1-2 raw egg whites to salty soups; they congeal and soak up the salt. Then put the liquid through a strainer or simply skim off the egg white.

If a soup or gravy contains too much
fat, let it cool and then skim off the fat.

Sauces, gravies and dressings

A good sauce or dressing can add colour, flavour and even texture to an uninspiring dish. Once considered an essential culinary accomplishment, rediscover this kitchen craft and create inspiring accompaniments for salads, pasta, rice and roasts.

Wine, plain flour, cornflour or puréed vegetables all contribute to good gravies, while oil and vinegar are the starting point for a tasty dressing.

DEGLAZING

Deglazing is the simplest way to get a good base for a meat gravy.
- After browning, remove the meat from the pan and add water, stock or wine.
- Bring to the boil, stirring constantly, to incorporate the meat dripping.

THE trick to thickening gravy

Thickening gives gravy the right consistency and improves the flavour. Try these tried and tested tricks.
- Purée vegetables cooked in the meat stock to thicken gravy and add taste.

When you deglaze meat juice with wine or stock it takes on a more concentrated taste.

- Add 2 teaspoons of cranberry sauce or redcurrant jelly to give gravy a fruity kick.
- When braising meat, add a few small pieces of bread to the pot: they will disintegrate and make the meat stock creamy.
- Plain flour and cornflour are good thickeners. Just stir them into a little water and add to the gravy before bringing it to the boil. If you use half plain flour and half cornflour, gravy turns out thinner and lighter.
- A roux is a mixture of equal amounts of fat and flour, which is then cooked and combined with liquid to make a base for sauces. For a béchamel sauce, add milk to the basic roux; for brown sauce, add some meat or dark vegetable stock.
- A mixture of flour and butter (*beurre manie*) is good for thickening sauces shortly before serving. Work them together in equal proportions, then flake off small pieces and whisk them into the boiling sauce or gravy. Mixed flour and butter keeps for two weeks when refrigerated.
- Shortly before serving, soured cream, crème fraîche or double cream can be beaten into a meat stock to thicken it.
- Mix a few spoons of gravy with an egg yolk and add to the rest of the hot gravy while stirring constantly. Do not let the liquid boil or the egg yolk will 'scramble' and make the gravy lumpy.

SUCCESS with gravies

- Season roux-based gravies only after thickening. Don't forget to taste them before and after seasoning.
- Boil a gravy thickened with roux for at least 5 minutes to eliminate the floury taste. Add herbs such as thyme or sage to give gravy extra flavour.
- Prevent a skin from forming on light gravies by drawing a piece of butter over them with a fork.
- To enrich a gravy with butter, remove the pan from the hob and add the butter one piece at a time, letting it melt while swirling the pan a little.
- Whisk gravies vigorously to ensure they are creamy and free of lumps.
- To test the consistency of gravy, stick a wooden spoon into it. The gravy should coat the spoon.

Making a roux

1 Melt or sauté 20g butter in a pan and add 20g flour, while stirring vigorously with a spoon.

2 Keep stirring until the flour blends with the butter. For a dark roux, lightly brown the mixture.

3 Add your liquid (stock, water, milk, etc) and bring to the boil, stirring constantly until the sauce thickens.

● You can make gravy go further and add another flavour by adding stock, soured cream, cream, crème fraîche, wine, sherry or even the water from cooking vegetables or pasta.

● For perfect smoothness, strain gravy before serving.

THE perfect dressing

Vinaigrette is still the classic salad dressing – although it tastes good with fish and meat dishes, too – but it requires high-quality ingredients in the right proportions to work its magic.

● If you remember this one simple rule, you will never have to consult a vinaigrette recipe again: use vinegar and oil in a 1:3 proportion. If you are adding mustard, use about 1 teaspoon per 4 tablespoons of salad dressing.

● Stir some salt and pepper into the vinegar, and whisk in the mustard. Gradually pour in the oil while stirring constantly, until the vinegar and oil combine to form an emulsion.

● If the ingredients start to separate, beat the vinaigrette vigorously again.

● If you are in a hurry, put the vinaigrette ingredients in a lidded jam jar and shake to combine.

● If the vinaigrette is too oily, add more vinegar, mustard and spices. If the vinaigrette tastes sour, add oil and salt. A little sugar won't hurt, either.

● You can make a vinaigrette in advance. It will keep in the refrigerator for around three weeks. Shake it thoroughly before use.

● You can vary the basic vinaigrette recipe by adding crushed garlic, mayonnaise, nuts, dried fruit, sesame seeds or herbs, depending on the dish to be served.

Use the correct proportions of oil and vinegar to make the perfect vinaigrette.

SEASONAL PRODUCE

Growing fruit and vegetables at home is satisfying, healthy and economical. However, you need to know each crop's best season – although geographical location and access to a greenhouse or polytunnel can affect this.

Follow these guides for harvesting produce – then turn to pages 170-1 for traditional methods of storage.

Fruit • Main availability

Apples • August to December

Apricots • July to September

Blackberries • End of July to beginning of October.

Blueberries • July to September

Cherries • End of May to July

Currants (black and red) • June to August

Damsons • September and October

Figs • August to October

Gooseberries • June to July

Grapes • August to October

Melons • July to September

Peaches/Nectarines • July to September

Pears • August to October

Plums • July to September

Quinces • September to January

Raspberries • May to October

Rhubarb • March to June

Strawberries • May to July

Vegetables · Main availability

Artichokes (globe) · April to September

Artichokes (Jerusalem) · October to March

Asparagus · April to June

Aubergine · July to September

Beetroot · May to December

Broccoli · July to October

Brussels sprouts · January and February; October to December

Butternut squash · August to September

Cabbage · All year

Cauliflower · October to April

Celery · June to October

Coriander · May to October

Courgettes · June to September

Cucumbers · August to October

Endive · September to November

Fennel (Florence) · July to October

French beans · May to October

Garlic · June to July

Kale · June to November

Lamb's lettuce (corn salad) · May to November

Leeks · July to February

Lettuce · May to October

Onions · July to September

Parsley · All year

Parsnips · November to March

Peas · June and July

Potatoes · July to December

Pumpkins · September to November

Radishes · April to October

Rocket · April to October

Spinach · April to October

Swede · October to March

Sweetcorn · July to October

Sweet pepper · June to October

Tomatoes · July to October

Turnips · June to October

Shopping

People who shop and store their food logically save a lot, both in terms of time and money. Making a detailed shopping list and sticking to it reduces the risk of buying too much – and succumbing to impulse purchases.

Fresh, high-quality and often well-priced goods are available direct from farm shops or a local farmers' market. Wherever you shop, if you stick to seasonal and regional products, you'll get fresher goods and protect your pocket as well as the environment.

Choose seasonal and regional produce to save money and protect the environment.

PLANNING purchases

- If you organise a supermarket list according to product categories, you'll find you get through the shopping much more quickly.
- Place the shopping list in a visible spot in the kitchen. That way, anyone who needs something can jot it down immediately. But don't forget to take the list with you when you shop.
- Making a meal plan for the coming week can help you to purchase wisely, buying only the necessary ingredients. But be flexible: try to incorporate any specials that are available in the store or supermarket.
- Never shop while hungry as the danger of impulse buys rises exponentially and you are likely to spend more money than you had planned.
- Opt for generic brands whenever possible. These tasty copycats are made by reputable companies and are much less expensive.
- You will often find good deals at weekly markets. Bear in mind that shortly before closing time the prices on fresh goods often drop significantly.

CAREFREE shopping

To protect groceries from damage, put durable items (such as cans and bottles) into the trolley first, then fresh foods on top. If you cannot refrigerate frozen produce quickly, use special insulating bags.

- It is usually cheaper to buy in bulk, especially if you can share groceries with neighbours or friends.
- To reduce rubbish, choose multi-use packaging or buy products in bulk.
- Beware of buy-one-get-one-free offers unless you are sure you will use the extra food, can freeze it or share the purchase.
- You will generally find the most economical items on the lower shelves in supermarkets.
- Look for supermarket freezer cabinets that are neatly organised and not coated with thick ice. If the goods show any signs of freezer burn, walk away.
- Dented cans are best left on the shelves.
- You can tell when fish is fresh by its clear, bright eyes, bright red gills and shiny, moist skin.
- Well hung beef will be deep red; pork should always be pink.

STORAGE

Store foods properly. Temperature, light and moisture can affect their appearance, taste and vitamin content. Foods that are stored improperly can easily become a health risk or go bad.

● Prepackaged goods with a long shelf life, such as dried foods, are best kept in a dark cupboard or a dry larder.

● Products kept in jars, bottles or clear plastic bags should be stored in the dark.

● Bread remains fresh for a long time inside a cloth bag, which takes up little room.

● Put new supplies behind those you already have to hand. Check use-by dates but don't be a slave to them. Use your eyes and nose to test for freshness. Yoghurt and dairy products with bulging tops are almost certainly inedible.

WELL cooled

● Foods that spoil quickly, such as fish, meat, sausages and prepared foods, belong on the middle shelves, but be careful that meat juices won't drip onto dairy produce.

● Store all dairy products and cheese directly above the meat drawer.

● Foods that need only minor cooling, such as butter and eggs, are best stored on the top shelf or in the door compartments provided for them. Opened jars of jam and chutney can also be stored here.

● Fruit and vegetables should go in the lower drawer, without packaging, if possible.

● Take thawing meat out of its packaging and put it in a bowl or on a plate in the refrigerator, covered with aluminium foil to defrost.

● Fresh fish will keep for up to a day in the fridge; if cooked, it will last for two to three days.

● Store eggs in the fridge with the rounded ends uppermost; that way, the air remains on top and the inner membrane doesn't dissolve.

● To stop mould from forming, store blue cheeses separately from other types of cheese. Wrap each piece of cheese separately. Old-fashioned waxed paper is excellent for cheese, or use greaseproof.

● During the cooler months of the year, less delicate fruits and vegetables can also be stored in a garage.

STORE FOODS in the refrigerator CORRECTLY to increase shelf life.

Soups and stocks

Stock is a traditional kitchen staple, the byproduct of other kitchen processes and the starting point for many soups. Boiling vegetables, meat, fish or poultry in water produces stock, as does boiling down leftover bones, aromatic vegetables and spices.

A clear soup can be eaten with added chunks of meat, chopped vegetables, egg, noodles or dumplings. More substantial soups use stocks as a base but are thickened with a roux, cream, crème fraîche, soured cream or puréed vegetables. Purée or liquidise the pan contents after cooking to create a smooth soup.

SOUPS and broth

Use as large a pan as possible, as the ingredients must be able to float.
● For a simple meat broth, you need about 400g beef or poultry, 2 carrots, 2 trimmed leeks, 1 onion and 1 stalk of celery, plus salt, pepper and a bay leaf. Simmer all the ingredients in 1.5 litres of water for 30-40 minutes.
● Cook noodles, barley or rice separately and add to the soup later or it will turn unappetisingly cloudy.
● Spices such as bay leaves, cloves and star anise are best wrapped in muslin, tied into a bag and hung in the pot while cooking so you won't have to take them out afterwards.
● If you want to eat the meat cooked in the soup, such as boiled beef or chicken, make sure the water is boiling before adding it.
● A little cognac or a dash of vinegar in the cooking water makes the meat in a soup particularly tender.

STOCKS

To make a stock, you also need to add meat bones or fish bones and carcasses, if you can get them. Or make a stock from a selection of vegetables.
● To make a light stock, for every litre of water in the pan add about 750g chopped bones, 1 onion, 1 bouquet garni, mirepoix (a chunk of celeriac, 1 carrot and 1 leek), a few peppercorns and 1 clove of garlic. Slowly bring to the boil, then simmer on low for 4-5 hours before straining.
● For a dark stock, roast the bones and vegetables in the oven first.
● Stocks should simmer but never boil. That way, the stock is richer tasting and clearer. In addition, boiling for a long time causes the stock's vitamins and minerals to go up in steam.
● Stock is ready when it has a concentrated aroma. If chicken bones were used, they should be coming apart.
● For a fish stock, mix 1.5 litres of water with about 500ml dry white wine and add vegetables, salt, pepper, parsley and about 1kg lean white fish trimmings. Bring to the boil, reduce the heat and let it simmer over a low heat for 20 minutes. If you cook the stock any longer, it will become bitter.

TIPS for soups & stocks

While cooking soup or stock you will notice foam from congealed protein forming on the surface. Skim it off regularly, along with excess fat.
● Remove fat by skimming it or letting it soak into some kitchen roll. You can also let the liquid cool, then remove solidified fat from the top with a spoon.
● Clarify a finished stock by boiling a beaten egg white in it. The egg white absorbs suspended solids as it congeals – just strain the stock before you use it.

Fresh herbs and spring onions garnish clear soups perfectly.

- You can also clarify stocks by straining them through a fine sieve or a piece of muslin.
- Add little or no salt to soups and stocks initially as they become more concentrated as liquid evaporates. It is best to add spices and seasoning carefully towards the end of cooking.
- Use only whole peppercorns in stocks. Ground pepper starts to taste bitter if it is cooked for a long time.
- Use parsley stems for stocks – the flavour is stronger than in the leaves, which tend to turn bitter after lengthy cooking.
- Stocks or soups usually keep in the refrigerator for two to three days and can last up to three months in the freezer.

PURÉED & cream soups

For puréed soups, root vegetables, legumes, potatoes and squash or pumpkin are the best choices. The starchy ingredients combine with one another, making the soup nice and creamy. Add some curry paste to a puréed soup for a more intense flavour.

- For a puréed soup, soften about 550g vegetables in a pan with a few tablespoons of olive oil, then cook them along with root vegetables, salt and pepper in 1 litre of water. Press the soup through a strainer or purée in a blender, then season to taste.

- After braising the vegetables, cream soups can be thickened with 1-2 tablespoons flour and finished off with 250ml cream, soured cream or crème fraîche.
- Thin out soups that are too thick with milk or a little water. Or thicken soups that are too thin with some cornflour made into a paste with milk or water.
- Puréed soups can be garnished with cream, butter, fresh herbs, croutons or crumbled bacon, or finish with a swirl of plain yoghurt.
- When preparing a cream soup, start by making the roux (see page 209), then whisk in the stock a little at a time so that no lumps form.
- To stop a cream soup from curdling, boil acidic ingredients such as tomatoes well. Then make sure you pour the soup into the cooking pan containing the cream, rather than the other way around.
- Finish a minestrone or other clear soup with freshly grated Parmesan and croutons.
- For a semi-smooth texture, process or liquidise a soup mixture with a series of quick pulses.

Meat stock

1 Cover the washed vegetables, spices and meat with cold water.

2 Bring the stock to the boil, then simmer for a while. Insert a spoon between the pot and the lid to let the steam out.

3 When the cooking is done, remove the meat and pour the contents of the pot through a strainer and into a second pan.

Spices

Pepper, ginger, saffron and nutmeg are an integral part of today's cuisine, adding taste and colour to many dishes. But all spices must be stored correctly or their flavours will fade.

Using spices correctly will enhance a meal, quite literally 'spicing it up'. Here is how you can keep them fresh and put them to good use.

Use a ceramic mortar as wood absorbs the flavour of spices.

Fresh curry powder

2 tablespoons coriander seeds
2 teaspoons cumin seeds
½ teaspoon mustard seeds
1 teaspoon black peppercorns
1 teaspoon fenugreek seeds
10 fresh curry leaves
½ teaspoon ground ginger
1 tablespoon ground turmeric
1 teaspoon chilli powder or
cayenne pepper

Dry roast the coriander seeds, cumin seeds, mustard seeds, peppercorns and fenugreek seeds over a medium heat until the seeds darken and become fragrant. Stir constantly to prevent burning. Leave to cool, then grind to a powder.

Dry roast the fresh curry leaves, grind and add to the mixture along with the ground ginger, ground turmeric and chilli powder or cayenne pepper.

STORING spices

● Store spices in airtight, opaque containers in a cool, dark location. They last longer when not exposed to sunlight.
● Since spices tend to give off and absorb flavours, always keep the same spice in a given container.
● Stored properly, spices keep for up to two years and many last even longer. But they are best bought in small quantities and used more quickly – ideally within six months.

PROPER use of spices

Clever use of herbs add subtle flavours that can help you to reduce the amount of salt you add to food.
● Spice meats and vegetables before cooking, when they are most absorbent. However, salt removes moisture from raw meat making it tougher, so it is wise to salt meat only after browning.
● Marinades season and tenderise meats.
● Spices should heighten the taste of the food, not overpower it. So use them in moderation.
● Many spices, such as paprika, chilli, garlic and curry, lose their flavour or become bitter when you add them to bubbling hot oil or butter. They release their aroma and taste better when lightly sautéed.

COMMON kitchen spices

● Use caraway to season bread, cabbage dishes, casseroles and curry.
● Cayenne pepper is dried, ground chilli. It adds pungency to sauces and stews, such as chilli con carne. Use it in moderation and taste between each addition to ensure that a dish is not too fiery.
● It is a good idea to wear rubber gloves when chopping fresh chilli peppers. The oils in the inner membranes and the seeds can burn your hands, eyes and lips. For milder dishes, discard the potent membranes and seeds. Dried chilli peppers can be ground as required in a mortar or spice mill.
● The aroma of cloves evokes thoughts of Christmas treats and mulled wine, but their spicy-sweet flavour also works well in marinades, apple dishes, pear compote and with red cabbage.

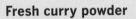

- Stick or ground cinnamon is used in Asian dishes, sauces, baking and many desserts.
- Coriander is used in Spanish, Middle Eastern, Asian and South American cuisine. The leaves should be used fresh and can be sprinkled like parsley on salads, soups and other cooked dishes. The seeds are best bought whole and ground as needed. Coriander is the base of most curries and can be used for meat, poultry, fish and vegetables.
- Cumin gives a bite to plain rice and is used in many stews, curries, grills and lamb and chicken dishes. It is much loved by Indian, Middle Eastern, Spanish and Portuguese cooks.
- Curry, a mixture of various spices, is used to add a savoury punch to curry or rice dishes, sauces, poultry and meat.
- Paprika ranges from sweet to spicy or smoky. It releases its colour and flavour when heated, but burns easily. So add paprika only when liquid ingredients are present and don't cook it for too long.
- Ginger adds a fresh, slightly pungent taste to baked goods, stir-fries, rice dishes and fruit.
- Raw horseradish is particularly pungent. It gives a kick to roasts and winter soups.
- No cook can get by without pepper. Unripe green peppercorns have a citrusy aroma. Black pepper is harvested shortly before ripening, then fermented. Red peppercorns, in contrast to white, are unshelled and have a more delicate flavour.
- Saffron adds taste and colour to rice dishes, seafood and lamb.
- Star anise, which has a mild aniseed taste, and is a key flavour in Chinese cuisine.

- Allspice or Jamaica pepper is a key ingredient of Caribbean jerk seasoning and an ideal ingredient for meat rubs of all kinds.

WORKING with spices

- Choose whole spices when possible (for example, peppercorns). They retain their flavour better.
- Grind spices with a mortar and pestle just before use for maximum flavour. Add a little salt for traction.
- Spice mills are useful for nutmeg, pepper and other hard spices. Make sure they are easy to clean. Or buy a coffee grinder specifically for the purpose.
- Soft spices, such as red peppercorns and juniper berries, can be crushed, cut or mashed with a knife.
- Reserve special graters for ginger or nutmeg, as their flavours can be transferred easily.

ASIAN CUISINE relies on a FRAGRANT cornucopia OF SPICES.

Sweets and treats

Refined sugar should be consumed in moderation, but we all need the occasional treat. Many homemade goodies are easy to make and – if you use fresh, quality ingredients – even better than shop-bought.

Homemade ice cream or sorbet on a hot day is delicious and refreshing, while truffles or candied fruit are a fitting finale to a meal with friends.

CHOCOLATE temptation

Ganache, a mixture of chopped chocolate and cream or butter, is the basis for many types of truffles and is simple to make. Use more chocolate in proportion to cream for a firmer ganache.

● While ganache is still warm in the pan, add liqueur, rum, sparkling wine or champagne, honey, coffee, peanut brittle, nougat, dried fruit, syrup or many other delicious ingredients.

● By blending ganache made from different types of chocolate, you can make marbled chocolates. For example, the contrast between a layer of white chocolate with vanilla and a layer of darker ganache with orange liqueur is both attractive and tasty.

● You can decorate truffles with many things, as long as they taste good and look pretty. Try icing, whole nuts, candied fruit or coloured decorations made from marzipan. You can roll the balls in cocoa powder, icing sugar or coconut, or use finely chopped nuts and chocolate crumbs.

ICE CREAM: ideal for hot days

Water is the basis for sorbets, while ice cream gets its creamy texture from milk and cream. You can flavour both with sugar, fruit and any number of other tasty ingredients.

● For sorbet, heat up about 250ml water and 200g sugar and boil for 2 minutes until the sugar dissolves. Stir in about 550g puréed fruit or 500ml fruit juice and a little lemon juice. Leave it to cool. Flavour with sugar or lemon and freeze.

● Ice cream is equally easy to make: beat 3 egg yolks with 85g caster sugar until foamy then stir in 225ml cream and 175ml milk. You could also add puréed fruit, nuts, chocolate or yoghurt.

● If you don't have an ice-cream maker, put the ice cream or sorbet mixture in a shallow metal bowl in the freezer. When nearly frozen, stir thoroughly and freeze again. Repeat this process several times.

TIPS for making ice cream

● Ice cream loses flavour when frozen, so the initial liquid should taste quite sweet and strong.

● If ice cream turns out too grainy, it may be because you have used too much water or too little sugar. Alternatively, the ice cream may have frozen too quickly or wasn't stirred frequently enough.

Making ganache

one In a small pan, quickly heat 150ml double cream. Bring to the boil, then take the pan off the hob to cool completely.

two In the meantime, chop up about 300g good-quality chocolate (at least 70 per cent chocolate solids) and melt it in a bain-marie.

three Stir the cream and chocolate together to produce a smooth liquid. If necessary, stir in a little more cold cream.

While melting chocolate, stir it continually to make sure it doesn't separate and go grainy.

3 Leave to steep for a day before removing and drying the fruit. Then boil up the sugar solution and pour it over the fruit again.

4 Repeat this process five times. The last time, boil the sugar solution down further and let the prepared fruits dry on a cooling rack.

● Strawberries can be crystallised but other berries aren't good candidates because they are too soft.

● For fruits such as grapes, which can be crystallised whole, it is best to poke a hole in them to allow the sugar solution to penetrate them. Crystallise orange and lemon slices with the pips removed and peel intact, using only organic fruit.

● To coat fruit with chocolate, cut it into bite-sized pieces and spear them with a wooden cocktail stick before dipping them in liquid chocolate.

INVENTIVE snacks for children

Children can often be persuaded to eat fruit and other nutritious foods if they look appealing.

● Fashion an edible caterpillar by skewering melon balls, grapes and cheese cubes on an uncooked spaghetti strand.

● Pour melted chocolate over a mixture of dried fruit, cereals and nuts to create a delicious fruit-and-nut cluster.

● Cut pineapple pieces into triangles and arrange on a plate in a circle to create the rays of the sun. Spoon vanilla yoghurt into the centre for dipping.

● For the classic 'ants on a log' treat, fill the centre of a stick of celery with peanut butter and sprinkle raisins (or 'ants') along the log.

● Create a flower children will want to devour by using segments of a clementine or tangerine for the petals, a strawberry for the centre and kiwis sliced into the shapes of leaves.

ICE CREAM on a stick IS EASY TO MAKE with lolly MOULDS.

CRYSTALLISED fruit

Fruits are crystallised by dipping them in a strong sugar concentrate to preserve them and give them a sweeter flavour. Alternatively, you can coat pineapple, apples, bananas, pears or strawberries with chocolate.

1 For the syrup, boil 1 litre water and 1kg sugar until strings form.

2 Suspend 550g fruit or pieces of fruit in a strainer in a glass bowl and pour the sugar solution over the fruit until it is completely covered.

Table and place settings

An imaginatively decorated table makes a dinner party a feast for the eyes as well as the stomach, and can be quite a talking point – just make sure the decorations add style without interrupting the conversation.

When preparing for a festive occasion, arrange the table so it is topical and attractive – but remember to keep things simple for both you and the guests.

TABLE linens

Choose table linens according to the occasion, making sure they don't clash with the dinner service, silverware or glasses.

● A tablecloth should hang over the edge of the table by about 25-30cm on all sides. Lay a table protector under it so your table isn't damaged by hot dishes and spills, and to keep the tablecloth from slipping.

● White tablecloths allow the greatest latitude in choosing the other decorations. To add some colour, place a runner or band of pretty fabric lengthways or widthways across the table.

● For rustic settings, use simple straw place mats.

● If you need several tablecloths for a long table, make sure that the edges overlap and that the tablecloths are all straight.

SET GLASSES above the SILVERWARE ON THE RIGHT-HAND side.

around the napkins with raffia, or transform silk ribbons into colourful napkin rings. At Christmas time, fasten small wooden toys around the napkins with red or green ribbons.

TABLE settings

● Plates should be placed about 1cm from the edge of the table. The distance between the centres of two place settings should be 60-80cm, so that you can fit a good number of guests at the table without anyone feeling cramped.

> ### GOOD TO KNOW ✓
>
> #### A seasonal table
> Generally, nature provides us with most of the things we need to create a beautiful table. In autumn, there are colourful leaves, berries and nuts for decorating. In spring and summer, make inexpensive table decorations from fresh green twigs, daffodils, tulips, daisies, grasses or moss. Evergreen branches, mistletoe and pine cones add a festive element in winter.

● Napkins can be tucked into wine glasses or folded and laid on the plates or to the left of them. Use napkin rings or fold the napkins into decorative shapes.

● With a little creativity, you can make one-of-a-kind napkin rings. Tie some grasses, flowers or leaves

- There should be no more than four pieces of silverware on the right, three on the left.
- Always place the dessert silverware horizontally above the plate. Point the handle of a fork to the left and the handle of a spoon to the right. If you are serving fresh fruit for dessert, place a dessert fork (instead of a spoon) at the top edge of the plate.
- For a formal dinner, add extra knives and forks for a first course outside the main course cutlery.
- Place a bread plate, along with a small butter knife, to the left of each setting if necessary.
- Set glasses above the silverware on the right-hand side in the order in which they will be used.
- White wine glasses should be positioned slightly further right.
- Red wine is generally served in large, wide goblets that taper slightly towards the top.
- Fill wine glasses only a third to a half full so that the wine has room to breathe.

THE final touch

Don't overdo table decorations. If your guests have to peer through the flowers to talk to each other, you may have gone too far. Table decor should encourage a festive atmosphere without hindering conversation.
- Place cards belong at the top of the place setting and can be decorated with drawings, photos or small flowers.
- For children's birthdays, spell the names of guests using alphabet biscuits, sweets or cupcakes.
- Menus are used mainly for official occasions, but at private parties they can add sophistication and

Long-stemmed glasses and individual gifts help to create a sense of celebration.

serve as souvenirs. Make them yourself for a personal touch: experiment with different shapes or decorate them to coordinate with the invitation.
- For large parties, you can also write the menu in chalk on a blackboard.

FLOWERS & candles

- Use floral decorations that are appropriate for the season and nature of the gathering. Remember to cut them short so they don't hinder eye contact across the table.
- Instead of using a large vase of flowers as a centrepiece, which usually has to be removed during the meal, scatter flower petals or arrange small, individual bouquets or pots on the table. Individual flowers and long-lasting greenery, such as sprigs of ivy, also dress up a festive table.
- Make sure candles are in secure candlesticks so they don't tip over.
- For a fairly large table, use small lanterns or tea lights in decorative glasses. Match the colour to the floral decoration.
- To make a beautiful table even more festive, scatter polished pieces of coloured glass, shiny confetti or coloured leaves over the tablecloth.
- For special celebrations, put a small chocolate on every plate, such as an Easter egg, Santa Claus or small heart – according to the occasion.

THRIFT IN THE KITCHEN

With a little planning, skill and creativity, you can save plenty of cash in the kitchen – without giving up on good food. Look for savings by altering household appliance usage, changing your shopping habits and making savvy use of leftovers.

MAKE FRUIT YOGHURT BY STIRRING JAM into plain yoghurt

Try these alternatives

- Instead of making a roux, thicken gravies or soups with mashed potato or another mashed vegetable, such as swede or parsnip.
- If you don't want to buy a whole carton of buttermilk for a recipe, add 1 tablespoon of vinegar or lemon to just under 250ml whole milk and leave it to stand for 10 minutes.
- If you run out of cream for making sauces, use condensed milk or coffee cream instead. However, these replacements won't froth up so can't be used as a substitute for whipped cream.
- Preserved fruit makes a delicious topping on a cake.
- Most nuts can be used for pesto, including walnuts and almonds.
- You can turn granulated sugar into icing sugar easily in a coffee grinder or spice mill.
- Make stale bread into breadcrumbs and freeze, ready for use in everything from stuffing to toppings for dishes such as fish pie or cauliflower cheese.
- Make a funnel from a plastic bag by cutting off one corner.

Save energy

- Household appliances typically use up about 30 per cent of your home's energy. It is worthwhile in the long run to replace energy-wasting appliances such as old refrigerators, freezers and electric cookers.
- A gas cooker is more economical than an electric one.
- Look for good-quality, thermo-conductive pots and pans.
- Always put pans on the heating element that is closest in size to their base, and use a tight-fitting lid while cooking.
- Use residual heat by turning off an electrical element about 5 minutes before whatever you are cooking is done.
- A pressure cooker can save up to 30 per cent of the energy used for hob cooking if the cooking time is more than 20 minutes.

Enhance flavour

- Brush dried-out sponge cake with milk and bake it briefly in the oven so it tastes fresh again, or use it up in a trifle.
- Purée uncooked mushrooms with a little stock, freeze them and use later to add flavour to sauces or soups.
- Turn a hollandaise sauce into béarnaise with a pinch of tarragon.
- Add chopped gherkins, chives and capers, plus some lemon juice, to mayonnaise for a flavour-packed tartare sauce.

Budget

- Plan your menu around the supermarket specials.
- Seasonal items are not only cheaper but also tastier and fresher because they have taken less time to travel from producer to consumer.
- By tracking household expenditure for a month, you can get a good idea of where your money is going and modify your spending if necessary. An easy way to do it is to keep an envelope in your handbag or car and use it to store receipts, then tally them at the end of the month. Alternatively, keep a notebook in your handbag and jot down what you spend.
- Stocking up sensibly saves pounds. When goods become expensive, just pull out the berries you froze in the middle of summer.
- Buy unsliced bread as it stays fresh longer. If a loaf looks too long, cut it in half and freeze one portion.
- Packaged foods always cost more. Prewashed salad, for example, will set you back an extra 50 per cent.

With a little planning, skill and creativity, it is easy to save cash in the kitchen.

Cut down on food costs

- Try not to let foods spoil. You can make purées, jam, smoothies or compotes from fruit. Freeze leftover gravy and sauces (in an ice cube tray, for example).
- Choose offal – liver and kidneys are inexpensive and tasty when cooked with onions and herbs.
- Shop thoughtfully: brisket, silverside, pork hock and other less expensive cuts of meat are 10 to 30 per cent cheaper but need longer marinating and/or cooking times to render them tender. Plan your meals in advance and take advantage of slow cookers to make the most of your money.
- After you have eaten a roast chicken or turkey, boil up the bones to make stock.
- Even leftover alcohol can be re-purposed: use wine that has gone sour instead of vinegar in a dressing or marinade.
- Precook: cooking twice the amount of potatoes or pasta saves time. A day or two later you can warm them up and serve, or use in a soup or casserole.

Style and comfort

Our surroundings should be havens, where we can relax, unwind and enjoy time with family and friends at the end of a hectic day. Here are some useful tips to help you to enhance individual rooms and create the comfortable refuge you want your home to be.

Balconies and patios

While the size and position of a balcony or patio are often beyond our control, with a little effort and the right accessories any outdoor space can be turned into a relaxing oasis.

When thinking about how best to utilise these outdoor spaces, try to imagine an outdoor room. Choose appropriately sized outdoor furniture and consider how to protect yourself from the sun.

OUTDOOR furniture

- For small balconies or patios it is a good idea to look at space-saving furniture, such as a wall-mounted folding table with a swivel-out table leg to stabilise it.
- Stackable stools and folding balcony chairs fit into a limited space, which makes them convenient to store. This is especially important for houses with small balconies or limited storage space.
- On a small balcony, chairs with adjustable backrests are more practical than lounge chairs.

- Multi-functional furniture makes maximum use of the space: watertight chests, for example, make convenient seats as well as providing storage for chair cushions and pillows.
- A canopied beach chair makes the most of a long, narrow balcony. Set up lengthways, it makes a cosy seating area as well as providing a screen from the wind. Protect it against rain and weather with a robust, weatherproof tarpaulin.
- Weather-resistant outdoor furniture is a sensible option to avoid covering up or carrying heavy furniture inside every time it rains. Garden furniture

AN ADJUSTABLE TABLE is IDEAL for the BALCONY OR PATIO.

made of woven synthetic fibres looks like wicker furniture but resists inclement weather and UV rays.
● Hammocks or hanging chairs are more appropriate for larger patios – you need room to swing and a stand or solid beam to hang them from.
● Big balconies or patios are ideal for sun loungers, deck chairs and clusters of sofas centred around an outdoor table.
● Large outdoor spaces can be structured into 'rooms' with plant pots, folding screens, raised flower beds, canvas curtains and outdoor furniture.

SUN protection

Most balconies and patios require some protection against sun and glare in the form of a parasol, shade sail, awning or trellis covered in greenery. Here are a few of the more common choices.
● Parasols come in many sizes, shapes, patterns and materials. Buy them in the autumn if you want a bargain. A parasol usually stands on a pedestal but you can get space-saving brackets that screw onto the railings or fixtures for the wall or ceiling, which are better for small balconies.
● Retractable awnings are fabric or vinyl shades on a permanently mounted frame. With the help of a crank or a motor they can be fully or partly extended, depending on your preference and the location of the sun. Installing an awning can be difficult because the frames tend to be heavy and they must be firmly mounted. If in doubt, employ a professional to install it.
● Shade sails are made of a tough, sturdy cloth and secured to fastening points with cords or snap hooks.

They come in standard shapes, but it is possible to have one custom-made to fit a patio or balcony.
● Plants can also be used to provide some shade. Tall tub plants, such as quick-growing bamboo, are a good choice. You can also plant grapevines and other greenery to climb a lattice frame or trellis, or train climbing plants up a pergola.

CREATE an outdoor room

Just like the interior of a house, an outdoor space needs to be well decorated if you want it to feel cosy and look complete. Small accessories can play a big role here, for example, by helping to create a maritime theme with colours or lanterns.
● Terracotta pots, ancient-looking sculptures or wrought-iron furniture create a Mediterranean mood.

Lushly planted pots and planters beautify any outdoor space.

● A small pond with water lilies or a large basin with floating candles helps to create a tranquil, relaxed ambience. The gentle splash of a fountain or small waterfall can have the same effect.
● Bright, colourful seat cushions and incense sticks will evoke the spirit of Asia, as will red and gold accents – and, of course, a bamboo plant or two.

Beautiful bedrooms

To create the right environment for a good night's sleep make your bedroom a relaxing haven, adding all the accoutrements of comfort and serenity.

Getting the correct amount of sleep is key to our health and well-being. Consider investing in a new bed, matress and bedding if slumberland eludes you.

CHOOSING a headboard

Not every bed has a headboard, but they can add the finishing touch to a bedroom's decor.
● MDF (medium density fibreboard) may be relatively inexpensive but it is less sturdy. Also, some MDF is made using toxic glues and binders that may affect the air quality in the bedroom. Check there is no chipboard beneath upholstered headboards.

SLATTED frames

These offer some real advantages in terms of bed height and weight.
● Rigid slatted frames vary considerably, depending on whether or not the frame has been reinforced in vulnerable spots. Some models allow you to adjust the supports to suit your sleeping position.

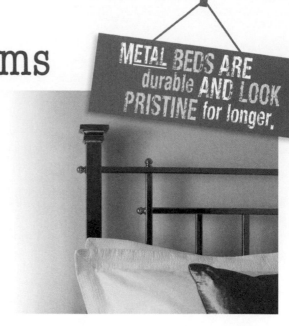

METAL BEDS ARE durable AND LOOK PRISTINE for longer.

● Adjustable slatted frames let you alter the height of the bed – an advantage if you want to read before going to sleep. But a well-made adjustable bed can be expensive.
● Futons sometimes come with simple, adjustable wood frames.
● Check the number of wooden slats to determine the quality of the frame – the more the better.
● Electric bed frames let you adjust the frame effortlessly with a remote control.
● If you and your partner have different needs, consider a double bed with its two halves joined – each tailored to your individual requirements.

To create the right environment for a good night's sleep make your bedroom a relaxing haven

CHOOSING a mattress

Selecting the right mattress means not only finding one that's the right size but also looking at the materials used and whether it suits your budget and needs. Here are the most popular mattress styles.

● Continuous coil mattresses are the cheapest, but the coils often wear out more quickly and because the springs move as one unit you are more likely to be disturbed if your partner moves during the night.

● Open coil mattresses are one step up, and are constructed of single springs fixed together by wire.

● Pocket spring mattresses are the most comfortable. As each spring is sewn into its own pocket, it moves independently and provides better support.

● Memory foam mattresses, or memory mattresses, are topped with a layer of temperature-sensitive viscoelastic material that moulds and remoulds to the body's contours. Although more expensive, they provide support and help to relieve pressure on painful joints. However, the foam is affected by room temperature so can be cold in winter and make you feel very hot in summer. You will find that the foam 'loses its memory' after about ten years.

● Latex or high-density foam mattresses offer a high degree of sleeping comfort as they conform perfectly to your body and sleep position while minimising pressure on your head, shoulders and hips. They are breathable, so you won't overheat, and are also a good option if you are allergic to house dust mites.

● Protect a mattress and maximise comfort with an under blanket or purpose-made cover.

THE right pillow

Always choose a pillow that conforms to your sleep position. If you sleep on your side, your pillow has to prop up your head so it should be thicker than the pillow for a back-sleeper.

● If you have problems with your neck, opt for a smaller pillow (about 80cm wide but only 40cm long). A larger pillow could shift during the night, resulting in neck and back pain in the morning.

● If you have selected a pillow carefully but your neck pain remains, consult a doctor. Neck pain can be due to serious issues that require treatment.

● Goose and duck down are the softest, lightest natural fillings but not good if you are prone to allergies as they harbour dust mites. They need regular shaking to avoid lumpiness. Good alternatives are synthetic down or foam.

● Test pillows for their springiness – they should quickly return to shape after being pressed – and replace them as necessary.

SIDE SLEEPERS need a PILLOW THAT SUPPORTS their HEAD.

DUVETS and sheets

You will probably want duvets and sheets made from different fabrics to cater for changing seasons.

● A duvet or continental quilt is the most popular form of bedding and, like pillows, may be filled with

GOOD TO KNOW ✓

The canopy bed

Many people dream of having a bed with a canopy. In former times, it was not only a symbol of wealth and nobility, it offered a degree of privacy – desirable when several people had to share a room. Thick brocade curtains have now gone out of fashion but a canopy bed still provides a feeling of safety and security, as well as a touch of old-fashioned romance.

feathers or synthetic material. Choose a duvet by its thermal resistance or tog rating; this will range from the warmest at 13 or 14 to the coolest at 4 or 5. For maximum flexibility through the year, choose a duvet made in two parts that fix together with poppers, one with a medium and the other a low tog rating.

● Sheets and duvet covers are made from all kinds of materials, such as cotton, polycotton, satin and silk. Flannel sheets will warm you in the winter but cotton or linen is the best choice in summer.

Blankets and throws

Whether you are relaxing on the sofa with a good book or spending a summer evening on the patio, there's nothing nicer than wrapping yourself up in a soft blanket or comfy throw.

For cool summer evenings or during late spring or early autumn, a pure wool blanket or throw is just right for wrapping around your shoulders or draping over your knees when you are feeling chilly.

WOOL blankets

● The traditional wool blanket – made of pure sheep's wool – still has a place in many households. It is wrinkle and odour resistant.
● Merino wool, lambswool, camel hair, alpaca or cashmere blankets are of especially high quality and can't be beaten when it comes to keeping you warm

It's comforting to wrap up in a cosy throw when you want to relax on the sofa.

GOOD TO KNOW ☑

Quilts: rich in tradition

Quilts are bed covers sewn in three sections, providing insulating layers, and they are easy enough to recreate. For the display side, use fabric remnants to create a patchwork. The middle layer consists of soft padding and the backing is usually plain. Quilts can be used in many other ways – as bed throws, tapestries, sofa blankets, tablemats or rugs.

● Don't wash wool blankets – or wash them rarely. Always check the laundering instructions. Instead, air them regularly and brush them occasionally.
● Blankets made of polyester or fleece are easier to launder regularly and much lighter in weight than most woollen blankets.
● Cotton blankets are easy to care for, durable and affordable. As light blankets they are suitable for every season and they are easy to care for: just put them in the washing machine and then the dryer.

THROWS

Throws are medium-sized decorative blankets usually used on the bed or sofa. They are made from a variety of materials and come in different sizes.
● A decorative two-sided, patterned or coloured throw on the sofa can quickly give a living room a fresh new look. Or use a larger throw on a bed to change your bedroom decor.
● Plain-coloured blankets or throws can have a balancing effect on colourfully patterned sofas.
● If space is at a premium, an attractive throw and colourful pillows can transform a bed into a comfortable place to sit.
● Throws need to be laundered frequently, so make sure you choose one that is machine washable – and safe to put in the dryer, too.

Book care

Books make wonderful companions, enthralling, educating and entertaining us. For that reason, every book collection should be lovingly cared for and properly organised. But root out volumes you no longer read and give them to charity.

Keeping books well organised and dust free will help to prolong their life and ensure that you and your family can enjoy reading them for years to come.

GOOD TO KNOW ✓

Vacuum cleaner accessories

Many vacuum cleaners have a small dusting brush tool that can help you to preserve your books and keep them dust-free. A small brush with soft bristles should allow you to dust whole rows of books gently and quickly without having to empty the shelf. Often, all you have to do is pull the books out about 5cm.

STORING books properly

- Choose bookcases with glass doors that will keep out dust.
- Display multi-volume dictionaries or antique books along walls with open shelves, so that they are easy to access.
- Try to install your bookshelves in a dry room or use a dehumidifier in damper climates or rooms. Moisture can damage valuable books.
- Don't pack books too tightly on the shelves, especially over a heating unit; it will cause the glue in the bindings to become brittle.

- To slide books out easily, make sure there is about 2.5cm of space between the top edges of the books and the bottom edge of the shelf above. This makes it much easier to take out the books you want to read.
- Arrange books according to authors, themes or special subject areas. This makes it easier to find what you are looking for and much simpler to replace a book afterwards.
- Store heavy volumes on the bottom shelf. This will make the bookcase more stable.
- Leave plenty of room between subject areas so you can add new books without having to reorganise the entire bookshelf each time.

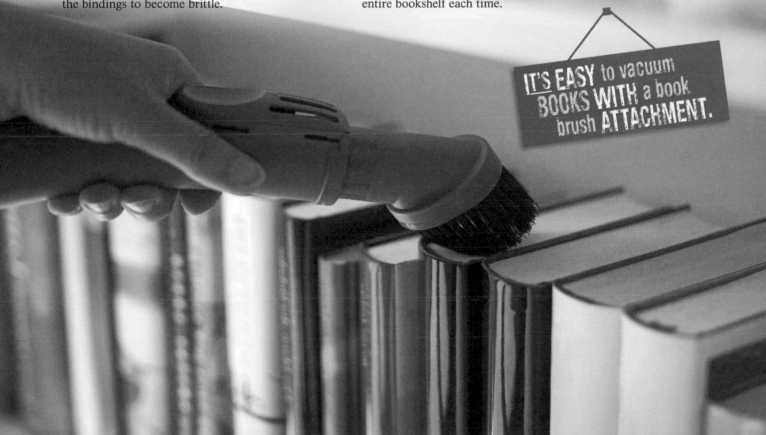

IT'S EASY to vacuum BOOKS WITH a book brush ATTACHMENT.

Candlelight

Candles, together with oil and tallow lamps, were once the home's only light source. Today, we use their warm, soft light more as a decorative element or for creating an intimate atmosphere.

The use of candles dates back to ancient civilisations and over the centuries they've been made with a variety of substances, including tallow, beeswax and palm, coconut and olive oils. Today, most candles are made from paraffin.

A FEW ARRANGED CANDLES CAN BE simply elegant.

MATERIAL

● Candles made from beeswax are of high quality but relatively expensive. Real beeswax candles give off a sweet fragrance when they burn.
● Petroleum-based paraffin is the material most frequently used in the manufacture of candles today. Paraffin candles are popular as they are an affordable option.
● Soya wax is made from hydrogenated soya bean oil. Because it is a natural substance, soya wax is a much more environmentally friendly choice than paraffin. Soya candles are also less sooty when they burn.
● Stearin, or stearic acid, is made from vegetable and animal fats. Adding stearin to paraffin candles makes them easier to manipulate during production. The candles also burn more slowly.

CREATIVE candlesticks

A special meal is enhanced by elegant silver or crystal candlesticks. But what about a cosy evening by the fire or a romantic getaway at the cottage? Here are some time-honoured solutions for adding candlelight to the occasion.
● Use an old wine bottle with a raffia-wrapped bottom, or any nicely shaped bottle. Just insert a candle in the top; the wax flowing down the side will give the bottle a vintage look. Use a plate or coaster to avoid staining the cloth or table beneath it.
● You can also paint bottles, cover them with découpage or fill them with coloured sand or ornamental stones.
● For a pretty dining-room table centrepiece, create an arrangement in a pot with an oasis of floral foam and seasonal blooms, then tape toothpicks to the ends of candles and insert them in the centre.

● Make candle holders by hollowing out spaces for votive candles in melons or in pumpkins, as you would for Halloween.

LANTERNS

Sometimes a lantern works better than a candle – outdoors, for instance, as candles can be too unstable or are blown out easily.
● Candles in jars can look like twinkling fairy lights on the patio. Use large-bellied jars with sand, stones or seashells filling their bases.
● Use glass-frosting spray on jars containing candles to create a calm, soft light.
● Drop tea lights into well-cleaned baby food jars and arrange them in small clusters on the table and sideboard. They'll look fabulous.
● Make tin cans into lanterns by punching holes in a pattern with a hammer and nail. To avoid denting the can, fill it with water and freeze it for several hours.

PLAYING with fire

Candles are useful, warm and romantic. But they can be dangerous. Stay safe with these three simple rules.
1 Never leave candles unattended, not even for a short time – this includes tea lights and lanterns.
2 Use a container in which the candle can stand up straight.
3 Trim the wick down to about 1cm.

Choosing carpets

Style but also practical issues will govern your choice of floor coverings. Where areas are likely to be well-trodden, it's sensible to opt for hard-wearing fabrics. Elsewhere colour and effect will play a greater role.

There's an enormous selection of carpets and runners available. Before deciding what to buy, decide how the carpet is going to be used, whether it will fit in with the room's furnishings, what the floor currently looks like and how much you are willing to pay.

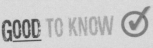

GOOD TO KNOW ✅

Carpet quality

The main indicators of carpet quality are its density (how closely individual strands are packed), its twist (how many times the carpet's fibres have been spun) and the height of its pile. When you bend it backwards, the less backing you can see the better the carpet. Look for individual carpet fibres that are tight and neat; loose or flared strands indicate substandard quality. As a general rule, the higher the pile the better the carpet will wear.

THE right choice

● On wood, laminate or stone floors, a rug's primary function is to create an accent, so the design is extremely important.
● A rug in a child's room should be easy to clean and care for. Choose one that is washable or whose pattern hides stains.
● In an office area with rolling desk chairs, carpets have to be durable.
● Different certificates provide helpful information. The Wool Mark, ensuring the quality of fibres is perhaps the best known. Others, such as the GoodWeave label, are ethical, helping oppose illegal child labour in the carpet industries of South Asia.

TYPES of carpets

Different carpets are suitable for different situations. You wouldn't want a luxurious, deep-pile carpet in the porch any more than you'd put AstroTurf in the living room. Here are some of the things to bear in mind when you are choosing a carpet.
● Sturdy carpets have short, tough fibres. They are often made wholly or at least largely from synthetic fibres.
● Carpets of higher quality are usually made from natural fibres such as wool with deep pile. They are softer but also get dirty more easily and require more maintenance.
● With loop-pile carpets, the threads of the pile, whether they are made of wool, synthetics or mixed fibres, are sewn into the basic weave. Since loop-pile carpets are relatively impervious to stains, they are a good choice for rooms with more traffic. However, beware if you have pets – their claws can catch in the loops and pull.
● Velour, plush or velvet carpets are cut-pile carpets in which the loop pile is sheared off for a softer texture. Because of its velvety surface, this type of carpet works well for bedrooms or living rooms. A plush carpet in which the threads have been twisted double under heat is particularly sturdy.
● Needle-felt carpets are high-tech and durable. They're produced by electrostatic attraction of individual fibres, which results in a highly resistant carpet. Ultra sturdy, they are most commonly installed by hotels and businesses rather than home owners.

Choose hardy, woven carpets or runners for busy areas of the house.

TIPS for carpets and runners

Here are some things to bear in mind when it comes to the safety, maintenance and finishing of any carpet, runner or rug you decide to lay in your home.

● Look seriously at carpet tiles. They are a sensible option for busy areas of the home. You can replace worn or dirty squares rather than an entire carpet.

● Install wood trim along the border instead of carpeting. Wood doesn't get nearly as dusty and it is far easier to clean than most carpeting.

● Use cornflour paste to stick down harzardous corners. When the paste dries, iron the corners flat (put a piece of wrapping paper between the iron and the rug just in case). Then remove any remnants of the cornflour with a small nail brush.

● Turn a runner 180 degrees occasionally to prevent excessive wear on one end, especially if that end is near the front door or another busy part of the house.

● Choose dry carpet-cleaning products rather than carpet foam for cleaning deep-pile carpets. It can take several days for a damp deep-pile carpet to dry completely.

● Hire a carpet steamer or cleaning machine once a year to clean all carpets thoroughly.

● Help deter moths and eliminate unpleasant smells by sprinkling carpets and runners occasionally with a little cedar or lavender oil.

● Renew faded colours by wiping down the carpet with a solution of water and vinegar, mixed in a

TO AVOID COMPRESSION marks REPOSITION FURNITURE regularly.

10:1 proportion. Just be sure to test an inconspicuous area for colour fastness first.

● Deal with indelible stains by covering them with a rug or, if you can afford to take the risk, by creating a carpet of many colours. Punch out the dirty spots with a metal hole punch. Then, acquire some discontinued carpet patterns (inexpensive) or samples (free). Punch out shapes to insert into the carpet's holes. If you get it right, you will end up with a creative design. If not, console yourself with the thought that you'd have had to replace it anyway.

Clocks

The comforting ticking and nostalgic charm of an old-time clock can still provide a striking element in today's home decor.

The earliest mechanical clocks ticked away in cloisters back in the Middle Ages, while pendulum clocks have been around since the 17th century. Since then, timepieces have become increasingly precise and compact, and the advent of the digital clock has changed the way we regard timekeeping. But many people continue to take pleasure in the steady rhythm of old-fashioned analogue clocks.

GOOD TO KNOW ✓

A radio-controlled clock

People who shy away from the maintenance requirements of an antique clock will appreciate the advantages of a radio-controlled model. This ingenious invention independently registers the signal beamed from a long-wave radio time signal broadcaster or the global positioning system (GPS). This way, it always displays the correct time and never needs to be adjusted by hand. These clocks also take care of the seasonal change from winter to summer time automatically.

ANTIQUE clocks

● Grandfather clocks are classic, weight-driven pendulum clocks. They fetch high prices at an antique shop, but you might be lucky and find one at a flea market or online – though bear in mind that it may need to be refurbished, which can be quite expensive.
● Smaller wall clocks are more compact and so easier to integrate into a space than a huge grandfather clock. Popular finds include quirky cuckoo clocks and pendulum clocks with a wooden case and glass on three sides.
● Mantel clocks come in many different forms. Some are protected by a glass cover, while others have wooden cases with glass fronts.
● Bakelite, and other Art Deco clocks from the 1930s are already sought-after collectibles. Look for them at flea markets or car-boot sales.

WINDING a clock

Mechanical clocks need to be wound regularly, most having an eight-day cycle. Choose a day and time to wind an antique clock as consistency is important to good maintenance.
● Wind a weight drive before the weights reach the bottom. When you are resetting the weights, lift them slightly to protect the bearings (and oil them sparingly occasionally).
● Winding a spring drive calls for particular care. Count the turns so you don't twist the key too far.

CLOCK maintenance

Here are some other points to bear in mind to ensure that an antique clock remains in good running order.
● Never move the hands on antique clocks backwards. You risk damaging the mechanism. Stop the clock temporarily instead.
● Treat the varnished case surfaces carefully with cleaning polish and then seal them with a good antique wax or beeswax.
● Remove the pendulum and immobilise the pendulum arm when transporting a pendulum clock. For short moves, hold the case in such a way that the pendulum can lean against the mechanism.
● Learn to live with a little imprecision if you are the owner of an antique clock. Continually adjusting the time can damage some delicate mechanisms.

Most longcase clocks chime the time on each hour and fraction of an hour.

Colours

The colours you surround yourself with at home can influence your mood and sense of well-being. In general, the function and style of a particular room determines whether you choose warm or cold, dark or light, and strong or subdued colours.

Different colours have different characteristics. They can give a home a modern or traditional feel and by combining different colours, can add visual interest to a room.

Colour charts displaying cold colours (green to purple) and warm colours (pink to yellow).

COLOUR considerations

● Bright colours (vibrant shades of green and blue, red yellow and orange) provide an expansive feeling. These are friendly, happy colours that encourage communication and are therefore especially welcome in the dining area and kitchen.

● Dark colours, such as deep red, purple, and dark shades of blue and green, can have a constricting and gloomy effect. But when applied in the right place or as accent elements, they can help convey comfort and security.

● Warm colours (orange, yellow and pink hues, for example) raise the perceived temperature of a room. For that reason, they are best used in rooms that face north. Because they inspire activity, avoid using them in rooms meant for relaxation, such as a bedroom.

However, yellow in a bedroom is very good for combating SAD (seasonal affective disorder) and early morning depression.

● Cold colours, such as icy blues and green, have a calming effect. They are especially suitable for bedrooms, helping you to relax in the evening and wake up refreshed the next morning. But they need to be used with care in north-facing rooms.

● Navy blue can create a cold atmosphere that discourages conversation; do not use it in living or dining areas.

● Red raises the energy level of a room but it may also make people more irritable and hostile – so it is not a good choice for a child's room. Use it as an accent rather than a base room colour.

● Grey should be avoided for the dining area and kitchen – unless you want to dampen your appetite.

HOW colours work in a room

The right choice of colour can make a room appear bigger or smaller and make ceilings look higher or lower. This visual effect can actually compensate for some of a room's flaws.

● Use bright light colours for small rooms; they will make them appear more spacious.

● Choose warm, dark colours, such as deep shades of red, for large rooms to make them look cosier.

● Use paler colours on low ceilings: the ceiling appears higher when it is painted in a lighter shade than the walls.

● Use darker colours for high ceilings. If you want to reduce the height of a room visually, opt for a dark-coloured ceiling. The ceiling will also appear lower if you paint the bottom area of the wall in a lighter shade that gradually darkens as it rises towards the ceiling. If you want the ceiling to appear higher, the colour should gradually become lighter as it rises from floor level.

● Opt for lively shades for narrower rooms; bright colours on the walls make a room appear wider to the eye.

● Darken wide rooms. If you paint two walls opposite each other in a darker shade, a wide room will look less cavernous.

THE right colour strategy

To ensure a harmonious design for a room, you will also need to choose attractive colour combinations for furniture, fabrics and accessories that complement the walls and ceiling.

● First, choose a basic colour that you like and make sure it's appropriate for the room. Use this colour for the walls, rugs and curtains, perhaps in varying intensities.

● Choose a consistent secondary colour for furniture and accessories. For a unified look, choose a complementary colour or design the room in colour coordinates (for example, different shades of the same colour). Or choose two secondary colours, but in that case the colours should appear next to each other in the colour spectrum.

● Be careful when combining two colours of different intensities. For example, placing strong

Pair neutral colours with fresh accents in the form of cushions, artwork, accessories and throws.

colours next to pastel shades forces the eye to jump back and forth between light and dark. This can create a visually disturbing effect and affect the room's atmosphere. All pastels mix well together.

● Pair neutral colours such as white or magnolia with some fresh accents in the form of cushions, artwork, accessories and throws in colours such as red, green, blue and even pink.

● If you are uncertain what colour to choose, ask for advice from a paint store or interior designer. Buy 'tester' pots and experiment before you commit.

INJECT a little colour

Want to make a few changes to liven up a room? Here are three easy do-it-yourself projects.

● Re-cover a lampshade. Trace and cut out the shape of the shade in wallpaper. Glue the ends with wallpaper paste for a slipcover. Summer calls for a fresh lime green against neutral furnishings. In winter, warm it up with a rich brown or red.

● Create a fabric wall hanging. Sew a hem across the top and bottom of a piece of attractive fabric and insert a wooden dowel at each end.

● Paint the wall behind book and other shelves an accent colour that coordinates with the room.

Decorating

Each season has its individual charm, and this can be reflected in your home and garden. Bring the rhythms of nature into the house by choosing seasonal fabrics, table decorations, flower arrangements and embellishments.

SPRING

Emphasise Easter by using colours such as bright and pastel yellows, pale and vivid greens, and eggshell blue.
● Decorate the windowsill with early flowering plants, such as hyacinths, tulips and daffodils, in small terracotta pots.
● Opt for sheer curtains, tablecloths and choose bedlinen that will lighten the look of a room. Look for plain, embroidered or printed designs that complement interiors.

Opt for sheer curtains to lighten the look of a room and complement interiors.

● Rework your accessories. Bring out floral china or clear glass, and switch dark-coloured sofa cushions for more cheerful tones.
● Reorient furniture that focuses on the fireplace and arrange it to take advantage of a garden scene.

SUMMER

A varied mix of bright colours can express the fullness of the season. Here are a few ways to add summertime touches.
● Laying a white linen slipcover over the sofa helps to create a summery mood, especially if you use colourful cushions as accents.
● Counteract summer's heat with cool hues: outdoor tables draped with blue and white are like a cool breeze in the garden.
● Create a maritime flair by filling glass tubes with sand of different colours and grain sizes, and setting them out on the balcony or patio.
● Swap a heavy, oriental carpet for a light and natural sisal rug.
● Replace fireplace logs with white candles of different heights and sizes.
● Arrange sunflowers in glass vases and light them up with lanterns at dusk to add a warm, decorative element to garden parties or barbecues on the patio.

AUTUMN

This season is characterised by nature's rich harvests. With just a little effort, you can decorate a house or apartment with foliage and fruit from field and forest – think pumpkins, squashes, red and golden leaves, and pine cones.
● Echo the changing colours of the leaves with accents in warm shades of orange and red. Dark green and harvest gold are also the perfect match for warm autumnal hues.
● Let autumn usher in soft, textured fabrics such as suede, velvet and felt. Satin in autumnal tones adds glamour to table decorations.
● Place decorative elements such as fresh or dried leaves, chestnuts, acorns or beechnuts in plain bowls to add interest to tables, sideboards and windowsills.

● Create autumnal room and table decorations from small squashes, dried wheat stalks, corn cobs and gourds: simply spread them out on the table.
● Bright red apples and juicy pears in a large glass bowl make an attractive – and edible– centrepiece.

WINTER

Even though it may be cold outside, you can create a cosy atmosphere indoors. Turn up the heat and spruce up your home for winter with these tips.
● Bring ice and snow to mind with white, turquoise and bluish-green. Of course, the true colour of winter is snow white, which can be beautifully combined with other seasonal colours.
● Alternatively, chase the winter chill out of your home with rich shades of purple, gold and red.

● Use winter decor fabrics such as soft wool, luxurious satin, chenille and lambswool for cushion covers, throws and drapes.
● Add warmth to a room (literally and figuratively) by swapping sheer curtains for heavy drapes.
● Use bright woollen rugs to warm up the living room or bedroom floor – it's quick and inexpensive.
● Make a colourful, cosy winter bolster by sewing two scarves together lengthwise, stuffing them with wadding and securing the ends with ribbons.
● Use decorative glassware such as vases or lanterns to contrast pleasantly with rustic-looking decorations made of wood.
● Opt for a vase full of berried holly or evergreen twigs, or a bowl filled with glittering Christmas bauble decorations, apples or oranges for an attractive winter touch.

Rich shades of purple, gold and red keep the winter chill at bay.

Displays

Almost everyone has a collection of cherished objects to display, whether it's porcelain, coins or trophies. When shown in a glass cabinet or display case with the proper lighting, collectibles are conversation pieces, too, that add interest and a personal touch to a room.

There are three fundamental factors to consider when it comes to displays: the case, the glass and the lighting.

A TASTEFUL DISPLAY CAN BECOME A focal point IN A ROOM.

Keeping display objects behind glass protects them and will prevent them from becoming dusty.

DISPLAY cases

It is easy to create display cases of comic figures or cars for a child or adolescent's room – or your own.

● Looking to display miniature items? With a bit of luck, you might find an original typesetter drawer or case (originally used to store printer's letters) at the flea market or on the internet.

● Bear in mind that presenting objects in open display cases means they can easily become dusty, which means more work for you and an increased risk of damaging the objects.

● Keep valuable collector's items behind glass. In the old days, almost every living room had a glass cabinet for displaying precious crystal, porcelain figurines, coins or pewter. These cabinets still work well today, and make for an interesting design element in themselves because they are rare and have a certain nostalgic appeal.

LIGHTING

The way you light treasured objects can play a big role in how attractive the display is. It may even be worthwhile to install a special lighting system to offset treasured collectibles properly. Here are a few pointers for displaying items in the best light possible.

● A lighted glass cabinet displays objects in the right light. You may also choose to highlight special pieces with small halogen spotlights.

● A light shining directly from above can be harsh and unappealing – ideally light should be projected at a 60 degree angle.

● Before choosing a lighting system, check that electric light won't damage any of the display objects. Not all lights are cool enough for use in a display case, and some collectibles are sensitive to both temperature and light.

Door wreaths

These are an individual and friendly way to welcome guests to your home. Choose the basic components of a wreath according to the season and occasion. During the Christmas season, decorate wreaths with pine cones and ribbons to spread Yuletide cheer.

The proper wreath can set the right mood for any get-together – simply fasten decorations to suit the occasion. Every season brings with it a new collection of flora that would go well in a wreath.

AS the seasons change

● A simple wreath made with willow twigs can be a lovely harbinger of spring.
● Decorative grasses and hay, as well as flowers, make pretty components for a summertime wreath.
● In the autumn, use tendrils of hops, ivy or grape leaves.
● In December, wreaths of entwined evergreen branches are a reminder that Christmas is coming.

SPECIAL occasions

● Use Santa Clauses, chubby-faced angels and wooden toys to herald Christmas, and colourful Easter eggs and chicks for Easter.
● Explore the shape possibilities. Door wreaths don't have to be round. Other basic shapes include hearts and ovals, but there is no need to stop there.
● Ribbons, strings of pearls and dried flowers can add a wonderful romantic touch. To create a cheerful mood for an anniversary party, choose a variety of bright and colourful ribbons.

INCORPORATE DRIED FRUIT INTO A festive Christmas wreath.

● Create unique designs with figurines, hearts and twinkling lights. For a wedding, weave a little bridal couple into a wreath of roses.
● Dried flowers give wreaths a lovely floral touch but they tend to be fragile, so don't use them on doors that are frequently opened and closed.
● To celebrate a child's birthday, display the boy's or girl's age in the centre of the wreath using small, brightly coloured flowers or beads.
● For a teen who has just passed his or her driving test, make a clever wreath that includes little cars or traffic signs.
● Fragrant decorations such as herbs and spices look lovely and smell even better when incorporated into a wreath.
● Spray evergreen foliage with gold or silver paint for an extra festive touch.

An autumn door wreath

one Deck a Styrofoam wreath with twigs, ivy tendrils or other greenery. Wrap it in place with florist's wire.

two Add decorations such as flowers, fruit and other embellishments. Attach items that lack stems with florist's wire.

three As a final touch, decorate the wreath with colourful ribbons.

Fireplaces and wood burners

The sound, smell and warmth of a crackling fire adds ambience and a traditional welcoming feel to a home. Fireplaces and wood-burning stoves can also help to reduce heating costs by lessening your dependence on more expensive fuels.

Attractive as they are, wood-burning fireplaces actually suck heat out of the home through the chimney, as well as releasing emissions into the environment. If you have an old wood burner, consider putting in a fireplace insert or replacing the current one with a newer, high-efficiency model. No matter what type of fireplace you have, it is a potential hazard, so make sure you observe the appropriate safety regulations. Also you need to be sure you are not breaking clean-air regulations. You need a wood burner described as 'clean burning' in a smoke-controlled area. If you are unsure, check with your local authority.

OPEN fireplace

A built-in fireplace can be a source of cosy warmth, but you will need to take into account certain factors to keep it safe.

- Use a fireplace screen to prevent sparks flying out.
- Install a fireproof glass panel for safety and an unobstructed view of the fire without fear of sparks.
- Make sure you have the right fireplace tools: bellows and tongs aid in stoking the fire. A small shovel is useful for removing ash.

A GAS FIRE WARMS UP A ROOM without producing ASH.

WOOD BURNERS and fireplace inserts

The contemporary variants of the open fireplace are wood burners and fireplace inserts. In both cases, the fire burns in a closed chamber and is visible through a glass panel. You get all the appeal of a wood-burning fire with none of the hassle, smell or soot. Here are a few things to consider.

● Install a wood-burning fireplace insert. It can help you to maintain a traditional look by turning an inefficient masonry fireplace into a wood burner with an efficiency rating of about 70 per cent. However, one disadvantage of wood burners and inserts is that dust particles land on the stove and burn up there, making the air dry although the fire is closed off.

● Wipe the glass panel regularly with window cleaner to remove soot. Just remember to spray the window cleaner on a cloth rather than directly onto the panel – and only when it is cold.

● If you are looking for something traditional, opt for a cast-iron stove.

● Find out the efficiency factor of a stove. It informs you how much energy is actually transferred to the surrounding air in a room.

GAS fireplaces and fireplace inserts

Gas fireplaces and fireplace inserts give you the look and warmth of a wood fire but with the added efficiency and ease of use of a modern convenience: you can turn them on or off with the touch of a remote control button or wall switch.

● Choose a gas fireplace to fit the decor of your home. A two-sided gas fireplace makes a lovely room divider, for example. But if you are fond of the antique mantelpiece, buy an insert to fit the fireplace rather than replacing the whole thing.

● Find a certified inspector to check the fireplace periodically for carbon monoxide and leaks. Many utility companies provide a yearly safety inspection service for furnaces and water heaters – ask if they will include the fireplace.

● Check the chimney flue for blockages, such as bird nests and leaves, before lighting the first fire in the autumn. Have the chimney swept every year, when it can also be checked for safety.

● Gas fireplaces don't produce smoke and other by-products that affect people with asthma or allergies. They are easy to maintain and don't require you to sweep up ashes or deal with stray sparks.

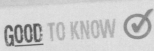

GOOD TO KNOW ✓

The right fuel

Modern wood burners burn logs, wood briquettes and wood pellets. Around 700kg of dry, seasoned firewood can replace the energy from about 210 litres of fuel or 200m³ of natural gas. Ashes of a light grey to grey colour are a sign of efficient combustion, containing no carbon remnants – you can even use them in the garden as a high-potash fertiliser. And add dried orange peel as you start a fire. It burns well and gives off a pleasant aroma.

A cast-iron stove adds a traditional touch as well as contributing heat to a room.

Floral decorations

A bouquet of flowers isn't just for young lovers or celebrations – the right floral decoration can bring a room to life. Take advantage of garden flowers to create beautiful displays in summer, and use dried flowers and arrangements to add an elegant, welcoming touch in winter.

Flowers are an attractive addition to any room. Whether placed in a corner or used as a centrepiece, they add colour and rich scents everyone can enjoy.

CUT flowers

A fresh bouquet of cut flowers looks lovely. Employ these traditional tips to prolong enjoyment of them for as long as possible.

● Gather cut flowers only during dry weather. Snip open blooms in the morning and those that haven't yet opened in the evening.

● Put flowers into a bucket of lukewarm water up to the petals immediately after picking. Before you arrange them in a vase, cut the stems again.

● Spray cut flowers with water occasionally to keep them fresh in the heat of summer.

● Lengthen the life of flowers with a milky sap (like euphorbias and poppies) by dipping the stems in boiling water for 10 seconds to cauterise the stalks.

● Gently remove all lower leaves from the stem so none will be immersed in the vase water.

Floating garden

Decorative bowl
Water
Flowers or leaves
Stones

Fill the bowl with water and add the flowers or leaves. To keep the stems under water, weigh them down with stones. Rub the big leaves with a little vegetable oil so they shine. You can even add a little aromatic oil to the water to enrich the aroma.

● When flowers begin to wilt, shorten the stems by about 2.5-4cm and put them in hot water for a while, using a paper towel to protect the blooms from the steam. Then move them to deep, cold water for 1 hour before arranging them in the vase again.

● Use wire or tape to provide extra support for the stems of bent flowers.

DRYING flowers

● Hang freshly cut flowers (tied together in small bouquets) upside down in a dry, well-ventilated room. The stems will thin as they dry, so you may have to tighten the strings. After about three weeks, use the flowers for a wreath or other arrangement.

● Dry flowers with glycerin to preserve their shape and suppleness, although their colour will probably change. Roses, peonies and anemones are good candidates. Pour one part glycerin and two parts

CUT STEMS ON A SLANT so more water is ABSORBED.

Creating floral arrangements

1 Get some floral foam from a local florist and soak it thoroughly. Cover it with moss and then wrap it tightly with wire.

2 Beginning at the centre, carefully insert the selected flowers. As a rule, flowers that appear together in nature tend to harmonise in floral arrangements as well.

3 Attach decorative elements such as Easter eggs, berries or fruit using florist's wire or toothpicks.

boiling water into a tall, narrow container. Cut the flowers on an angle and place them 8-10cm deep in the hot solution. Store the container in a cool place until little glycerin drops appear on the leaves.

● Dry flowers such as French marigolds, peonies and carnations with silica gel crystals from a craft store. Shorten the stems to about 5cm and stick them onto a blunt wire. Spread the silica gel on the bottom of an airtight container. Lay the flowers on top and cover them with 2-3cm of silica gel crystals. Place the closed container in a warm place for two days.

● Use silica sand to preserve the shape and colour of zinnias, marigolds and orchids. Place the flowers in a tightly sealed container with a 5cm layer of sand. Take care that the flowers don't touch each other – once dry, they can't be separated. Close the container tightly, taping it shut, and set it aside for about ten days. You can reuse the sand as long as you let it dry at moderate heat (100-120°C) in the oven.

FLORAL arrangements and aids

Beautiful floral arrangements can be fashioned out of both fresh and dried flowers, but never combine more than five different types of flower.

1 Add the greenery first, then the tallest flowers (for structure) in the centre and at the sides. Use odd numbers of flowers for the most pleasing effect.

2 Next, add the shorter filler flowers; distribute them evenly. Opt for a simple dome-shaped arrangement that will look beautiful from all sides.

3 Add more visual interest with trailing greenery that droops over the edge of the vase or container.

● Use moss with fresh flower arrangements, since you will need to moisten it regularly.

● Pebbles or marbles can add a decorative touch to an arrangement in a tall vase.

● To keep wild flowers from wilting, insert the stems into plastic bags filled with water and tie them up.

VASES and vessels

When you are adding flower displays, it's important to consider the vessel in which you are placing them.

● Use an attractive bottle for a single, tall flower.

● A wooden frame turns into a living picture if you drill holes in the bottom for test tubes; fill the tubes with a continually replenished display of fresh flowers.

● Wildflowers, cornflowers and other informal arrangements look good in woven baskets lined with watertight plastic.

● Terracotta pots make rustic but pretty containers for many types of flowers and branches. Line them so that moisture doesn't leach through. It's even better to place a small vase inside that's hidden from view.

● Pewter harmonises with peach-coloured petals.

Flowers and context will determine the type of vase.

Fragrances in the home

Many of our fondest childhood memories are associated with certain smells, such as the aroma of biscuits baking or lavender in a cupboard. Bring back treasured memories – and create new ones – with flowers, herbs, potpourris or fragrance sachets.

You don't need artificial products such as air fresheners or fragrance oils to produce pleasant home scents. There are many natural solutions if you are willing to explore a little bit, and the rewards are sweet smells without harsh chemicals.

NATURAL fragrances

Remember the basic rule: don't overdo it. It is easy to become accustomed to certain fragrances and use too much of them. Here are some tips and tricks.

● Place a big bouquet of fresh flowers, particularly lilacs, roses or lilies, on a table in the corner. It will fill a living or dining room with a wonderful scent.
● Hang herbs such as rosemary, peppermint or thyme to dry from a ceiling or balcony. They give off a pleasant aroma.
● Fill an attractive bowl with fruit and spices, such as oranges, limes, green apples, cinnamon and cloves. It will look decorative and fill the room with intriguing smells.
● Make a potpourri using dried flowers and leaves from aromatic plants. Rose petals, jasmine, lilies, lavender, rosemary, cinnamon sticks and vanilla pods are wonderful additions. Set them out in open bowls or fill a sachet; just make sure all ingredients are thoroughly dried beforehand. To intensify the fragrance, add a drop or two of essential oil.

Fragrant sachet

one To make a fragrance sachet, cut two pieces of material each about 5 x 10cm. Sew them together, right sides in, along the length of two sides and the width of one. Linen, cotton and cambric work well for sachets, but transparent fabrics such as organza look particularly pretty when filled with colourful petals.

two Turn your bag right-side out. Sew a seam around the open end of the sachet and pull a string through so that it can be closed nicely, or simply tie it closed with a ribbon after filling it.

three Fill the cloth bag with flowers or herbs and tie it off.

● Assemble your own potpourris from dried flower petals, fruit, herbs and spices. A sachet or small cushion filled with them makes a special gift.
● In winter, place orange and lemon slices near a radiator and let them dry there.
● Use scented candles to create a soft, warm or romantic atmosphere. They don't have to smell of roses or vanilla – even plain beeswax candles emit a distinctive fragrance.
● Experiment with incense. Using smoke to send prayers to the gods is one of the oldest known forms of ceremony, used in everything from the censers of the Catholic church to pagan bonfire rituals. But incense sticks don't have to be about religion – they are a good way to spread scent through a room.
● Or opt for a fragrance lamp. They are available in many shapes, but the principle is always the same. A candle is placed under a plate or small bowl, which is

Fill a fragrance lamp with water and any type of oil that suits your mood.

filled with water and a few drops of essential oil. The flame beneath warms the water-oil mixture and spreads the fragrance throughout the room.

● Make a scent ball: spread glue on a Styrofoam ball and fasten on dried flowers from a potpourri.

● Stud an orange with cloves and pieces of cinnamon stick. Tie with ribbon and hang up as a pomander.

● Create a simple fragrance dispenser: fill a lidded jar with rose petals and sprinkle each layer with salt to prevent them from rotting. When you wish to add fragrance to a room, unscrew the jar for a while.

HOW fragrances work

The sense of smell is more important than most people realise. Advocates of aromatherapy believe the power of scent can even have a healing effect. Here are some of the more common scents.

● Eucalyptus is a bright, fresh scent that most people find refreshing. It clears sinuses when your nose is blocked due to a cold, and some feel it heightens concentration. It can also repel certain insects.

● The bewitching fragrance of lilac is stimulating but can be a little strong. For this reason, it is best to use it sparingly or in conjunction with other scents. In some cultures, lilac is a symbol of love.

● The fragrance of honey can help you relax and dispel nervousness. Due to its healing properties, the ancient Maya considered the bee a sacred animal.

● Lavender is thought to have a relaxing and refreshing effect and can act as a counterbalance to nervous tension and depression, and promote sleep. For this reason, it is often found in balms and creams, as well as in aromatherapy products.

● The fragrance of tangerines and oranges awakens memories of summer and so has a well-deserved reputation for chasing away winter depression. This may be why, in Western cultures, it is not unusual to put a tangerine in Christmas stockings.

● When suffering from a cold, use the fragrance of peppermint to clear congestion. Its strong, exhilarating smell can also enhance your powers of concentration.

● The beguiling fragrance of roses has a practical role to play in reducing pain and producing euphoria. It can be helpful in stressful and hectic situations.

● Ancient uses for vanilla included as an aphrodisiac and for fever reduction. Recent research suggests that its smell can reduce anxiety.

● The fragrance of cinnamon has a warming and relaxing effect on most people.

● Cloves smell spicy and sweet, and have a mildly sedative effect.

● Patchouli has a strong musky aroma and is thought to have aphrodisiac qualities. It is now more popularly used for rejuvenation.

● Sage has a refreshing, calming odour.

● Rosemary is believed to strengthen the memory.

● Cedar has a warm, grounding aroma and is often used as an aid to meditation.

The scent of roses reduces pain and produces euphoria.

Fresh, bright bathrooms

The bathroom is the perfect place to relax at both the beginning and end of the day, so it is worth spending a little time creating a serene, spa-like feel. A few small changes are often all it takes to add a comforting ambience to even an older bathroom.

New mirrors and lighting can help freshen up a tired-looking bathroom, and changing the fabrics, towels and tiles sets a new tone. But ventilation and hygiene should remain a top priority as mildew will damage the room's appearance and can be harmful to health.

THE right set-up

● Use a softer bulb as dim lights can help create the right ambience for a relaxing bath. Also, artfully placed candles can add to the spa feel and provide flattering lighting for almost any skin type.
● Use light colours such as white, pale blue and beige or light-coloured wood to make smaller bathrooms appear larger. If you can't replace tiles or repaint the room, choose accessories in lighter shades.

● Install a new shower partition. This can be done without much trouble and it works wonders, especially in older bathrooms.
● Clear out the clutter: the only things on display should be those in daily use.
● Store towels in a bathroom cupboard if space is limited. Or, for a chic yet practical solution, roll them up, secure each with a ribbon and place them standing up in a basket by the bath.
● A larger bathroom is ideal for exotic, humidity-loving bonsai trees and tropical plants.
● Add suitably upholstered furniture to create a stylish, inviting place to relax. Stools with terry towelling seats are practical and provide comfortable seating while you paint toenails or moisturise skin.
● Bring in soothing scents such as lavender, orange and vanilla in the form of aromatherapy candles, lotions or scent diffusers.

MODERN fabrics and fixtures HARMONISE WITH OLDER TILES.

Replacing damaged tiles

Damaged tiles can cause bigger problems in a damp environment. If you discover a cracked or chipped tile, follow these simple steps as soon as possible to limit the damage. If you can't match tiles exactly, use contrasting replacements.

one Remove the grout around the damaged tile and use a dry-cut saw drill attachment to slice diagonally through the tile, or make several holes with a drill.

two Chip off fragments of the tile with a hammer and flat chisel until the whole thing comes loose. Remove as much of the old adhesive as possible.

three Cover the back of the replacement tile with bonding material and press it onto the wall, using tile spacers to position and fit it correctly.

four Let the glue dry overnight (or as directed by the manufacturer), then apply grout and clean.

- Keep a waterproof clock in the bathroom – one that can be attached to the tiles with suction cups is perfect – so you can keep an eye on the time while getting ready.
- Get a water-resistant radio if you like to start the day with music.

BEST fabrics

Well-chosen fabrics can turn the bathroom into a comfortable, attractive space. They should also be kind to skin.

- Choose a shower curtain that complements the rest of the bathroom furnishings. Waterproof fabric curtains are more chic and attractive than plastic ones but need regular cleaning, so make sure they are machine washable.
- Opt for matching towels in different sizes to give the bathroom a coordinated look. Thick, high-quality towels will make drying off after a bath an absolute joy and will last for years.

PROBLEM tiles

If old, cracked tiles, greying grout or 1970s decor are cramping your style, don't begin demolition work immediately. Take a cue from earlier generations and fix rather than discard.

- Try thoroughly cleaning the existing tiles, which injects new life at once. For stubborn stains, use water mixed with ammonia or alcohol, and remove limescale with vinegar. Rub old tiles with a little linseed oil to make them shine again.
- Repair small cracks in the tiles with matching paint from an art supply store. Mix a small amount of the paint with grout and apply to the tiles to fill hairline cracks. Rub smooth.
- Refresh older tiles by repainting the grout (special grout-colouring kits are available). But remember that grout paint can't be used on non-enamel tiles or on top of water-resistant joint sealer.
- Visually enlarge the bathroom with adhesive mirror tiles. They are a good choice when replacements for old tiles are no longer available. Or paint over unwanted tiles with tile primer and lacquer to cover them.
- One simple but effective solution is to apply a new layer of tiles over unsalvageable existing ones. This is cheaper and easier than removing the old layer and resurfacing the wall.
- Even easier is to paint over tiles with tile paint – but only if tiles aren't going to get wet regularly.

THE mildew menace

The best remedy for mildew is good air circulation, which gets rid of moisture so mildew can't take hold. Turning on the bathroom fan is the first step, but try some of these tips, too.

- Wipe condensation from the shower wall and tiles with a squeegee after each shower.
- Use vinegar to clean the corners between the shower or bath and the tiled wall regularly.
- Paint non-tiled wall surfaces and the ceiling with mildew-resistant paint.
- Remove the grout where mildew has taken hold, clean thoroughly and seal the cleaned edges with new grout.

An elegant unit of shower shelves keeps the bathroom tidy.

Furnishing for comfort

Plush, upholstered furniture can add style or period appeal to a living room. Remember, however, that its most important qualities are hidden beneath the fabric covering.

Sofas or armchairs are likely to be used on a daily basis, so as well as looking attractive and harmonising with the existing furnishings they need to be practical, sturdy and hard-wearing both in terms of construction and fabric covering.

COMFORT COMES first WHEN CHOOSING A SETTEE

SELECTION criteria

● For a seat in a lighter or plain dark colour, check whether it has removable covers that can be washed or dry-cleaned.
● See if the upholstery fabric has a water and stain-resistant Teflon coating. This will help to keep furniture looking like new.
● Consider buying an easy-to-clean leather sofa, particularly if you have children who are more likely to spill things or a pet that likes to be curled up beside you on the sofa.
● Keep a new armchair or sofa away from direct sunlight to prevent cover fabrics from fading.

● Alternatively, use a cover fabric that may fade but can be easily removed and stored when guests arrive.
● You can buy high-quality upholstered furniture with different degrees of firmness. The heavier you are, the firmer the cushions should be for comfort.
● Consider the height of the seat and back rest as well. It is easier to pull yourself up from the higher seats of taller furniture than from a low-slung couch.
● If you have weak legs, consider an armchair with a mechanical or electric power lift. They help you stand by raising and tilting the seat in an ergonomically appropriate way.
● Measure it out. You won't really know whether a sofa is a good fit until you get it home, but you can get a sense of whether the size is right by taking the measurements of a potential purchase and then laying out squares of newspaper in that exact size in the living room.
● Give yourself plenty of time before deciding what to buy. However stylish, a sofa and chairs need to stay comfortable even after a long evening spent in front of the television.
● Stand about 1m back from the sofa you are considering and look at it with a critical eye. Does the pattern line up? Does the sofa look symmetrical? Do the cushions fit together in a straight line? Does everything look right?
● Don't buy a sofa, chair or love seat without first sitting in it just as you would do at home – even if this includes lying down or slouching. Noticing the flaws in a piece of furniture only when you have got it home can be an expensive mistake.

Heating and ventilation

By managing a home's heating and ventilation systems properly it is possible to save on energy costs, create a healthy interior climate, reduce greenhouse gas emissions and banish mildew before it sets in.

Most people feel comfortable in a room with an average temperature of between 18°C and 22°C. However, different areas of a house require specific temperatures to ensure comfort and prevent mildew.

HEATING sensibly

- An average of 18°C to 20°C is usually sufficient in the bedroom, where you can always add an extra blanket on a chilly evening.
- Keep the bathroom well heated. In the morning and evening especially, the bathroom should be pleasantly warm – the towels, too. A temperature of around 18°C might be a bit too cool when you are stepping out of the bath or shower.
- Never let rooms cool off too much: the energy expended to reheat cold walls and floors is comparatively high.
- Under floor heating can offer an added element of comfort while adequately heating the entire home.
- Save energy by keeping doors that connect to uninsulated rooms (such as the garage or a cellar) closed and make sure they fit tightly.

THE right ventilation

- For quick air circulation, open two windows that face each other to create a draught.
- Open the windows three times a day for 3-10 minutes for effective circulation without cooling your home down excessively. Experts recommend short, periodic bursts of ventilation to maintain air quality.
- Air a house for longer periods during warmer months, unless you suffer from hay fever. If so, stick with short periods of ventilating, even in the summer.

PREVENTING mildew

Too much moisture creates a breeding ground for mildew. In living areas you may find mildew on wood, carpets, wallpaper, drywall or masonry. Improper ventilation is usually the cause.

- Air the room immediately if steam rises while you are cooking or bathing. Just open a window or turn on a fan.

A CEILING FAN KEEPS AIR MOVING during hot and cold SPELLS.

- Keep doors closed to prevent moist air from spreading to the other rooms.
- Open the windows when drying laundry inside the house to allow damp air to escape.
- Leave about 2-5cm clearance, if possible, between big wardrobes and outer walls. Mildew can easily develop behind the wardrobe without being noticed.
- Regular ventilation is especially important if you have newer double-glazed windows. These are virtually airtight so prevent fresh air from entering the home.

Adequate ventilation provides the right humidity level, which is generally about 50 per cent relative humidity.

Home safety

By taking care and using commonsense when planning a renovation or arranging furniture, you can make life more comfortable and reduce the risk of accidents in the home.

Safety issues can affect every room and area of the house, including the bathroom, balcony and back garden.

STABLE HAND RAILS PROVIDE security AND can add style.

GOOD TO KNOW ✓

Avoid accidents

Most falls occur indoors, so look for problem areas in the home. Are there obstructions such as raised thresholds? Or electrical cords you could trip over? Are there rugs or runners that could slip or cause someone to fall? Is the lighting in each room adequate?

USEFUL aids

● Ensure there is a light switch right next to the front door so you won't have to stumble around in a dark house when you come home at night.
● Choose stylish light switches that contrast with the wallpaper or paint to make them more visible.
● Plug in a small night-light in the hall to ensure safe passage to the bathroom in the dark.
● Opt for shatterproof glass patio doors to prevent accidents. Use decorative aids, such as adhesive designs or stickers, to make the glass visible.
● Use sound-absorbent curtains or floors to reduce echoes and other noise.
● Choose dining-room chairs with arm rests for comfort. They also make it easier to stand up after a meal.
● Sitting down and getting up are much easier if you choose sofas and chairs that have raised seats. Similarly, make sure that a new bed is a comfortable height before you buy it.
● Make sure you don't forget your keys or mobile phone by designating a storage area next to the front door that is invisible from the outside.
● Attach timers to table lamps so that they turn on automatically once or twice a day to deter burglars. It also ensures you come home to a lit house at night.

BATHROOM and kitchen

The bathroom or kitchen probably offer the most potential for disaster in a home. Here are some preventive secrets.
● Install sturdy bath rails to help you get in and out of the tub.

- Stick water-resistant, adhesive anti-slip stickers to the bottom of the bath. They offer better protection against slips and falls than a bath mat.
- Buying a shower or bath seat is not only for older people – it can simplify bathing children, too.
- Install a single-handled tap. It is a safe and easy-to-operate fixture that allows even little ones to regulate water temperature, protecting against scalding.
- Take a look at a space-saving, collapsible step ladder. It will allow you to reach the uppermost cabinets in the kitchen and high book shelves without risk, and they can be stored away neatly.

BALCONY and patio

A patio is a relaxing place to unwind after a hard week. Just make sure that it provides a safe environment as well.
- Flat surfaces as you enter or exit a balcony or patio reduce the risk of tripping. For a balcony you step down to, it may be worth using a wooden structure to raise the level.
- A screen that protects against the wind enables you to enjoy an outdoor space even on cooler days.

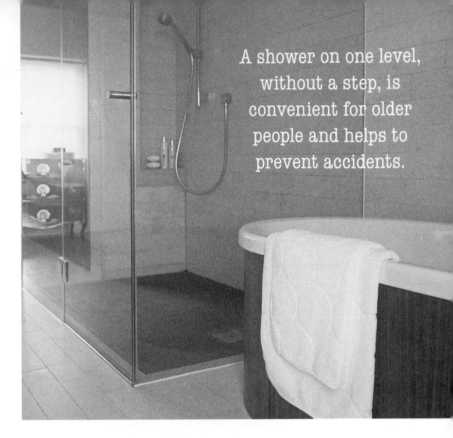

A shower on one level, without a step, is convenient for older people and helps to prevent accidents.

- Make sure the height of the balustrade on the balcony complies with building regulations. Take special care to make sure children are constantly supervised if they are able to access balconies.

GARDEN and entrances

The steps from the patio to the garden can be an attractive feature but they need to be safe. Install a handrail and make sure there is no loose mortar that could make a step unstable and cause a fall. Here are a few other things to consider.
- Repair uneven pathways and keep them clear of leaves, broken branches and overgrown plants.
- Illuminate the entrance to the garden and steps with solar glare-free lights to avoid tripping in the dark.
- Install motion detectors on lights outside the house and in the garden so that when someone approaches the surroundings light up. This is also a useful security feature.
- Lay non-slip doormats that are flush with the floor to make entrance areas safer.
- Install an easily accessible doorbell and a legible house name or number to help visitors find the right home. The postman and parcel carriers will also appreciate a house number that is clearly identifiable from the street and a large, visible letterbox well placed for easy access.
- To help keep mosquitoes and other insects away, install a pump in an outdoor water feature. Mosquitoes can only breed in standing water.

A tasteful outdoor wall light adds a touch of class as well as security.

Indoor plants

Plants add atmosphere to a home. Tall specimens such as dragon trees or palms act as accents, while smaller ones can decorate a windowsill. When choosing a plant, consider carefully where you are going to put it in order to enhance the decor and ensure its survival.

When properly cultivated, orchids will flower over and over again for years.

Armed with a few tips and patience, anyone can care for a variety of plants that will help enliven a living space. And remember, a room with plants is known to help lift the spirits.

GOOD TO KNOW ☑

The right location

Besides water, plants need light to thrive. A window facing south that gets plenty of sun is ideal for cacti and succulents. Ferns, orchids and spider plants prefer a north-facing window or relatively dark areas of a room.

A healthy room climate

Plants are not only beautiful, they increase the humidity level and oxygen content of a room during the daytime. This is good for the climate and, ultimately, for your health. This is especially important in winter, when the central heating can dry the atmosphere.

● Certain plants are particularly effective at counteracting the release of volatile and potentially harmful chemicals from various substances (such as insulation, plywood or varnishes) at normal atmospheric pressure – and contributing to balanced internal humidity. These include: areca palms, dwarf date palms, Boston ferns, rubber plants and peace lilies. Plants kept in water (such as bamboo), with no need for soil, can increase this effect.

● Don't fill a bedroom with plants as they use up oxygen and produce carbon dioxide at night.

● Bear in mind that big plants such as dragon trees (*Dracaena*) produce more humidity than smaller ones.

WATERING plants correctly

Many plants do well when watered from above, however, there are myriad factors to consider depending on the size and species of plant. One general guideline to follow is that you should always use water at room temperature. Another is always to water plants with furry leaves, such as African violets, from below to avoid damage.

● Put bulbs and other plants with sensitive roots, such as poinsettias, in a large bowl and water them from below to keep the bulbs or roots from rotting. Also, check planters about an hour after watering and drain off excess water.

● Keep rainwater and stale mineral water handy as both are good for watering plants, as is cooled, salt-free water that has been used for boiling eggs.

● Leave tap water to sit overnight before you water plants with it. Fluoride and other minerals dissipate or settle when the water is left to sit.

● Immersion can be beneficial to many plants, including cacti and citrus species. Immerse the root ball in its pot in a large container of water. It is saturated when no more air bubbles rise.

● Spray plants occasionally. Use lime-free water to keep limescale from building up. Pamper plants (especially cacti and those with large leaves) with a cleansing shower twice yearly: put them out in the rain or spray them from a shower head. Tip the pots of cacti to prevent water running directly into them.

REGULAR care

With a little tender care, plants will reward you with healthy foliage and gorgeous flowers.

● Check new plants thoroughly for pests before buying. Indoor plants can be susceptible to spider mites, aphids and other troublesome creatures that spread easily.

- Dust plants carefully from time to time to let the leaves breathe. The undersides need particularly close attention.
- Wipe plants with larger leaves with diluted beer to make them shine.
- Remove wilted flowers and leaves immediately to keep plants looking good and to help prevent them attracting bugs.
- Prevent a harmful accumulation of water by occasionally loosening the soil. When repotting, ensure the new pot's drain holes are clear. Line the bottoms with shells or curved shards from clay pots to keep the holes open but prevent soil escaping.
- A mixture of sieved compost, garden soil and sand makes good potting soil. Also, egg shells (for calcium), coffee grounds and black tea make excellent natural fertilisers.
- Leaving plants exposed to heat or draughts is the equivalent of leaving outdoor plants unprotected from the elements.
- Water plants more sparingly during the cold months, as winter is a time of rest. Also, keep them in rooms with cooler temperatures if possible.

CUTTING back

Trim fast-growing plants regularly during the summer. Careful pruning controls plants and encourages growth. It will help them grow stronger and is a good way to shape certain species.

A garden in a bottle

A terrarium is the perfect way to bring a little greenery to a cramped space. This is how to assemble one.

one Line the bottom of a large, big-bellied bottle with gravel. Then, use a lolly stick to extend the handle of a spoon and a fork (use some tape to attach them).

two Put a layer of soil on top of the gravel and smooth it out. Use a cardboard funnel to keep the sides of the bottle clean.

three Insert suitable plants, such as ferns, small trailing figs and ivy, using the spoon and the fork as tools. Don't plant them too closely as they need room to grow.

four Water the plants carefully with cooled boiled water, and keep the bottle tightly sealed. If the bottle fogs up, you have overwatered it.

five Don't place the bottle garden in direct sunlight or on or near a heating unit.

- Prune woody plants once a year, but when depends on the type of plant. Plants that flower before midsummer are best pruned immediately after flowering.
- Increase the number of shoots and flowers by cutting back the main shoot. This will cause the plant to sprout new shoots.
- Make the cut clean, no matter why or where you prune. Use a sharp knife or secateurs.

PLANTS HELP to give ROOMS a PLEASANT FEEL.

Light conditions

Make the most of natural light in your home to safeguard the
environment and reduce electricity use during the day. Opening
the curtains wide to let the morning sun into your home
provides a positive energy boost at the start of the day.

It makes sense to situate the kitchen and living room
where they get the maximum amount of daylight. But
unless you have designed the house yourself, that's
not always possible. Even so, there are plenty of
tricks to help you make a dark home (or an overly
bright one) more appealing.

PROVIDING the proper lighting

● A dark and windowless hallway can create a
gloomy first impression. Illuminate it evenly, perhaps
with ceiling spotlights along its length.
● If natural sunlight only penetrates the part of the
room that is close to a window, capture that elusive
light with a mirror and reflect it wherever you wish.
● Very pale-coloured or well-illuminated walls give a
room an expansive feeling because they appear to
reflect the sunlight. They can also make a room
appear larger and friendlier.
● By designing windowsills and patios to look like a
continuation of the interior furnishings, you can
cleverly enlarge a room. The natural light will evoke
the same colours inside as well as outside.

● Dark walls absorb light and therefore seem
visually closer to the observer – an optical trick that
you can use for large or long, stretched-out rooms.
● To reduce glare, place the TV out of the path of
direct sunlight and reflected light.

PROTECTION from sunlight and glare

Although bright, sunlit rooms are generally pleasant,
too much sunlight can be stifling in the heat of
summer. Here are some ways to cool things down.
● Shutters deflect sunlight when placed on the
outsides of windows.
● Equally important is protection from glare.
Anti-glare devices such as louvred blinds and interior
shutters go on the inside of windows and usually
have adjustable wooden or plastic slats that direct the
light up or down, or otherwise control it.
● Panel curtains mute the direct light from the
outside yet still allow light to enter, keeping all areas
of the room free from glare.
● Anti-glare screens reduce eye strain when using a
computer monitor on a sunny day.

To prevent eye strain,
reduce the contrast
between a computer
monitor and a room's
ambient light.

Lighting

Lighting has an obvious function, but it is important to light a room properly. Harsh, glaring lights in the study can cause headaches or eye strain, while insufficient lighting in the kitchen or workshop could be dangerous.

A CLASSIC floor lamp ADDS ELEGANCE AND DIFFUSES soft light.

To keep you and your family safe, observe the manufacturer's guidelines when hanging lamps, chandeliers and wall lamps and follow lighting tips for each room. Seek professional advice on all electrical matters.

BASIC lighting

● The light source should illuminate the room fully and evenly.
● Light projected from the ceiling should not diminish the effect of nearby floor or table lamps.

GOOD TO KNOW ✓

The end of incandescence

In 1880, Thomas Alva Edison patented the incandescent lightbulb. Today, the bulb's fate is sealed: many countries have either banned incandescent bulbs or will begin phasing them out by 2012. Energy-saving lightbulbs (such as CFLs or LEDs) use a fraction of the energy of traditional lightbulbs and come in many different forms, colours and temperatures. By making the switch you can reduce lighting energy consumption by up to 80 per cent.

HANGING lamps

A hanging light fixture is a common sight in a modern home, but not all are properly hung. Follow these suggestions when installing one.
● Pendant lights are perfect for casting light on dining-room tables. Hang them about 76-86cm above the table (for a 2.5m ceiling) so that they provide enough light without disrupting sight lines or blinding guests.
● If the ceiling is higher and you plan to use large, dramatic table centrepieces, hang a pendant light a little higher so guests can see one another.

CHANDELIERS

Glittering crystal chandeliers are the most elegant forms of hanging lamps. They have become much more affordable through the use of glass-like synthetic materials. However, a chandelier won't show a dining room to its best advantage if it is draped with cobwebs and a layer of dust, so it needs to be cleaned regularly. (*See Good Housekeeping, p.143.*)

WALL lamps

Wall lamps supplement the main light source in a room and play an important role in mood lighting. Here are some suggestions for using them to their best advantage.
● A beam of light shining up or down on the wall can visually separate different areas of activity in a room from each other.
● If you want wall lamps to help spread the light evenly in a room, make sure that more light is shining upwards than downwards.

FLOOR lamps

Portable floor lamps come in many different attractive styles and will always liven up dark areas of a living room or den.
● Floor lamps create a cosy atmosphere, lighten dead corners and provide reading light.
● Torchères (lamps on tall stands) can add a striking accent and also be used as reading lamps.

TABLE lamps

Table and desk lamps can be both decorative and useful.
● They can look pretty on an occasional table or they may be essential for bedtime reading or for close work in a study.
● You can make an imaginative lamp base from a clay pot, clear vase or ceramic urn and a lamp kit from a DIY store. Be sure to only use materials that aren't combustible, as the lightbulb will emit heat.

KITCHEN

Basic lighting is provided by a ceiling light, but halogen spotlights or fluorescent tubes installed under hanging cabinets can illuminate important working areas much better.

● Recessed downlights over the sink or stove can create good task lighting for cooking, baking or scouring pots and pans.

● Kitchen islands and breakfast bars can be effectively lit and highlighted with a series of pendant lights.

● A pendant over the kitchen table provides lighting for doing homework, tackling paperwork, undertaking hobbies and many more things.

LIVING area

The living room is less about function and more about ambience, so tall torchère lamps can provide basic lighting. But torchères will reveal any bumps on the ceiling and walls, so place them carefully. There are several other things to consider.

● Islands of light created by several small table lamps accentuate certain spaces.

● Coloured lights can create a fun effect, depending on the colour and position.

● To relieve eyestrain, illuminate the wall behind the TV with a soft light source.

● Indirect lighting, mounted, for example, along skirting boards, in glass cabinets or behind a curtain rod can provide a pleasing lighting effect.

DINING area

● Choose a hanging lamp with a pleasant and glare-free light to install over the dining table. A dimmer switch that allows you to adjust light conditions is ideal: turn the light up for a family dinner, down for an intimate meal for two.

BEDROOM

● A ceiling light that can be turned on and off at the door or from the bed generally provides the basic light source in a bedroom. However, it is also important to invest in a good bedside reading lamp if you enjoy reading in bed.

BATHROOM

● Halogen lamps or fluorescent lights provide basic lighting in many bathrooms these days. For a more focused illumination of the vanity area, you can install additional lights that don't cause glare or shadows above or on both sides of the mirror.

THE DINING AREA should be STYLISHLY LIT and FREE from glare.

Looking after your home

By employing a few tried-and-tested tricks you can keep new furniture or a freshly varnished floor looking pristine and your sofa and armchairs in showroom condition.

Slipcovers, armrest covers, throws and furniture cups or felt pads all work well to keep furniture and floors free from damage.

SEATING

Living room chairs and sofas experience heavy wear, but a few well-chosen covers will reduce the need for repairs and increase their life expectancy.
● Slipcovers prolong the life of upholstered furniture. They come as either custom-made or loosely draped coverings that protect the sofa from small, sticky fingers and four-legged friends. When slipcovers get dirty, just remove them and wash them. They can also be changed with the seasons.
● Slipcovers also serve a decorative purpose: they enhance seating arrangements and ensure that sofas and different types of chairs are well coordinated.
● The armrests of chairs and sofas wear out the quickest. Use armrest covers to protect against abrasion and sweat.
● Throws that cover an entire sofa have a modernising effect on older or outdated furniture, giving the living room a new look.
● A decorative blanket draped over the seat of a chair or sofa can save seating from wear.

If you can't get the stains out of upholstered fabrics, opt for an elegant, custom-made slipcover.

● It may be possible to have removable covers dyed if you want a change of colour.
● Discourage your cat from damaging furniture by investing in a scratching post and placing it near where it sleeps or eats. Or try a cat repellent (available at pet stores) to keep it off furniture.

FLOOR

A new floor enhances any house. Although wear and tear may be unavoidable, you can reduce damage by taking a few simple precautions.
● Always put felt pads under tables and chairs. Self-adhesive pads often come loose, leaving ugly remnants of glue. Opt for more durable, screw-on pads instead.
● Put a protective mat under the desk chair. Even casters specifically made for hardwood floors can mark them.
● Prevent dirt and stones from being trailed through the house by asking people to remove their shoes at the door. You could perhaps provide some welcoming warm slippers or woolly socks for visitors to wear.
● Use area rugs to protect any section of the floor that might be susceptible to spills or extra wear.

Wearing slippers indoors will help to protect floors.

NATURAL MATERIALS

Go green when painting, decorating and choosing home decor, avoiding chemicals and pollutants that damage the planet. Natural, eco-friendly materials tend to be good for your health and the environment.

Tile or stone floors are long-lasting, stylish and free of toxic materials.

Floors

Today, beautiful, eco-friendly and stylish flooring is well within the reach of anyone who is renovating their home or having a new house built.

- Cork flooring is warm and springy underfoot. You can even sand down a cork floor and apply a new polyurethane finish. But check first with the manufacturer that it is safe to do so.
- Linoleum is also a natural product. Due to its noise-reducing, antibacterial and antistatic qualities, it is currently undergoing something of a renaissance.
- Naturally oiled wooden floors are an alternative to engineered hardwood, laminates or glued parquet floors. You can sand off scratches and repair them with oil as they occur.
- Tile or stone floors are extremely long-lasting and free of toxic materials unlike many types of synthetic flooring. Cared for properly, they look good and enhance most types of property.

Walls

From plaster to paint, walls play a huge role in the quality of a home's interior environment. It is not difficult to find materials that won't emit harmful fumes. These are some of the choices available.

- Lime plaster (not to be confused with lime-based render) is made of finely sifted sand and lime. It is good for controlling humidity and offers tremendous durability.
- Clay plaster has a similarly positive effect on the climate of a room. It can be used as a base, top coat, brush plaster or textured exterior plaster.
- Clay paint consists of clay, chalk, water and pigments. It is non-toxic and affordable.
- Whitewash is ideal because it resists mildew. It is best applied to lime and clay plaster, but will also stick to oil paint or wallpaper.
- Milk paints are good for allergy sufferers. Based on a centuries-old, non-toxic formula made from casein (a milk protein), water, limestone, clay and natural pigments, they come as a fine powder that can be mixed at home. They stick only to unfinished, porous surfaces such as wood and plaster.

Choose natural materials for your bedding.

SOLID WOOD can influence THE CLIMATE OF a ROOM.

Beds

For maximum comfort and sound health, it is a good idea to stick to natural materials when it comes to your bed. A solid wood, wrought-iron or stainless-steel bedframe is a good choice.

- Opt for eco-friendly materials when buying a mattress. In terms of energy consumption, all-natural latex (free of synthetic latex) rates much better than synthetic latex made from petroleum, even when you factor in harvesting and transport.
- Before buying a futon, check the stuffing to see if it has been enriched with breathable, warming, natural fibres such as horsehair, virgin wool, coconut fibres or natural latex.
- Choose down, feathers, cotton and linen for blankets and bedding unless you are an allergy sufferer. If so, opt for blankets that can be washed in water up to 60°C, or hotter, and pillows and duvets that are made of synthetic fibres.

Furniture

Furniture made from natural substances and finished with chemical-free stains or varnishes not only looks beautiful, it is also non-toxic.

- Solid wood furniture is preferable to pieces made from MDF or veneer, which can emit formaldehyde. A bonus: by absorbing and releasing humidity, solid wood polished with beeswax improves the atmosphere of a room.
- Wicker furniture made of cane, bamboo, rattan or willow provides a natural ambience. Just moisten the furniture with a damp cloth to freshen it up.
- Wrought-iron or stainless-steel furniture is a good alternative, especially for allergy sufferers.
- When buying upholstered furniture, seek out pieces made with natural materials – soya or vegetable-blend cushions, for example, as well as cover fabrics such as hemp and organic cotton.

Household fabrics

When choosing the fabrics and linens for your home, going green means looking a little more closely at the labels before you buy.

- Look for blankets made of wool or organic cotton.
- Look at organic cotton, bamboo and linen for bedding and sheets. The absorbency of the natural fibres contributes to a dry, warm sleep environment. They are also cool in summer.
- Consider felt-backed bedroom curtains for allergy sufferers. Not only does this have an insulating effect, but such curtains tend to absorb the vapours and scents in the air.

Parties

Part of the fun of hosting a party is decorating the house beforehand. With a little bit of ingenuity, it's easy to create the perfect party space, whether it is for a family event, a barbecue with friends or a child's birthday.

Flowers, candles and party favours set a welcoming mood for a summer party outdoors.

There's always something to celebrate but each occasion requires a different approach. Before sending out the invitations, settle on the theme and type of party. It could be an evening gala with a theme such as 'film stars', a casual brunch or a garden party with a barbecue and games.

FAMILY parties

● Invitations are usually written, although you might send an email for a casual event. Make the invitation original, personal and tailored to the occasion.
● Using place cards? Make sure they suit the context. Be imaginative: hand-lettered stones, homemade cupcakes or little silhouettes in miniature picture frames can spell out where people are to sit just as well as cardboard cards with ornamental calligraphy.
● Give your event a restaurant feel by posting the menu on a blackboard or printing it on decorative paper, then rolling and tying each copy with a ribbon and placing one on each guest's plate.
● Ensure the floral decorations suit the occasion: a bouquet of wild or informal-looking natural flowers works well for a barbecue, while a single amaryllis in a slim silver vase adds elegance to a dinner table.

OUTDOOR parties

Tables set for summer need fresh flowers and lots of greenery. For example, a garland of variegated ivy looks festive on a brightly decorated table. Deck out the table any way you want, but try to stick with one or two bright colours, such as a cool blue and white or sunflower yellow.
● Once the table decor has been decided, add a playful touch to cocktails or ice-cream sundaes with cheerful paper umbrellas. Straw hats with colourful flower wreaths hung around the terrace contribute to a rustic decor. Also, it is a good idea to make sure there are plenty of flickering lanterns spread along the tables at an evening barbecue, while strings of lights are a must. Torches can also be used to create a romantic light in the garden, but keep them well away from trees and other plants to avoid accidents.
● A well-decorated party space doesn't have to be just about looks. Scented candles can serve a double

purpose, contributing to the ambience as well as keeping insects away. The scent of citronella, mint or eucalyptus work best.

A CHILD'S birthday

Children look forward to their birthdays all year long. Fortunately, there are lots of simple yet effective ways of preparing for their special day.

● First, let the children make their own party invitations. They can pull together fanciful creations with stickers, crayons and coloured paper, or hone their design skills on the computer. It's fun for them and gives the invites a special, personalised touch.

● On the day, use bright paper plates, napkins and confetti to add colour to the table.

● Make the birthday girl or boy's seat of honour stand out from the rest: crown it with streamers and garlands or a bouquet of balloons.

● Use paper tablecloths and put cups of crayons on the table so young guests can practise their artwork and be entertained while they wait for their food.

● Blow up plenty of colourful balloons. They can also serve as place cards if you write a child's name on each and tie it to his or her assigned chair.

● Opt for a sweet and scrumptious wreath as the final touch: use toothpicks to stick liquorice, jelly sweets and candied fruits onto a Styrofoam form.

CHRISTMAS

You really don't have to spend a fortune to come up with tasteful holiday decorations. In fact, traditional items like red velvet bows, nutcrackers and handcrafted nativity scenes or angels can often decorate a home more festively than any life-sized Santa Claus climbing the wall, or rows and rows of lights flashing on the roof.

● Buying a real tree? If you have to trim off the lower branches, keep them to use for wreaths, floral arrangements or garlands. Then, mirror the tree's festive green hue with embroidered napkins, place mats or tablecloths and add a splash of red – such as ribbons or smaller accessories – to set it off.

● Place nuts, pine cones or glass marbles sprayed with metallic paint in a silver bowl and decorate it with ivy tendrils. Or create a pyramid of oranges, then tuck whole walnuts, brazil nuts and hazel nuts into the crevices and place the pyramid on a bed of pine boughs to create a beautiful but edible holiday centrepiece.

● Create an inviting Christmas atmosphere with candles in pretty candlesticks or lanterns spaced

Flying balloons

Paper cups
Strings (about 1m long)
Balloons (filled with helium)

Half fill around ten cups with water. Dip the end of each string in a cup and put everything in the freezer. After freezing, tie the balloons to the strings and distribute the cups around the room. As soon as the ice melts, the balloons float up one after the other. One warning, however: don't do this outside, as the material they are made of could be harmful if swallowed by birds or other wild animals.

throughout the room. (Be sure that no candles are too close to the Christmas tree.) Strings of lights, lanterns or a variety of glittery ornaments can be used to decorate the windows, terrace, balcony or trees around your property.

FESTIVE candles, berries AND FOLIAGE all add to the CHRISTMAS SPIRIT.

Pictures

Devise a tasteful arrangement and use the right lighting and suitable frames to draw attention to oil paintings, family portraits or photos from your last holiday.

A little preparation will help ensure that pictures are hung properly, arranged stylishly and shown in the best lighting conditions.

A neat, geometric arrangement looks cohesive and clear

ARRANGING pictures

The basic principle when arranging pictures is that the horizontal centre of each picture or group of pictures should line up at eye level on the wall. You can drop the centreline 30-60cm in areas where the admirers will mostly be seated, such as a living room. Beyond that, you are free to arrange them as you like, however it helps to observe certain ground rules.

● Hang pictures that are the same size and shape next to or underneath each other in strict geometrical order.

● Create a harmonious arrangement by organising pictures of different sizes according to imaginary lines that go through the middle and the lower or upper border of the specific picture.

● Be creative: arrange a larger group of pictures in a square, oval or circle. Alternatively, orientate them according to an imaginary cross in the middle of the arrangement.

PLACING pictures in the best light

● Even in a well-illuminated room a picture light can accentuate a piece of art or group of photos.

● Well-placed lamps either below or above can cast a mellow light on pictures.

● To bring out the glow in a photo, illuminate it from behind. Press a photo to a piece of clingfilm, glue it onto matt glass, and install a light behind it so that it shines through the picture.

HANGING pictures properly

Besides the good old picture hook in the wall, there are many other options for hanging pictures almost invisibly.

● Hang pictures from a rail or moulding using nylon line that matches the colour of the ceiling and is almost invisible. You can move them back and forth and adjust them for height, too.

● Hang artwork on the wall over an art shelf, or set it directly onto the shelf.

● If a picture always hangs askew, use a little Blu-Tack or masking tape behind the corners of the frame to hold it in place.

Pillows and cushions

Bedroom pillows and scatter cushions provide a fresh, creative and inexpensive way to decorate as they come in a wide variety of shapes and sizes, colours and fabrics, textures and patterns. They can enliven any room, create a dramatic effect or provide the perfect finishing touch.

One of a pillow or cushion's main jobs is to ensure comfort and support. Follow a few simple guidelines to take full advantage of their potential benefits.

FOR everyday use

● Choose a pillow or cushion filled with down and/or feathers if you prefer a soft, warm pillow or cushion.

● Take a look at speciality pillows filled with buckwheat, spelt or memory foam that offer better support for your neck.

● If you suffer from neck pain, consider a neck roll pillow that can relieve neck strain. Inflatable neck pillows are ideal for travel and don't take up much room in luggage.

● For back pain, try adding a relatively firm wedge cushion to the chair. The tapered angle of the wedge encourages correct posture, providing relief for your spinal column.

● Why not get some general-purpose seat cushions for hard-surfaced chairs, benches or stools? The cushions will increase comfort considerably.

● Choose fabrics that will harmonise with the rest of the decor.

● Floor cushions are both portable and comfy. They make wonderful spots for sleeping pets or napping children to curl up, while beanbag chairs filled with Styrofoam or sand provide remarkably comfortable yet movable seats.

DECORATING with cushions

Providing comfort isn't a cushion's only function. When covered in attractive and appropriate fabrics, cushions and pillows can also play a central role in a home's decor.

● Deck out a sofa, armchair, futon or bench with accent or matching cushions to create a connection between the different kinds of seats or to complement the overall design and colour scheme of the room.

● Arrange cushions of different shapes and fabrics on a plain sofa. Vary patterns but stick to the same colour for a pleasing effect.

Clusters of cushions add a splash of colour to a plain sofa and can tie in with a room's decor.

● Opt for floral patterns, paisley or brocade for a country look. By contrast, simple upholstered furniture and a contemporary ambience tend to call for cushions with geometric patterns or brightly coloured silk and velvet fabrics.

Decorating cushions

Cardboard stencil
Plain-coloured cushion cover
Fabric paint

First, cut your stencil to the desired shape (or buy one ready made from a craft store) and stick it onto the fabric with masking tape. Apply the fabric paint over the stencil carefully: only the cut-out pattern will be coloured.

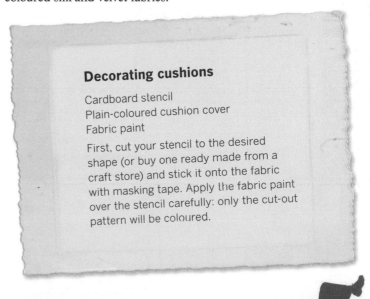

Space saving

When the cupboards are bursting, bookshelves are overflowing and the computer is sitting on the dining room table, it's time to start finding ways to save space. Every home has a few empty nooks and crannies that can be pressed into service.

Most parts of the house have the potential to provide extra storage space, including the kitchen and breakfast room, dining room, bedrooms, hallways and corridors. Look at the potential space-saving areas available before deciding how to proceed.

ADDITIONAL cupboards and shelves

● Create additional storage space under a pitched roof or a sloping staircase by adding a moveable wardrobe on casters or a custom-built cupboard door.

● Turn a nook into a cupboard by installing a rod for clothing. The space between the clothes rod and the back wall should be the width of half your widest hanger. If there's no room for a door, install a roller blind or a curtain.

CREATE space in the bedroom

Bedrooms, especially children's bedrooms, can be a perennial problem but are not difficult to organise effectively. Here are a few ideas for making the best possible use of the space available.

● Use the space under the bed. Tuck away extra blankets or pillows for guests in drawers, low chests or sturdy boxes that are low enough to fit under the bed. A chest of drawers with casters and handles or loops is particularly easy to move in and out. This works best with a bed with a taller frame.

● If there is no space under the bed, hang shelves above it instead. If you want to store important but not particularly attractive objects on open shelves, use a set of matching decorative boxes to hide them away while keeping them in view.

● If you have a high-ceilinged bedroom, consider installing a platform for the bed to turn it into a loft-style bed. You will have created more and can put the area beneath it to good use. What works best in this case will be a chest of drawers, a trunk or a set of shelves covered with cabinet doors.

USING hallways and corridors

Hallways and corridors can make up a large percentage of a home, yet they are rarely used as storage spaces. Why not put every nook and corner to work with custom-fitted shelves? Home-improvement stores offer modular, multi-use shelving systems that can stretch the available storage space without stretching your budget. But there are all sorts of other solutions as well.

There is plenty of extra room beneath a loft-style bed for cupboards and shelves.

● Install a shelf or shelves over the front door or along one wall in the hall, along with some well-placed hooks. This will make it easy to organise coats, boots and shoes, scarves and sporting equipment. You can use matching pull-out baskets to make everything look tidy and attractive. Or you might want to conceal it all behind a curtain or blind.

● Or don't hide the shelf away, but showcase it instead. If you equip the shelf with halogen spotlights, it will create a welcoming island of light in the hallway.

● Another great idea is to build a small custom-made office or chest of drawers to go beneath a staircase in the hallway. A desk made exactly to fit and one or two shelves is perfect for using this potentially dead space.

STORAGE in the kitchen and dining room

A cosy kitchen will quickly become the hub of a home, so it is essential that there is enough storage space to keep it neat and uncluttered.

● Store larger items such as pots, pans and small appliances in the kitchen's floor-level cupboards. For higher up, choose overhanging cabinets that are the right height. The topmost shelves, which you may only be able to reach with a stepladder, should contain any appliances or other items that you do not use frequently.

● Alternatively, hang pans and utensils from a custom-made hanger attached to the kitchen ceiling.

● Use any unused nooks or corners to hold the rubbish and recycling bins or a kitchen towel rack.

● Install tall cabinets with vertical slide-out drawers that are accessible from two sides. They are versatile space-savers, while a corner cabinet with a carousel makes good use of a space that often goes to waste.

● Remove large, rarely used items, such as catering-size pans or dishes, and keep them in the loft. Keep a list of what you have put away.

MORE space in children's rooms

Children's rooms serve as the place where they play, learn and sleep, so they have to perform miracles when it comes to providing adequate storage space. Over time, most children's bedrooms accumulate a mind-boggling assortment of toys, clothes, books and crafts that all have to be stored somewhere.

● Here, again, a loft bed can provide valuable extra space. Put a desk or seats underneath it.

● It there is no shelf space, hang a bag made from sturdy material or a mesh net from a ceiling hook for storing stuffed animals, balls and other toys.

● Use colourful baskets, boxes and chests to store toys in the evening. Make it easy for children to sort and put away their own toys by gluing a magazine picture of the contents (for example, cars, blocks, dolls' clothes) on each container.

A WHEELED cabinet AND SHELVES TURN A nook into AN OFFICE.

Storage solutions

Cupboards, wardrobes, dressers and shelves can easily become dusty and untidy. Organise your belongings in neat and decorative ways with functional furniture and a system that can help you create order from chaos.

In times past, people managed with much less space. Granted, they probably had fewer belongings but even today it is possible to organise storage to make better use of the room available.

A TIDY cupboard

● Never overload clothes hangers, shelves and drawers. Clothing should be easy to remove and to put back.

STORE SELDOM-USED ITEMS in neat boxes ON higher shelves.

● Select multi-armed hangers for use in a smaller cupboard.
● Make the best possible use of space by installing two clothing poles in the cupboard: one for longer clothes such as dresses and another for shorter ones such as shirts.
● Keep accessories such as scarves, ties, belts, stockings and gloves in a drawer. Alternatively, hang handbags, belts and ties from hooks on the inside of the cupboard door.
● Use drawer inserts for keeping items organised and accessible.
● When you pack away seasonal clothes, dust your shelves. Dust mites are bad for fabric and people with allergies, and can make clothing appear dirty.
● Donate garments to charityif they no longer fit. Even if you do manage to get back into them, the chances are they will be out of fashion.

CHEST of drawers

Drawers of different depths can be useful. Store undergarments in shallow drawers and sweaters in deeper ones, for example. But consider the following when looking for a new chest of drawers.
● Check that the drawers slide smoothly.
● When stowing clothing in a drawer, leave about 2.5cm of clearance between the top edge and the contents so it can be opened and closed smoothly.
● Arrange the chest so you have enough room to put items in and take them out easily.
● Avoid storing heavy items in drawers as you risk breaking the bases.

SHELVES and glass cabinets

These are an obvious solution for storing and organising, which is why they are ubiquitous. But there are a few things to consider.
● Open, decorative shelves make excellent room dividers – tastefully separating the dining area from the living room, for example – and they also provide storage. But they do need to be dusted regularly.
● Opt for a cabinet with a glass door if you are not fond of the duster.

- When buying shelves, make sure they can accommodate the weight of whatever it is you are planning to store. Lightweight shelves may buckle or even break under a load of heavy books or tools.

CHESTS, Welsh dressers and sideboards

Nice as they are, a shelving unit or glass cabinet won't fit in every room. Sometimes a sideboard or Welsh dresser is the best solution.
- You can use a lovely old chest with no partitions to accommodate either large items of clothing or children's toys.
- Put a Welsh dresser or sideboard in the dining room or a large kitchen for a decorative storage place for dishes, glasses and silverware, as well as offering extra space to lay out food and drinks during a dinner party. If you don't have room for an attractive but heavy old antique, there are many streamlined models available today.

PRACTICAL aids

With a little creative flair it's possible to find a storage solution for practically every problem.
- Use decorative boxes to store off-season clothing or footwear, as well as seasonal decorations. Store them inconspicuously on the top shelf of the wardrobe, under the bed or in the loft. By adding peepholes covered with plastic wrap or a photo of the contents, you can find what you need at a glance.

A partitioned box with a transparent cover lets you see at a glance what's inside. Just store it under the bed when you don't need it.

- Store clean cashmere and woollens in zip-up plastic bags to prevent attack by clothes moths.
- Wheeled trolleys allow you to store diverse items in an organised manner. A tool trolley is a classic example – it can be rolled out of its niche as needed.
- Magazine racks are practical, too – just don't forget to empty them regularly.
- Put up a peg board with hooks to keep workshop tools neatly organised. Traced and painted outlines of the tools make organising easy.
- Hooks are great for kitchen utensils and magnetic strips work well for knives.
- Sturdy plastic boxes are good for storing general household items.
- Paper is a perpetual problem, from bills to school notices cluttering counters. Keep an expandable file folder in the kitchen with sections for significant categories. Tuck it away out of sight when not in use.
- Store all your instruction manuals in a box or binder in the kitchen.
- Store toys in see-through stackable containers that are clearly labelled. That way, kids can put them away themselves easily and favourite toys won't get buried at the bottom of a large box.
- Store bath toys in a mesh bag and then hang it from the tap or showerhead until it is fully drained.

Do-it-yourself decorative boxes

If you need some boxes for storing small household items, you don't need to spend a fortune. Shoe boxes can be smartened up nicely to do the trick with inexpensive art supplies.

one Select a suitably sized shoe box for the items to be stowed.

two Select wrapping paper or wallpaper that matches the decor and cut it to size.

three Spread wallpaper paste on the box and glue on the decorative paper.

four Glue the paper to the box. When it is dry, stow items away.

Table linen and kitchen fabrics

The choice of fabric and colour does much to set the tone of a kitchen or dining room. Choose table linen, napkins and tea towels that add personality to the room and provide a warm welcome for both family and guests.

Vary your choice of tablecloths and napkins according to the occasion – be it a formal celebration, a family dinner or child's party. Buy tea towels in a variety of colours and fabrics, so you can easily find one or more to fit in with your decor, taste or requirement.

IN the dining room

● Buy table linens that match your dinner service.
● Opt for a tablecloth in tried-and-tested linen or cotton. Wash at the highest possible temperature safe for the fabric to ensure that any stains come out.

● Check that coloured tablecloths are colourfast or they will fade when washed.
● Make sure a new lace tablecloth is washable up to about 60°C.
● Before adding lace trim to a tablecloth, preshrink cotton lace by washing it. Otherwise, you risk having to remove it and sew it back on again.
● For everyday family use, consider a plastic tablecloth. They are extremely practical – and not only for an outdoor table. There are plenty of attractive and contemporary designs and colours.
● For a one-off occasion, consider paper. Paper tablecloths come in many sizes, colours and patterns,

HARMONISE LINENS and your DINNER service for AN attractive table.

but most can be used only once. Thicker ones can be wiped and used again if not stained.

- Use a table runner and placemats to protect a tablecloth. They can also be used to cover stains (though you should always have a fresh tablecloth to hand for emergencies).
- Placemats are made in many different materials, but for households with children wipeable plastic mats are ideal.
- You can lay a table runner across the table to add colour and style as well as to protect the table from scratches, scuff marks and stains.
- Get some small decorative metal or heat-resistant wooden trivets so you can put pots directly onto the table safely.
- Choose cloth napkins to add a note of elegance to a dinner party. For everyday use and less formal occasions, pick paper napkins that match the decor.

IN the kitchen

The tea towel is an essential kitchen item. It is used countless times for drying dishes and handling pots. But as well as soaking up water and spills, tea towels also harbour bacteria so must be washed regularly. However, the characteristics of each fabric differ – some are much more absorbent and therefore dry more effectively than others.

- Cotton tea towels are the cheapest option but they tend not to be the best choice. They don't dry especially well and often start to smell after a while. Even several washings may not eliminate the odour, so consider the cotton tea towel as a last choice.
- Linen or half-linen tea towels are absorbent and they dry items well without shedding lint onto sparkling glasses. They tend to be durable, although

Linen tea towels dry items well without shedding lint onto sparkling glasses.

they need to be washed a few times before first use to ensure they lose their stiffness and work their best.

- Microfibre tea towels are a more modern solution, absorbing up to five times as much water as an ordinary tea towel. They also dry rapidly after use. Microfibre is made from a blend of tiny nylon and polyester strands that are woven together. The only disadvantage is that people with very dry skin or eczema might find that microfibre tea towels can be annoyingly clingy and irritating to their skin.
- Always wash tea towels in the hottest possible water to remove all germs.
- Avoid chemical fabric softeners as these leave a coating on the towel that keeps it from drying dishes properly. Try using regular white vinegar as a softener instead. It has the additional advantage of eliminating unpleasant odours, even from cotton tea towels.

Folding a napkin into a lily

1 Fold a napkin into a triangle, then turn the right and left corners up to the upper tip to form a square again.

2 Now fold up two-thirds of the lower tip and then back down again to the bottom line. The top must not be longer than the bottom piece or else the folded napkin may tip over.

3 Finally, fold the napkin together by folding back the left and the right side to the rear, and tucking one tip into the fold of the other. Round out the shape by hand.

4 Turn down both upper corners (these form the leaves of the lily).

Wallpaper

Wallpaper is making a comeback. Not only can it look richer than paint, it gives rooms more dimension and can mask a wall that is uneven. Whether you opt for a muted colour or bright patterns, wallpaper should create a unified look with the fabrics and the style of furnishings in a room without overwhelming the space.

Wallpapering is easier that you think. A few tips, some basic instructions and a little imagination are all you need to change the atmosphere of a room.

The colour of this wallpaper works well with the sofa and accessories.

- For a striking wallpaper pattern to show itself off to best advantage, the furnishings should be low-key.
- Vertical lines on the wall make low ceilings appear higher, particularly if the ceiling is painted white.
- Before you start, get an idea of how a chosen wallpaper will look by holding a large piece of it against the wall.
- Textured wallpaper covers up small holes or lumps.
- Make a note of the serial or model number of the wallpaper you are buying so you can find it again if you need more for a perfect match.
- To ensure you have enough, buy an extra roll or two – especially if you can return it if unused.

THE right choice

Your choice of wallpaper will depend on personal preference and where you will be using it, as well as on the skill of the installer.

THE effect on a room

Wallpaper can have a significant impact on a room. It can subdivide and structure a room, as well as making it appear larger. The trend these days is to wallpaper only one or two walls rather than the entire space.

- Large patterns tend to dominate a room, making then ideal in a big room. In a small room or one full of nooks and crannies it can be overwhelming.
- Cover small rooms with a bright, monochromatic colour or small patterns.

Making fabric wall panels

By creating fabric panels for walls, you get all the drama of wallpaper but can easily change the patterns and colours.

one Pick up wood frames for stretching canvas from a craft store (they can be different or uniform sizes) and cut the fabric so it is slightly bigger than the frames.

two Iron the fabric, position it on the frame and tack it down on one long edge with a staple gun.

three Stretch the fabric taut over the frame and staple the opposite long edge, followed by the short sides.

four Finish the corners by folding down the fabric as if wrapping a present, then staple it down. Now you have a dramatic wall feature for little money or effort.

New styles and designs have made wallpaper popular.

- Vinyl wallpaper is easy to handle and stands up to scrubbing and moisture fairly well, so it can be used in bathrooms, kitchens and children's bedrooms.
- Flocked wallpaper has raised 'velvet' patterns and works well for creating decorative highlights and for formal areas such as dining rooms. It is washable, but can be damaged by rubbing and scrubbing.
- Grass-cloth, made from a weave of grasses, is better suited to areas that sustain little wear and tear.
- Fabric wallpaper is made of silk, cotton or linen, which is sometimes laminated to regular paper. It's not the easiest surface to keep clean and is fairly difficult to work with but looks stunning in a formal setting, such as a dining room.
- Foil wallpaper is made of patterned metal foil. It can add an interesting touch and reflects light well but is unforgiving if it gets wrinkled or folded.

WALLPAPERING made easy

Once you have selected the wallpaper, roll up those sleeves and get to work.
- To remove old wallpaper, pierce it with a scoring tool and moisten it with water. A little white vinegar added to the water acts as a glue solvent.
- Turn off the electricity at the fuse box before moistening wallpaper as plug points and light switches can pose a danger.
- Fill in holes and cracks in the walls with polyfilla, then sand them until smooth.
- Check the undercoat by sticking a piece of tape onto the wall and pulling it off with a jerk. If the paint remains on the wall, the wallpaper will stick.

Draw a vertical line before positioning.

- Sand down painted walls to help the glue to stick and eliminate unsightly bumps.
- Apply a coat of primer on dry plaster before beginning so the wallpaper will stick.
- For a perfect finish, apply lining paper before adding the final paper.
- Repaint the room's trim and ceilings before you begin wallpapering, and protect furniture and rugs from drips with dust sheets.
- Always begin wallpapering at the edge of a window.
- Using a level or plumb line, draw a vertical line from floor to ceiling as a guide to keep the paper straight. Repeat for each strip.
- Cut lengths of wallpaper about 10cm longer than the height from ceiling to skirting board.
- Fill a tray with lukewarm water and dip each length of pre-pasted wallpaper for 30 seconds or more. Change the water every six to eight lengths.
- Match the edge of the first length to the plumb line you have drawn on the wall, leaving 5cm of extra paper at the top of the ceiling.
- Match patterns exactly, and allow for them when calculating wallpaper amounts.
- Be especially thorough when gluing down the edges and corners of wallpaper seams.
- Trim excess paper at the top and bottom of the wall with a utility knife and ruler.
- Wipe wallpaper and skirting boards with a damp sponge to remove excess glue. Rinse the sponge well or streaks will appear as the paper dries.

Wall treatments

Colourful and creatively decorated walls accentuate a home's individuality and create a cheerful mood. You can add a special touch by using templates, sponges, wall stamps or other decorative elements to achieve the look you want.

Stencilling, cork tiles, mouldings, embellishments and wall treatments are some of the easiest ways to make an impact on any room.

INNOVATIVE wall treatments

Wall treatments have come a long way since the clumsy wall-sponging that was popular in the 1980s. Today, a stylish wall treatment can be all you need to transform a room.

● To achieve the appearance of a high-end wallpaper without the price tag, create it yourself. Paint the wall with a dark-coloured base coat, for example, and stencil or stamp the design in a lighter paint. You can get wallpaper stencils in small or large sheets, so you just roll the pattern and repeat. Another advantage: when you are ready for a change, just paint over it.

● Tongue and groove can lend a charming, slightly retro appeal to walls. You can buy ready-made panels that are simple to install. Just run them halfway up the wall and top them off with some complementary trim.

● Paint stripes either in contrasting colours or two shades of the same colour. They are easy to manage: paint a base coat of colour, then apply masking tape the width of the stripes and paint the second colour between the strips.

● Use wall stickers to decorate a child's room. They can be removed easily when your toddler outgrows his or her current passion. Similarly, wall graphics cost little and can add a note of drama.

● Paint one wall with blackboard paint for a dramatic and interactive finish.

PAINTING with stencils

Many historic buildings have splendid patterns that were created using stencils. Painting with stencils continues to be a way to give interior walls a strikingly individual design.

● You can buy prefabricated stencils at craft stores or you can make your own by tracing shapes onto poster board or stencil plastic, then cutting them out. This allows you to create any image you want.

● Limit yourself to two or three colours and opt for simple patterns if you don't have much experience using a paintbrush.

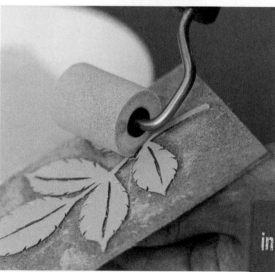

STAMP THE DESIGN in a lighter COLOUR ON DARK WALLS.

- Use a stiff, dry brush to ensure that the pattern shows up; the smoother the base coat, the easier it is to apply the pattern. But even materials such as textured or grained wallpaper can be embellished.
- You can use any fairly thick paint or opt for special stencil paints. Both will work well.
- Use stencilling to beautify furniture, doors, and fabrics (with fabric paint), as well as walls. Using the same stencil on everything helps unify the look of a room.

CORK tiles and wallpaper

Cork can add a warm and pleasant atmosphere. The bonus: it is a sturdy, water-resistant material that is immune to the effects of temperature.
- You can apply cork tiles to all stable surfaces, provided they are firm, clean and dry. Thoroughly remove old paint and smooth out imperfections on the wall with Polyfilla before you begin.
- Take the cork tiles out of their packaging two days before installing them and stack them in piles of no more than five pieces in the room where they are to be applied. This will prevent the tiles from contracting or expanding on the wall.
- Cork wallpaper can be combined with normal wallpaper, so is perfect for creating room accents.

Use wall stickers to decorate a child's room. They can be removed easily when outgrown.

- Let cork wallpaper soften for a short time after putting it up. Use a special seam roller to press down the seams.

MOULDINGS and embellishments

Decorative elements allow you to create stylish, classical or avant-garde designs – there need be no limit to your creativity.
- Use mouldings to embellish junctions where walls meet the ceiling or the floor. Off-the-shelf crown mouldings add grace and elegance to a wall. They will also cover up imperfections at the edges of the wallpaper, hide cables and offset plain curtain rods. They are inexpensive to install, particularly if you do it yourself.
- Install decorative trims, cornices and skirting board mouldings to cover up bumps and small tears in the plaster, and to hide small pipes.
- Opt for a ceiling rose to create an elegant effect in an older home that has rooms with high ceilings. Nowadays you can buy them in many different shapes and designs.
- Use decorative elements such as these to help visually correct and even out imperfections in a room.

You can buy prefabricated stencils at craft stores.

Window dressing

Coordinate curtains, drapes, blinds and accessories with the overall decor and proportions of a room. Choose window fabrics and decorative elements only after you have decided on the overall feel you want to achieve.

Window coverings, however, have to be functional as well as attractive. They protect a home from draughts, extremes of temperature and bright light, as well as from prying eyes.

CURTAINS for every style

There are several different kinds of window dressings, each with certain advantages.
● Voile or sheer curtains should limit the view outside as little as possible but, at the same time, they offer occupants their privacy.

● Drapes can keep out blinding light and act as a privacy screen.
● Blinds are often a better alternative to sheer curtains in contemporary-style homes.
● For a classic contemporary look, use distinct shapes, geometrical patterns and bright colours.
● To create a luxurious effect, combine light sheer curtains with drapes of silk, cotton or damask.
● Heavy valences of tufted cord, brocade or velvet look sumptuous. They can be combined with curtains of tulle and gauze. But this striking combination requires a room with a high ceiling or it can look stifling.
● Pleated pull-up sheers or drapes made from light fabrics with flowery patterns set a romantic and playful theme.
● Use decorative tie-backs to add interest and maximise light levels in a room.

SHORT or floor-length?

● Short curtains look rustic and suit short and square or wide, low windows. They stop just above the windowsill or a little below it.
● Arched sheer curtains are a good choice for windows with flowerpots or windowboxes, because the floral contents remain in full view.
● Floor-length curtains enhance the high, narrow windows often found in older apartments and houses. They should reach to about 5cm above the floor so that you can vacuum underneath easily.

THE effect on a room

● Choose floor-length curtains to make a room's windows look longer and the ceiling look higher.
● Install the curtain rod a little higher than the frame to make the window look taller.

BLINDS are ADJUSTABLE AND infinitely variable.

- Use wide curtains to make a narrow, high window look broader.
- Create uniformity between windows of differing heights by hanging all curtains at the same height.

CURTAIN rods

These days, many windows are covered by venetian blinds or roll-up panels of paper or reeds, but the good old curtain and rod still has its place in many homes. Of course, hanging curtains isn't quite as simple as hanging the blinds that cover so many windows but, if done correctly, they can bring a comfortable, old-world allure to any room. Here are a few things to consider.

- Display window treatments to their best advantage by installing curtain rods not less than 15cm above the window frame, and extending at least the same distance beyond the sides of the window.
- If a windowsill sticks out a lot further than the wall, it is a good idea to install wood moulding on the wall around the window first. Then, mount the brackets for the curtain rod on the moulding.
- Install the rod so that it is almost touching the ceiling, or is attached to it, if you want the ceilings to appear higher.

DECORATION

Window shades, blinds and curtains aren't the only way to dress up a window. With a little creativity, you can use all sorts of items to add visual interest.

- Install stained glass windows to give a room an individual appearance and shield you from prying eyes outside, too.
- Use the windowsill as a ready-made shelf for flowering or green plants. Or, add accessories such

WINDOW TREATMENTS GIVE free rein to YOUR CREATIVITY.

as pretty stones, elegant vases, ornaments or artistic flower arrangements.

- Install a glass shelf across the centre of the window, then place small plants on the shelf. This is especially good for bringing a touch of green to the kitchen or any area that doesn't require the same level of privacy as, say, the bedroom. Just make sure that the plants are not too heavy.

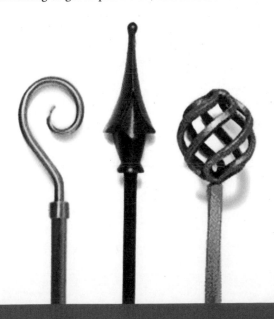

Curtain rod ends come in many shapes and materials.

Gardening
with nature

Generations of gardeners combined art, nature and horticultural know-how to create flourishing gardens that were both productive and beautiful – long before high-tech gadgets and chemical compounds became available.

Berries and currants

Small berry and currant bushes still deserve a place in every garden. Easy to grow and with fruit packed with vitamins and antioxidants, they tolerate a wide range of soil conditions and will thrive in a partly sunny garden sheltered from the wind.

Berry and currant plants can be grown as bushes or cordons, depending on the space you have available, but will do best in a moist, well-drained soil.

CURRANTS ARE AN EXCELLENT SOURCE of vitamin C.

FOR a good start

● Plant bare-root berry bushes such as raspberries in autumn or as soon as you can work the ground in spring so that the plants are well-established before hot weather arrives. Plant on a cloudy day or late in the afternoon.

● Plant potted berry plants at any time, as long as you can work the soil. However, bushes root better when they are planted in the autumn or spring.

● Buy varieties of blackberries and loganberries without thorns so that even children can pick their fill.

● Beautify an arbour or trellis with thorn-free, climbing blackberry varieties.

● Remove perennial weeds before you plant – they can rob the soil of essential water and nutrients.

● Plant the rootball of a currant bush around 7-8cm below the surface of the ground to encourage strong growth at the base of the plant.

● Put a little rock dust in the hole along with the compost for a red currant bush to add essential trace elements and boost pest resistance. You can buy it at good garden centres or online. The sprigs should be bare of leaves to a height of about 15cm above the surface of the ground.

● Leave enough space between plants to assure that they will thrive without competition. This also prevents grey mould during rainy summers.

● When planting raspberries, don't choose a location where raspberries have been grown in the preceding five years.

● Blueberries need acid soil to thrive. If the soil is alkaline, it is better to grow them in pots using ericaceous compost. Incorporate organic matter into the soil before planting to ensure strong growth. Good options include peat moss, well-rotted manure, straw, compost or aged sawdust. If possible, use only rainwater for watering.

PRUNING bushes

Prune berry and currant bushes before the onset of winter or right after the harvest. Early fruiting raspberry bushes bear on one-year-old shoots from the previous season; gooseberries and red currants on one to two-year-old shoots; and, blueberries after three to eight years. A good harvest requires proper pruning to provide the bush with light and air. Remove damaged canes or shoots, as well as sick shoots near the ground.

● After planting, cut raspberry shoots back to about 5cm above the ground.

● With blackberries, cut off brown, dead shoots right down to the soil and remove them from the bush in autumn or spring.

● In autumn or early spring, remove dead wood from blackberries and loganberries and cut back fruited stems of early raspberries. Trim stems of autumn raspberries to ground level.

● Cut back fruited blackcurrant stems after harvesting. New wood will then be produced from

Save supermarket containers for gathering your berries.

which next year's fruit will develop. Redcurrants, which fruit on the previous year's wood, should be cut back by about a third, removing any central stems that are cluttering the bush.

● Keep gooseberry bushes from having more than five strong shoots at the base or it is difficult for them to develop new shoots. Cut shoots back by about a third, and ensure that no shoots are more than six years old.

PROPER care

● Since most berry bushes are shallow-rooting plants, hoe them carefully and close to the surface, if at all, to avoid damaging the roots.
● Plant garlic, lily of the valley and yarrow next to gooseberries to help keep the bushes healthy and increase their yield.
● Plant wild garlic, marigold and forget-me-nots near raspberries to help keep pests away.
● Help many varieties of berry bushes with a layer of mulch consisting of dried grass cuttings, straw or leaves. It keeps weeds from growing and the soil

remains fine, crumbly and damp. If the berry varieties in the garden are not mildew-resistant, a layer of bark mulch will help.

AT harvest time

● Gather berries in fairly small containers to prevent them being crushed. Harvest with both hands by hanging a container with a handle over your arm. Line them with paper towels to absorb any juice.
● If the gooseberry harvest looks promising, pick some of the fruits early and preserve them. The remaining berries will grow bigger and better.
● As soon as berries begin to ripen, protect them from birds by hanging nets above the bushes – but make sure you can walk under them easily.
● Wait for whole sprigs to ripen then pick them off. You can then strip them easily with a fork.
● If you have a large space devoted to fruit, consider investing in a permanent fruit cage.

After picking the fruit, trim early raspberry canes close to the ground.

Planting berry bushes

1 Dig a hole for the plant about 50cm deep and as wide as the rootball. Loosen up the soil thoroughly to prevent waterlogging.

2 Add a little rotted compost to the hole to provide nutrients for the plant, then carefully insert the well-watered bush.

3 Fill in the hole with soil and compost, tamp it down carefully taking care not to damage the roots, and be sure to water thoroughly.

Bulbs and tubers

Crocuses, snowdrops and daffodils are among the first messengers of spring, whereas dahlias, gladioli and colchicums flower into late autumn. These traditional garden plants don't require a lot of care, and can easily be used to add colour to a balcony or patio area if grown in pots.

The saying 'the bigger, the better' really does apply to flowering bulbs, because the biggest bulbs are generally healthier and flower better.

MAKING the right selection

● Buy bulbs with undamaged outer skins and firm cores. If they are sprouting or exhibit decay, throw them out.
● Store bulbs that you can't plant immediately after purchasing in a cool, dark and airy place. Store rare and valuable varieties in the vegetable drawer of the refrigerator until they are ready to be planted.
● Plant lilies, lilies of the valley or glories of the snow (*Chionodoxa*) in shady areas; wood anemones (*Anemone blanda*) flourish in full shade. Daffodils will tolerate partial shade.

Plant spring-flowering plants such as irises, crocuses, grape hyacinths or tulips in the autumn.

PROPER planting

● Flowering bulbs grow best in loose, porous soil. If the soil is on the heavy side, fill planting holes with a layer of sand about 1-2.5cm deep. This keeps the bulbs from rotting when the weather is wet and cold.
● Remember that the depth for planting should always be at least twice the height of the bulb.
● Follow the instructions on the packaging of lily bulbs carefully as there are various considerations that will determine at what depth they are planted.
● Plant bulbs at different depths to get a nice, thick flowering. This produces better results, especially in restricted spaces such as a clay pot.
● Lightly water daffodils and fritillaries after planting. Their roots will grow more quickly in moist soil.
● Plant bulbs in plastic baskets, nets or wire mesh if you plan to take them out later. They will be easier to remove and will also be protected from rodents.
● Plant the tubers of summer-flowering dahlias, gladiolus or crocosmia (falling stars) in well-loosened and aerated soil.

CULTIVATING bulb plants

Most bulb plants flourish beautifully without special care. They need little watering, which makes them ideal for areas with less rainfall or for gardeners who have little time.
● Put a stake in the ground at the same time as you plant tall-growing plants, such as gladiolus or dahlias, that will need support later on. If you wait until the root system is already developed before driving a stake in, you risk damaging the plant.
● In damp regions, remove tulip bulbs after their leaves have wilted and store them until the autumn in a bed of sand, peat moss or sawdust in a dark place.
● Snip off the flowers of bulbs after they wilt in order to save them the energy needed to form seeds. Don't touch the leaves, however – the plant needs them to store nutrients for the winter. Keep them well watered, too, until leaves have died down.
● If possible, plant bulbs and tubers in different places each year to avoid the build-up of diseases caused by fungi or bacteria.

- Divide dahlia tubers with a knife before planting them in the spring. Each division should have a budding sprout.

PROPAGATING lilies through bulb scales

- Dig up the bulbs in the autumn and gently pull off four to six of the fleshy outer scales. Dust the wound on the parent plant with charcoal powder for protection and replant the bulbs.
- Place the scales to half their depth in a mixture of potting soil and sand. Keep them moist at room temperature by covering them with cling film. Don't expose them to direct sunlight.
- When small bulbs with delicate roots form (within about eight weeks), plant them individually in small pots so that only the uppermost tips poke out of the soil. Store them in a dry and dark place at about 5°C.
- Carefully plant bulbs as soon as the first delicate leaves sprout in the spring, choosing a day when it is not too cold.

PROPAGATING bulbs and tubers

The propagation of bulb plants is easy. Many form small bulbs or sprout tubers by themselves. These simply have to be detached from the parent plant and stored in a cool and sheltered place over the winter.

- Dig up bearded iris rhizomes every two to three years after plants have finished flowering. Divide them into several pieces and trim the leaves. Dispose of the oldest parts of the plants.
- Divide and move snowdrops immediately after flowering when they are still 'in the green'.

A multi-layered effect is created by planting low-flowering bulb plants in front of high-stemmed lilies, narcissi or tulips.

Climbing plants

Climbers such as clematis, wisteria or honeysuckle make glorious focal points when grown over an arbour or trellis. Their beauty and trailing habit are also ideal for concealing unattractive sheds, fences or bare walls.

Different species of climbing plants require different locations, so check what growing conditions they prefer before planting. They should be situated a certain distance from walls or climbing aids and provided with organic material to help them grow.

WHERE and how to plant

● Evergreen plants like shady or partly shaded north or northeast walls. These include ivy, evergreen honeysuckle and winter jasmine. This is also a comfortable spot for climbing hydrangeas.
● Deciduous climbers such as clematis, wisteria (*Wisteria sinensis*) and honeysuckle (*Lonicera*) thrive along sunny southwestern or southeastern walls.
● Plant climbing plants in spring so that they can become well-established before autumn.
● Cover the rootballs of the plants with at least 5cm of soil.
● Put a layer of coarse gravel or mulch over the planting site to keep the wall of the house from getting dirty when the plants are watered or when it rains hard. It also protects the soil against excessive drying out. Before adding the gravel, wait until the soil has compacted as a result of rainfall or waterings.
● Remember that clematis is fussy about keeping its feet cool – in other words, it likes a base in the shade. But as it also needs sun, shade only the base with mulch, a circle of stone, a brick or even a low plant. Other climbers also appreciate these conditions.

SUPPORT for climbers

Climbing plants hold onto their supports in various ways. Some, like ivy and virginia creeper, can cling by themselves without too much trouble. Others, such as clematis and wisteria, need support. When choosing a support, consider the characteristics of the plant you will be tying to it.
● Use vertical supports such as trellises for climbing roses and winter jasmine. Roses can also use their thorns to get a grip.
● Use either vertical or horizontal supports for clematis and perennial sweet peas.
● Hang supports from hooks on a wall. This makes it easy to take them off once plants have been trained.

Annual climbing plants such as black-eyed Susan can be raised in a pot and then placed on the patio.

REMOVE WILTED FLOWERS so plants PRODUCE new buds.

● Make wooden supports and trellises yourself. When choosing and working with wood, remember that the plants will add weight to the structure. Wind will tug at the vines and snow will provide an additional load, so plan for stability.
● Tie individual shoots to the climbing aid carefully and fairly loosely with soft string. They must be able to move in the wind and increase in diameter.

CUTTING climbing plants

If you cut back climbing plants regularly, they will become stronger and send out new shoots. Once the climbers have reached the desired height, simply prune them lightly every month. Shorten the tips of the shoots by a third to encourage new growth.
● Cut back early-flowering clematis immediately after it flowers. If you have a late-flowering variety, cut back the shoots from the previous year to two buds above the base.
● Prune species clematis, virginia creeper and climbing hydrangea only every couple of years.
● Cut back the shoots of climbing plants that are several years old significantly in spring.
● Cut back older plants radically once they have stopped sending out shoots the way they should.

● Protect bricks and gutters by keeping a close eye on powerful climbers such as Russian vine (Fallopia) and wisteria. Cut them back as required, otherwise serious damage to the wall or gutters can result.

CLEVER combinations

● Plant some ferns or other undemanding, shallow-rooting perennials around the roots of climbing plants to keep the ground below them from appearing too bare.
● Use a tree as a climbing aid: place a climber near the trunk of a tree with a narrow crown. For a tree with an expansive crown, plant the climber under the edge of the crown and allow it to climb up a rope to a sturdy branch.
● Combine two varieties of clematis that flower at different times, or an early flowering clematis and a climbing rose, for an attractive two-tone effect.
● Plant alternating varieties of climbers on the posts of a balcony railing.

Climbing roses need sun and can grow up to 5 metres high.

Companion planting

A traditional means of repelling pests, companion planting has been used for many years to protect prized vegetables. Aromatic plants such as garlic, marigolds and peppermint are all reputed to send a signal to unwanted insects to go elsewhere.

Companion planting is a kind of botanical buddy system. It works on the principle that plants that grow together interact and influence each other. Apart from removing certain nutrients from the soil, individual plants give off substances and fragrances that can be good or bad for their botanical neighbours.

WHERE and when to companion plant
● To implement companion planting, start by adding plants in the centre of the bed. Harmonious partners generally have similar seeding or planting times.

GOOD TO KNOW ✓

Harmonious partners

Make sure that plant varieties tolerate one another when introducing companion planting. Tomatoes go well with lettuce, cabbage, carrots, radishes, beetroot, celery, spinach and parsley, but not with potatoes, cucumbers, fennel and peas. Potatoes go well with horseradish but not with sunflower, tomato or cucumber. Improve spinach crops by combining them with peas and beans.

Root excretions from French marigolds help to keep ravenous pests away from broccoli.

● Companion planting tends to require less fertiliser because the plants absorb different nutrients from different soil depths, as well as excreting beneficial elements into the soil.
● Many plants enhance their neighbour's flavour. Caraway and coriander improve the taste of early potatoes, and dill intensifies the taste of carrots.
● Pests are less drawn to gardens where companion planting is employed because some plants form a protective shield for others by means of their scent.
● Garlic can prevent mildew and kills many fungi when planted near fruit trees.
● Nematodes will help keep slugs at bay but can't stand the root excretions from lilies or marigolds, while tomatoes drive away asparagus beetles.
● With carrots and onions you can kill two flies with one blow: they help protect one another from both carrot-fly and onion-fly infestations.
● Grow fennel to attract aphid-eating ladybirds.
● Ants will lead aphids onto sunflowers, so protecting a whole range of garden plants.
● Choose the right companion plants to reduce, if not eliminate, the need for pesticides. Eliminating pests will result in higher yields.

ADVANTAGES of companion planting
● Companion planting uses far less water. Since the companion plants grow close together and shade the ground more effectively, less moisture evaporates.
● Because the plants are closer together, you get a higher yield per surface area.

Compost

There is no need to spend large sums on modern plant feeds – the simplest and least expensive fertiliser is compost from your own garden. When well mixed over the course of a year, kitchen and garden waste turns into nutrient-rich humus, one of the best fertilisers you can find.

With a compost heap you are basically recycling nature, and the greater the variety of materials used the richer the subsequent compost will be. After as little as three to four months it may be possible to have raw or fresh compost, which is best for mulching or autumn fertilising of the vegetable patch. However, compost needs about a year to ripen fully. You will know when it is ready because it becomes fine, crumbly and dark, like fresh soil.

MATERIALS for the compost heap

All healthy, organic materials that rot within a year can be added to the compost heap. The following things can all be added.

● Kitchen waste, including: vegetable and fruit peelings and cores (as long as they haven't been sprayed with pesticides); spoiled or dried-out foods; coffee grounds, and tea bags and tea leaves; paper filters; and crushed eggshells.

● Paper from napkins, paper towels, uncoated paper and paper bags for disposal of biodegradable waste. Best shredded before adding.

● Garden waste such as shredded tree, hedge and shrub cuttings; residue from flowers and perennials; roots (but not those of perennial weeds); and, weeds without seeds.

● Grass cuttings can be added in thin layers, preferably mixed with coarser material so that enough air still gets into the compost pile. It also prevents grass going slimy in the bin.

● Potting soil, cut flowers, potted plants with soil and small pieces of untreated scrap wood.

Keep animals out of the compost by securing the opening with chicken wire.

SPECIAL considerations

- A partially shaded location is ideal for a compost heap. In bright sunlight the compost will dry out and in the shade it may rot. However, give an enclosed plastic bin full sun as it helps contents to rot quickly.
- Compost needs a certain amount of moisture and should feel like a sponge that has been squeezed out. A greyish-white coating or lots of ants mean that the compost material is too dry and must be watered.
- Good ventilation is also crucial. Use a pole to poke ventilation holes down to ground level.
- Turn the compost every six months or so.
- Surround a fairly small compost heap with a trellis on which nasturtium, vetches or other climbers can grow. Sunflowers are also good for concealing a compost heap.
- Fragrant plants can be used to combat the smell of rot but they are often unnecessary. Compost usually smells like fresh forest soil.

COMPOST must be ACCESSIBLE SO IT CAN be turned regularly.

Cress test for readiness

ONE Plant a little cress in a bowlful of material from the compost pile.

TWO If it sprouts in two to three days and develops green leaves, the compost is ripe and okay to use even for sensitive plants.

three If nothing sprouts or the cress develops yellow leaves, the compost is still too fresh for using in the garden – you will have to wait.

MAKING a compost heap

- A compost heap should be formed right on the top of the soil, never on stones or concrete. It needs contact with the earth so creatures that contribute to rotting can get in without hindrance and the resulting seepage can run off easily.
- Use the three-heap method for large gardens. Use the first heap for garden waste, which can be moved into a second heap in the winter. This can be left to ripen undisturbed. By the following winter, the ripe compost can be sifted and moved to the third pile for use in the garden.
- A composter with sliding panels at the bottom enables you to remove compost from the base. Removable sides are also helpful.
- A wire-bin composter with a hinged front provides easy access.

- Lay coarse wood cuttings to make up the bottommost layer of the compost. Place a little soil on top, then pile on the compost material to a depth of about 1m.
- Add a judicious sprinkling of lime sand or rock dust on each layer to provide the compost with important minerals and trace elements.
- Collect leaves separately in a covered circle made of wire mesh. They'll decompose more quickly and their acidic qualities will promote the growth of blackberries, raspberries and rhododendrons. Or pack leaves into large plastic bags and leave for two years to rot.

ENCOURAGING the composting process

A few shovels of garden soil or a herbal slurry (see *Fertilising and revitalising*, page 290.) can be used to speed up the rotting process. Or use a biologically formulated compost accelerator. Organic ones are available.

- Occasionally spray the compost with an undiluted slurry made from dandelion or stinging nettle, or a thinned-down borage extract.
- The chemical allantoin, which is contained in comfrey, encourages the rotting of straw as well as other plant remains that contain cellulose.

Crop rotation

Farmers have known the benefits of three-field crop rotation for centuries. Using the rotation principle in a vegetable and fruit garden will result in higher yields without excessive use of fertilisers, and it helps to prevent garden pests and fungal diseases, too.

If you plant the same varieties in the same spots every year, you will deplete the soil because the plants will always remove the same nutrients from the ground. Sooner or later, those nutrients will run out and crops will cease to thrive. More importantly, diseases will build up in the soil.

FORTIFYING rest

A planting cycle gives soil a well-needed rest so it can regenerate.

● The more time that passes before you put the same plant in the same location the better: a four-year cycle is ideal.

● Crop rotation often significantly reduces the need for fertilising, as some plants replace nutrients in the soil that other plants need. The roots of bean and pea plants release plenty of nitrogen when they rot, for example, which encourages growth in cabbage plants in the following year.

● Rotating crops also prevents pests and fungal diseases that spread in the soil. If the fungus spores and insect larvae can't locate the proper host plants, they die. For that reason, return cucumbers, tomatoes and peas to the same spots only after two years or more, and onions after three years.

● If a variety of cabbage is afflicted with club root, wait seven to eight years before replanting cabbages in the same bed.

● You can even rotate crops in a small garden with just one vegetable bed. Simply rotate the plant varieties by rows in subsequent years.

PLANNING crop rotation

When planning for crop rotation, consider the nutrient requirements of each plant. Some crops grow abundantly and quickly and use up plenty of nutrients – these plants need strong fertiliser. Other plants require fewer nutrients, grow more slowly and need more moderate fertilising. Plants with a low nutrient requirement just need a little compost.

● Follow a plant that uses lots of nutrients with a moderate nutrient consumer and finally, in year three, by one with low nutritional needs.

● Use a journal to plan crop rotation, keeping a record of precisely what, when and where plantings were done.

● For a well-planned crop rotation sequence, begin planting with vegetables in the cabbage family, followed by legumes, tubers and plants from the potato family (such as aubergines, chilli peppers and tomatoes) and, finally, by bulb plants such as leeks, garlic or onions.

START YOUR CROP ROTATION cycle by PLANTING cabbages.

Fertilising and revitalising

For centuries, gardeners relied on natural nutrient sources such as horse manure. Then chemical fertilisers became popular because they were inexpensive and easy to handle. But as concern for the environment grows, it is time to go back to organic methods to improve the soil and grow lush, healthy plants.

Green fertilising involves improving soil quality by planting specific crops with roots that will penetrate the earth to provide thorough aeration. Generally, they are dug into the soil before they flower to enrich it. The results may not be immediate but, in the long run, you will be rewarded with a healthier garden.

Mix compost and manure into new garden beds with a fork.

GREEN manure

● Leave small quantities of grass clippings and leaves spread out across the garden and lawn to decompose and enrich the soil.
● Use green fertilisation for empty areas and places with some permanent crops such as strawberries, rhubarb, roses and asparagus.
● Green-fertilising plants that are sown in the spring should be mowed or cut down shortly before they flower, then chopped up and raked in to enrich the soil.
● Leave green-manure plants sown in the autumn to stand through the winter and then mow or dig them in during the spring.

● Prevent diseases by ensuring that green-fertilising plants belong to a different plant family than that of the subsequent vegetable crop.
● Sow seeds of phacelia in any kind of soil in spring and summer. It grows quickly, is easy to care for and attracts bees.
● Dig green-manure plants in just before they flower, when they are most nutrient rich.
● Opt for yellow trefoil and hairy vetch. They are winter hardy and also a good bet for medium-heavy to heavy or light to medium-light soils.
● Green manure can also be cut and put on the compost heap.
● Use mustard for medium-heavy soils, but don't plant any vegetables from the cabbage family in that soil for the next two to three years.

NATURAL fertilisers

Organic fertilisers spread easily and continue working for a long time.
● Use compost that consists of waste from the garden and kitchen. It is a good source of nitrogen for the soil when added to the garden in the spring. Work it in to a depth of 2.5cm.
● Horse manure has to be well rotted or composted or it will burn the plants. If you buy it fresh, first pile it in a heap, water it thoroughly and cover it with a sheet of plastic so that it stays moist and rots. Spread it in the autumn and work it into the soil well. You can plant cabbages, lettuces and pumpkins immediately, but wait a year before planting beans, peas, carrots and radishes.
● Spread commercially available dried manure on the beds as a top dressing or use it as an effective composting accelerator.
● Spread bone and blood meal – which contain phosphorus and calcium – in springtime. Mix bone meal into the soil to encourage healthy growth in young shrubs and dose potted plants on occasion, too.
● Hoof and horn meal is easily absorbed by the soil and offers a quick supply of nitrogen, phosphorus and potassium. It can be spread

throughout the year, especially before sowing or planting. Use 60-90g per square metre.

● Fertilise potatoes, carrots, tomatoes, celery and roses with wood ash from a bonfire or indoor wood-burner. In springtime, spread the fine powder thinly in planting furrows or holes and then lightly work it into the soil.

● An occasional milk fertilisation benefits ferns, roses and tomatoes. Mix milk or whey with water in a 1:3 ratio. The plant roots will drink up the amino acids from the milk. The solution helps prevent mildew, too.

● Sandy soil needs fertiliser and added nitrogen as rain sweeps away many of its nutrients. It also needs compost to improve its texture and water retention.

● Crushed eggshells are a good source of calcium and they are especially useful in acidic soil as they raise the pH level. Another benefit: if you scatter them around tender young plants, their jagged edges

SPREAD THE COMPOST CAREFULLY, then work it in with A RAKE.

Stinging nettle slurry

About 1kg fresh stinging nettles
10 litres water

Crush the stinging nettles in a bowl. Add the water and thoroughly immerse the plants. Cover with cling film or a lid and leave to ferment for three weeks, stirring daily, then strain. Stinging nettle slurry strengthens resistance but should be used only on overcast days.

● Put coffee grounds directly onto the flowerbed to fertilise plants and keep slugs and snails away. You are unlikely to generate enough to fertilise the entire garden, so will have to use other fertilisers as well.

FEED your garden a slurry or infusion

Use herbal preparations to fortify plants against pests, diseases and fungus. Prepare them with diluted rainwater or tap water that has been allowed to sit.

● Create an extract or infusion by putting 1kg of herbs into 10 litres of cold water for at least 24 hours. Strain and use diluted. Stinging nettle extract keeps pests away; garlic extract with a little added liquid soap combats aphids and prevents and treats fungal diseases.

● Make a herbal solution with equal parts of the extract and cold water. After letting the mixture sit for about a day, boil it gently for about 30 minutes.

● Spray common horsetail infusion to ward off many diseases, including leaf drop in berries, leaf spot, peach leaf curl, powdery mildew, brown rot, rust, bottom rot in lettuce and scab. Mixed with a little soft soap it is effective against aphids and spider mites.

● Combat snails by spraying an infusion made from bracken, fir cones or strong coffee.

● Basil tea (8 teaspoons of dried leaves in 1 litre water) is an excellent spray weapon against aphids and spider mites.

● To make a tea, snip off fresh herbs and place them in a bowl or large cup. Pour boiling water over the herbs and let the liquid steep for about 10 minutes.

● Make a herbal slurry by soaking about 1kg of fresh or 150g of dried herbs in 1 litre of water for 10-20 days, stirring daily. Use diluted 1:10 with water.

● Use a birch leaf slurry with fresh leaves as preventative medicine for fungus on leaves and fruits; a dandelion leaf slurry promotes a better-quality yield from most berry and fruit trees.

Flowerbeds and borders

When planning and planting a flowerbed, the soil conditions and position of the bed each play an important role. Ideally, you should choose plants that flower at different times to ensure your garden always looks colourful and fresh.

If you select plants carefully, your garden will reward you with an almost constant display throughout the seasons, whether in sun or shade.

THE right plants

● Make a seasonal bed that positions early-flowering aliums next to colourful perennials. French marigolds, geraniums and pansies all flower for a long time.
● Shade-loving plants include cyclamen, anemones, epimediums, lady's mantle, lilies of the valley and bergenia.
● Damp, shade-tolerant flowers include Christmas roses, foxgloves, hostas and periwinkle, plus primroses and snowdrops.
● In the summer, plant ornamental grasses to create an attractive visual island within a flowerbed.

These grasses are eye-catchers until late autumn and can also be used to add a romantic note to summer flower bouquets.
● Plant some poppies, lupins and delphiniums for an early summer flowerbed. Peonies and multicoloured hollyhocks are also a good choice.
● Prevent flowerbeds from appearing bald out of season by leaving seed heads intact for interest in autumn and winter. Grasses, sedums and echinops are all good subjects. And consider some stone garden ornaments.
● Create flowerbed variety by using plants with attractive leaf shapes or shrubs with colourful bark or fruit. Try dogwoods, berberis or mahonia.
● Look for lots of buds, good branching and a well-developed, moist rootball when you are choosing plants to buy.

POPPIES AND LUPINS ADD vibrant colour in early SUMMER.

PLANNING the layout of the flowerbed

Use a detailed plan to make the work much easier. It will reduce costs as well, because you will end up with the right plants in the right quantities. If you are a novice gardener, start with a few fairly easy-to-care-for varieties.

● First prepare a sketch complete with a planting diagram. Take into account the size of the plants when grown and sunny or shady areas.

● Bear in mind that flowerbeds should never be too narrow. Many plants like to spread out, so leave enough space between them.

● Use a vertical planting scheme for a tiered effect: set out taller plants in the centre of the garden or at the back edge; place the medium-sized plants in front of them; and, put the lowest plants in front.

A set of miniature tools comes in handy for smaller jobs.

EDGING and maintaining the bed

A flowerbed doesn't have to be angular. You can create precise outlines in other shapes, too.

● Map out straight edges by driving two stakes into the planned ends and stretching a string between them.

● Lay out round, oval and elliptical beds by using a garden hose as a guide.

● Create irregularly shaped flower borders by outlining them in sand or stones.

● Use stepping-stones to make it easier to access hard-to-reach spots in the garden without treading on

Draw everything in the garden to scale, including intended plantings.

the plants. They also keep the soil looser as you won't compact it by walking on it.

● Add a layer of mulch between plants to keep soil damp and weeds at bay.

● For a more formal or Asian-style flowerbed, use white gravel or ballast stones for the same purpose.

PROBLEM locations

Use a little forethought and effort to establish flowerbeds in some of the problem areas of the garden.

● Protect new plants from sliding down a slope in the rain by embedding plastic netting in the soil. Or terrace a slope to create flat areas for planting.

● Put plants that tolerate more moisture at the foot of a hill where run-off accumulates.

● Situate plants that tolerate dry soil in front of a wall. The soil there warms up more, especially if it receives plenty of direct sun.

● Lay drought-tolerant plants at the corners of a garden, because these areas often dry out quickly.

● Create shade for sensitive plants in sunny gardens by using shrubs and small trees.

Fruit trees

Fruit trees serve a dual purpose: they make an attractive centrepiece in a well-planned garden as well as helping to feed the family. They need plenty of sun and dry soil in order to produce abundant fruit.

Always consider tree maintenance before deciding where to plant fruit trees. If space is limited, an espalier tree may be a good choice. Placing any type of tree in a protected location helps reduce the chance of damage from frost.

Choose a tree with a straight trunk and strong roots.

PLANNING and planting

When choosing a fruit tree, look for a straight trunk, strong, well-distributed side branches and strong roots. Fruit trees prefer permeable soil; the roots can't survive in heavy, damp soils.

● During transport and up to the time of planting, keep the roots moist. It is a good idea to wrap the rootball in a sheet of plastic.

● Water bare-root plants about 4 hours before planting, then remove any rotten or damaged root parts. Cut back the roots a little so they draw water better after planting.

● When planting in stony ground, dig a larger hole and give the seedling an ample amount of good earth and compost. Remove as many stones as possible before planting.

● Give a tree the right support and a good foothold. Drive a stake into the soil on the west side of the tree. It should reach into the crown. Old tights are great for securing the tree as they stretch easily and are soft enough that they won't harm the bark.

● Wrap the trunks of young trees with straw, reeds or jute strips to prevent drying out and cracking due to sun or frost.

FRUIT-tree maintenance

● Even resistant fruit-tree varieties must be checked for pests continually. For the first couple of years, keep the ground around a young tree free from plants right out to the drip line of the crown. That way its roots won't have to compete for nutrients.

● Keep about a 16cm weed-free ring around the trunk of all fruit trees in order to prevent crown rot. In the winter, cover the trunk on the southern and southwestern sides with boards or hessian sacks to protect it. A thick covering of well-rotted horse manure around the trunk helps keep the roots warm and prevents them from drying out.

● In spring, brush apple trees, which are vulnerable to woolly aphids, with an aloe tree paint.

Apple tree paint

1g aloe resin
1 litre hot water

Dissolve the aloe resin (available online and from some herbalists) in the water. Allow to cool. Use a large paintbrush to apply the paint to the trunk and branches.

- Break off the blossom of fruit trees in the first year to encourage new shoots to grow.
- Prune fruit trees on sunny, frost-free days in winter when plants are dormant. Remove crossing branches and open up the centre of the tree.
- Cut off diseased and damaged parts, as well as all branches that are growing towards the inside or straight up.
- Codling moths time their egg laying so that developing caterpillars burrow into and devour precious apples, pears and plums. Protect crops by buying and hanging up pheromone traps containing synthetic sex hormones that attract male moths and reduce the chance of mating. Set the traps in place from April to August.

ESPALIER trees

Espalier is the art of training trees to branch in formal patterns, usually along a wall or on a trellis. It is a technique that first became popular several hundreds of years ago in Europe's medieval gardens. Espalier trees make it possible to raise and harvest fruit such as apples and pears in a limited space. In addition to one and two-armed trees which grow flat against a wall, the U-shape is also a traditional espalier configuration.

- Espalier trees are high maintenance. In summer, they must be trimmed and tied in continually.
- Use wood or metal brackets to keep espaliers about 20-30cm away from the house wall so that the branches open and no heat can build up.
- An espalier tree should have paired side branches off the main shoot. It's easier to shape a tree when the branches are young and easy to bend.
- Shorten the shoots that develop from side buds some time in late spring and bend the main shoots in the desired direction 10-20cm before the tips.
- Cut repeatedly and prune continually until you achieve the right growth direction.

AN ESPALIER SYSTEM suits pears, APPLES and stone fruits.

OLD varieties of fruit

With the industrialisation of horticulture, many of the numerous varieties of fruit tree once grown in Britain have been lost – or at least are no longer grown commercially. However, there is now growing interest in these historical varieties and a garden is the ideal place to reintroduce heirloom fruit trees.

APPLES

Apples have been grown in the UK as a cultivated crop since Roman times. While apple growing in Britain is a much smaller industry today, many of our traditional varieties survive. Here are some you could choose to grow.
- **Old:** Egrement Russet, James Grieve, Laxton's Superb and Lord Lambourne.
- **Good for storage:** Bramley's Seedling, Cox's Orange Pippin, Worcester Pearmain.
- **Early:** Discovery, Merton Knave, George Cave and Grenadier.
- **Espalier growing:** choose apples grown on dwarf rootstocks that are described as MM106 or M26. Ideally they should fruit on spurs, not tips of shoots. Good dessert varieties to grow in this way are Sunset and Spartan.
- **Cider making:** Brown's Apple, Morgan's Sweet and Somerset Redstreak.
- **Baking:** Bramley, Golden Noble and Lane's Prince Albert.

SEVERAL OF THE OLD VARIETIES OF plum preserve well.

Fruit designated for storage should have no bruises. An apple picker prevents the fruits from falling onto the ground.

PEARS
- **Hardy, undemanding:** Conference, Beurre Hardy and Doyenne du Comice.
- **Store well:** Conference, Seckel and Winter Nelis.
- **Good cooking varieties:** Conference, William's Bon Chrétien.

STONE fruits

There are old varieties of cherries, plums and damsons, too. They make few soil demands and are generally hardy. These are some you might want to try growing.
- Good plums for preserving are the Damson, Mirabelle, Greengage and Victoria.
- With some plum varieties, the stone is easier to remove. Just use a teaspoon to pry the stone from a halved plum.
- Cherries with particularly split-resistant skins include the Stella and Early Rivers. Both are also excellent for preserving, as is the sour cherry Morello.

Garden birds

It is hard to imagine a garden without birdsong. But as well as serenading us, birds are a traditional method of keeping garden pests in check, so it is important to do what we can to encourage them.

If you offer birds favourable conditions (a place to sleep, nest and feel secure), you will soon enjoy many different sights and sounds in the garden.

FAVOURABLE conditions

● Plant as many shrubs and hedges as possible. These offer birds a place to nest as well as providing materials for nest building, secure places to sleep and, above all, shelter from wind and weather.

Bird feed

275g unsalted suet (or unhydrogenated coconut oil or bacon fat)
500g seeds, raisins and oat flakes

Cut the fat into little pieces and melt it in a pan. Stir in seeds, raisins and oats until a doughy mass forms. Leave it to cool, shape into little dumplings and put into a purpose-made feeder.

● Introduce climbing plants, too. Climbers are an ideal habitat for birds, especially when grown along masonry, walls or fences.
● Flycatchers, redstarts and wrens love small hollows in old trees.
● Plant berry bushes – flowering shrubs and trees that bear berries in the autumn supply birds with nourishment before winter. They especially enjoy blackberries, rosehips and blackthorn, or the fruits from hedgerow plants such as elderberry, holly and hawthorn.
● Plant the beautiful flowering honeysuckle. It offers birds an excellent place to nest. Its berries, however, are poisonous for humans. The same goes for the spindle tree.

HOSPITALITY

● Build a birdhouse or nesting box by using rough pine or thick plywood about 2.5cm thick. Remember to leave a fairly wide exit so birds can still get out easily. Check the books to see what each type of bird prefers and vary the design accordingly.
● Hang a nesting box in a sheltered and not-too-sunny place at least 3m high to give the nesting birds security, and locate the entrance hole so that it faces southeast. Put a generous layer of wood chips or shavings in the box.
● Install a bird bath on a stone pedestal that cats can't climb. Keep it filled with fresh water.
● In the winter, turn over the top layer of the compost heap. Birds will find plenty of insects on its underside.

Feed your garden birds and they will help keep plant pests away.

Gardening tools

You don't have to spend a fortune filling the toolshed with every gadget on the market, but you will need some basic gardening equipment to keep your garden looking its best.

'You get what you pay for,' is a traditional adage, and you will indeed pay a premium for high-quality gardening tools. But well-made, sturdy tools – a shovel, rake, watering can, garden hose, shears and trowel – will last a long time, so it is worth the extra expense in the long run.

LOOKING for quality

● Opt for spades and forks made from stainless or chrome-plated steel. They are expensive but are the most durable and don't rust.
● Choose small garden tools in bright neon colours – they'll be easier to find in the grass or beneath weeds. The handles themselves can also be painted.
● Avoid buying gardening tools from a catalogue or online. It is a good idea to try them out and see how they feel in your hands – and to check that they are not too heavy.
● Where possible or appropriate, select tools that have extendable handles to make garden work much easier on the back.

Avoid damage to gardening tools by cleaning them thoroughly, then drying the metal and wood parts.

CARE extends the service life

● Remove soil, grass cuttings or dirt immediately after using a gardening tool.
● Clean stubborn dirt from the cutting edges of garden clippers with white spirit.
● Prevent rust on metal surfaces by wiping them with an oil-moistened rag after use.
● Wrap uncomfortable handles with a little foam material for extra cushioning.
● If there is no room in the toolshed for a wheelbarrow, stand it upright with the handles against a wall so it won't collect water and rust.
● Use a medium wire brush to remove encrusted grass clippings from under the lawnmower.

PREPARATION for winter storage

● Sharpen the large blades of scythes or sickles with a whetstone and smaller blades with a file. Store indoors, safe from rain or snow – but away from small fingers.
● Oil wooden handles and shafts with linseed oil to keep them smooth.
● Clean metal parts with fine steel wool. Remove a light coat of rust by sanding it with medium coarse sandpaper.
● Don't keep metal tools on the floor. To prevent them from becoming damp and rusting, place them on a shelf or hang them from hooks screwed to the wall. Store the lawnmower on a wooden plank.
● Do not leave any wood or metal tools in the garden over the winter months. Bring everything indoors for protection against the weather.

Greenhouses

A greenhouse extends the growing season, enabling you to start tender plants from seed in spring, increase variety and ensure a continuous supply of produce. But unless it is well-heated, you will need to bring tender plants inside during winter.

Erecting a greenhouse is an investment, but the savings that will result in the long run should pay off the initial costs – and you will always be able to enjoy the sanctuary it offers.

LIGHT and air

- Make a greenhouse as bright and as large as possible to allow air to circulate.
- Orientate the greenhouse in an east-west direction so that it gets adequate light in the winter.
- Know that different plants have varying light requirements: delicate, soft-leaved plants, for example, need to be close to the glass.

- If space is tight add shelves and racks, putting plants with tough leaves beneath them.
- Remove or keep plants at a safe distance when using cleaning agents on the glass panes. Alternately, cover them temporarily with a sheet of plastic.
- In spring, paint the outside of the greenhouse with a whitewash to keep it from getting too hot in the summer sun, then wash it off in the autumn.
- Ensure good ventilation as you cannot always open windows. It is also important to create shady areas or to provide cover on particularly hot summer days. Blinds are ideal.
- On hot summer days, spray the inside of the greenhouse with water to keep it moist.

WHEN WATERING, don't let water DRIP ONTO PLANTS below.

Hedge care

A dense hedge is a good and natural substitute for a garden fence, especially if you choose evergreen plants. It will provide privacy, wind protection and a habitat for birds and other useful creatures.

The first task is to dig a planting trench for the hedge that is the correct length and width. The exact dimensions will depend on the purpose of the hedge, where it will be located and the type of hedging plants you have chosen. As a general rule, the trench should be at least 30cm wider than the plant's rootball and of equal depth.

Keep a row of hedging straight by marking out the bed with stakes and twine.

PLANNING the hedge layout

● Position one or two plants every 1m for a less dense hedge. For a dense hedge, place two to five plants in the same space.
● Stagger individual plants in a double-row hedge so that they can spread out better and make the hedge thicker.
● Plant a mixed hedge or tapestry hedge with different coloured leaves or needles for an attractive effect. You might combine golden privet with hawthorn and copper beech.

● Choose yew or holly if the hedge is to serve as a windbreak; cherry laurel is a particularly good choice for noise protection.
● Plant berberis, holly, hawthorn or pyracantha to preserve privacy and keep out intruders.

PLANTING the hedge

● Hedge plants grow in a narrow space, so it is essential that the soil be loose and enhanced with compost. For heavy soil, mix in some sharp sand or clay granulate for maximum effect.
● Put black plastic sheeting over the bed and slit a hole for each of the plants; the plastic will help keep the weeds down, reduce evaporation in the summer and protect the roots from frost in the winter. You can also spread mulch such as bark between the plants.
● Distribute the plants so that the sideshoots just touch one another. If you put them too close together you will restrict their growth.
● Avoid hollow spaces by jiggling the plants thoroughly so that the soil gets between the roots.

TRIMMING the hedge

A hedge's first 'haircut' should take place shortly after planting. Cut back deciduous bushes by half as they grow quickly. Trim plants with needles to a consistent height.
● Once a deciduous hedge is well-established, always prune back the new wood in the summer to encourage branching. In the winter, cut back into the old wood to encourage the formation of new, strong shoots.
● Trim evergreen hedges regularly but avoid cutting into the old wood or the branches are likely to remain bald.
● Shaped hedges should be trimmed three times in the course of the spring and summer.
● Trim the sides of hedges at an angle, so there is less on the bottom than the top, to make them especially bushy and thick.
● Spread a sheet of plastic or a tarpaulin at the foot of the hedge to make it easier to pick up the trimmings.

HEDGE maintenance

Hedges are relatively easy to maintain. If you prepare the ground well at planting, it is unlikely that you will need to fertilise any more.

● Fertilise evergreen hedges in the spring with hoof and horn meal. Otherwise, all you have to do is keep the soil loose.

● Keep a new or young hedge watered, especially during dry periods in both winter and summer. Evergreen hedges need less water than deciduous or flowering hedges.

● In late autumn, remove dead wood and weeds and, if necessary, restore the mulch layer around the hedge.

● Cover up bald spots in a hedge by planting a flowering climber such as honeysuckle. Train the shoots over the empty spots.

STRETCH A STRING ACROSS TO help cut the hedge STRAIGHT.

GOOD TO KNOW ✓

A fragrant hedge

Fragrant hedges of *Rosa rugosa*, box lilac or lavender awaken memories of days gone by. These hedges are low-maintenance, they don't need yearly trimming and they attract desirable guests such as bees, butterflies and birds.

Slow-growing box with its dense, evergreen leaves is the ideal choice for low pathway hedging.

Herb gardens

Herbs were once a standard feature in a kitchen garden, grown both for flavouring and for their medicinal properties. A ready supply of fresh herbs is just as useful today, and home-grown plants look good in the garden and are often tastier and cheaper than shop-bought packs.

Most herbs like a sunny area with loose, permanent soil, although some, such as parsley and chervil, do best in partial shade. Raised spiral herb gardens built up with rocks and soil have been used for many generations to get maximum productivity out of small spaces. They also tend to suffer from fewer pests and the garden is accessible from all sides. The basic design calls for a spiral or knot of rocks, enclosing soil in which many species of herbs are planted. The rocks warm the soil and the design allows for a wide variety of soil conditions.

ESTABLISHED herb beds

● Plant a herb spiral in spring or autumn on an area of at least 3m². Sketch out the shape in advance.
● Add a small pool about 80cm deep at the beginning of the spiral. Line it with pond liner and reinforce with stones.

● Dig out the remainder of the herb bed to the depth of a shovel and fill it with an 80:20 mixture of soil and well-rotted compost or horse manure. Create a small mound about 1m high in the centre of the bed and reinforce it in a spiral shape with stones.
● Fill the upward-spiralling bed with different types of soils to create the following areas.
1 In and around the pond at the base of the spiral, the loamy soil should stay moist. Watercress and water mint will thrive here.
2 The next level provides a sunny, compost-rich moist zone. Plant chervil, dill, garlic mustard, parsley, peppermint, sorrel, chives, garlic chives and wild rocket.
3 The normal area will be partly shaded, with rather dry humus soil creating the best conditions for fennel, coriander, lovage, tarragon, oregano, marigold and hyssop.
4 Plant herbs from warmer climates in the dry zone atop the spiral. The soil is permeable and water will drain naturally to the lower levels. This is where savoury, lavender, marjoram, sage and thyme grow best.

Plants find ideal conditions in the limited space of a herb spiral, which is less likely to attract pests than an ordinary garden bed.

Pick leaves and flowers by hand, but cut off tougher stems with gardening scissors.

● Unless the soil is chalky, add garden lime to give herbs an additional source of calcium and magnesium. Varieties that thrive with its help include savoury, tarragon, caraway, marjoram, mint, parsley, rosemary, chives and thyme.
● Plant a hedge of box, lavender or hyssop to protect herbs from wind and frost.
● Cut back bushy herbs like lavender and thyme in the autumn to prevent frost damage.
● Keep herbs from spreading out too much by cutting them regularly. Thin out perennials like oregano each autumn.
● Plant mint and lemon balm in clay pots to restrict their growth. They can form strong runners that quickly take over the entire bed.
● Seed annual herbs in a different location every year. This form of herbal crop rotation will help to keep the soil from becoming too depleted.
● Many herbs need room: lovage and fennel have large roots that can damage adjacent plants.

MAINTAINING the herb garden

● Give tough-leaved herbs such as bay and rosemary only a little water – they don't usually require much.
● Make sure water is able to run off effectively through a drainage layer of gravel or topsoil.
● Hoe the ground between herbs regularly to keep it loose and let the water run off without soaking in.

HARVESTING herbs and seeds

● Gather herbs and seeds in late morning or at midday for best results.
● Collect the seeds of dill, fennel, coriander and lovage when they turn from green to brown.
● Dig up garlic before plants flower and before the leaves dry up – that way the bulbs keep better.
● Harvest most herbs before they flower, otherwise their flavour fades or (in the case of sage) disappears. Lavender, thyme and oregano can, however, be harvested at flowering time.
● Pick lemon balm in the afternoon when the leaves develop their greatest intensity before or just after the flowers open.

HERBS CAN SPEND the winter ON THE kitchen WINDOWSILL.

Kitchen gardens

A kitchen garden is a long-standing part of horticultural tradition and garden design. Part of its charm is that it combines the beauty of flowers with useful plants like herbs and vegetables, which you can pick fresh and enjoy while they are still bursting with flavour.

We can thank medieval monks for developing gardening culture. As they became aware of the benefits of mixed plantings for protecting gardens from pests, these monks increasingly grew medicinal plants and flowers along with their vegetables and fruits. As early as the 16th century, the rural population of Europe used these cloister gardens as a model for their own gardens.

THE structure of a kitchen garden

● A kitchen garden can offer so much more varietythan a simple vegetable patch. A kitchen garden should be laid out on flat ground so that it is sheltered from the wind but receives plenty of sun. If possible, the minimum size should be about 25m².

● Plant carefully. It is important to consider the height and width of growth to ensure that one crop doesn't shade another from the sun. And also consider which vegetables should and shouldn't be grown together (see *Companion planting*, page 286).

Select varieties that will mature or ripen at different times during the summer and autumn to guarantee a continuous supply of fresh produce.

● Make a sketch before you cultivate the patch. The sketch should show the locations of the beds, pathways, possible sitting areas and desired plantings. This will make seeding and harvesting much easier.

● For a traditional kitchen garden, use a four-square design based on the intersection of two major paths within a symmetrical, enclosed area. When laying out the pathways, take into account the width of the plants that will grow along their edges.

● Use intersecting pathways in the plan so you can reach garden beds to plant, care for and harvest fruits, vegetables and flowers easily.

● Delineate pathways with ease by lining them with box hedges that grow no taller than 50cm.

● Save space in a smaller kitchen garden by edging individual beds with marigolds, other perennials and even chives and herbs.

Cultivate your kitchen garden in a sheltered but sunny area.

DELINEATE PATHWAYS by lining THEM WITH box hedges.

● Start laying out a kitchen garden in the autumn by planting the trees and shrubs, keeping in mind their eventual height and the shade they will create.

CENTRE and edging

● If you have enough room, consider planting a circular flowerbed in the centre, perhaps with a fountain or standard rosebush as a focal point.

● Place a wooden or cement basin in the centre for collecting rainwater. Disguise a less-attractive water butt with climbing plants.

● Install a small gazebo or bench under an arch covered with climbing plants in the centre of the garden.

● For larger properties, enclose the kitchen garden with a hedge or a fence. If the garden is fairly large, plant shrubs, hedges or hawthorn. Their solid green foliage provides an attractive backdrop to the multicoloured plantings and keeps the kitchen garden from becoming too busy.

● Incorporate trellis plants if the kitchen garden abuts a wall of the house. Hops are a good choice.

PLANTS for a kitchen garden

In a kitchen garden, plants are generally laid out close to one another in mixed plantings. The benefits are that soil doesn't get depleted, vermin and diseases don't spread, you don't have to fertilise too much and you get a large yield with little effort. It is fairly easy to selectplants as well.

● Include heirloom varieties of vegetable plants such as parsnips, runner beans and tomato varieties. It is also fun to grow orange beetroot and purple potatoes and carrots, which are old varieties.

● Plant medicinal and seasoning herbs such as valerian, savoury, dill, oregano, chamomile, garlic and peppermint between the vegetables.

● Introduce aromatic plants such as sweet pea, phlox, sage and centifolia roses. They attract butterflies and bees, which are important for pollination.

● Ensure you always have a variety of plants by including self-propagating perennials such as columbine and fennel, and biennials such as foxgloves and evening primroses.

● Plant marigold as it flowers abundantly from June until the first frost. The flowers can be eaten or used as a medicinal herb to prevent infections. In addition, they help ward off aphids.

● Place tall perennials such as delphiniums and hollyhocks against a fence or hedge where their long stalks can get support.

● Crop rotation reduces the need for fertiliser and helps prevent pests and diseases (see page 289).

Lawns

A well-kept lawn is a huge asset to a property, and can even add value. But whether it consists of decorative grass, a play area or simply ground cover, it needs to be laid out and maintained properly.

There are a few important considerations before seeding a lawn, including soil quality and what type of seed would grow best. For routine care you will need a lawnmower, rake, fork, hose and sprinkler.

DELINEATION BETWEEN lawn and beds IS AN AID to mowing.

PLANNING a new lawn

● Check the soil quality (see *Soil*, page 324). Is it heavy, light, rocky, acidic or alkaline? You will need to know in order to prepare the ground properly.
● Choose turfs or grass seed according to how you plan to use the lawn. If children or animals are going to romp around on it, opt for a heavy-duty type.

PREPARING the soil and getting the seed

The first step is to remove all roots, rocks and any construction waste, and thoroughly dig up the ground at least two weeks before sowing a lawn.

1 If the soil is very heavy, mix in some sand for better aeration and permeability. Enhance humus-poor soil with compost until it turns dark in colour.
2 At the same time, level out uneven areas with a rake and, on fairly large surfaces, pack it down with a light lawn roller. Then leave the soil to settle. Be careful not to over compact it.
3 Sow seed in spring while the ground is damp for best results.
4 Spread the seeds evenly by first practising with sand, or make a spreader by drilling small holes in the bottom of an old tin can.

Lay the grass turfs close together on well-prepared level soil.

Water can soak in more effectively through aeration holes poked into the lawn with a fork.

5 Choose a day with as little wind as possible. Spread half the seeds in one direction and the other half in the opposite direction.

6 Sow by sections and, on larger surfaces, mark off individual sections with string. Sow more heavily at the edge so the lawn comes in thicker.

7 Rake the seeds into the soil just a little so they don't blow away and can sprout properly. Pack down the seeds with the lawn roller or opt for a low-tech solution: footboards tied to the bottom of your shoes.

8 Finally, sprinkle water on the seeds but avoid washing them away. Don't apply mineral fertiliser during the first two weeks – it will retard rather than enhance germination.

MOWING and watering the lawn

How often the lawn requires mowing depends on the thickness of the turf. The cut grass should be no shorter than about 2cm. Shorter lawns burn easily in the summer.

● Mow a new lawn for the first time when the blades of grass are around 5-6cm tall.

● Shorten the lawn by no more than about 3cm per mowing. In the first mowing of the spring, remove only the tips of the grass.

● Leave the cuttings where they fall to provide nutrients to the soil and shade the lawn's roots. Excess cuttings can be composted.

● Water thoroughly once a week during summer dry spells. The soil should be moist down to about 15cm so the grass can form deep roots.

● Minimise evaporation by watering in early morning or after 6pm.

LAWN care

Scarifying or aerating will help to get old or untidy lawns back into shape.

● Scarifying involves using a rake to pull out the thick mat of dead grass and roots that accumulate under the living green blades of the lawn. Normally, you should scarify the lawn in spring or early autumn, every one or two years.

● Aeration helps combat waterlogging. Poke holes about 7-9cm deep with a fork.

● Spread fine sand over the surface of the lawn to deal with minor waterlogging. Worms will bury it when it rains.

● Use nitrogen fertilisers in the spring; in the autumn, opt for a fertiliser that contains more phosphate to stimulate root formation. Apply fertilisers high in nitrogen only in wet weather.

● Keep off the grass in frosty weather or the blades could break and leave unappealing tracks.

● If a section at the edge of the lawn is damaged, cut it out vertically and put it into the resulting hole upside down. Then even out the spot with a little soil, sprinkle some seeds onto it and carefully water the area.

Patching bald spots on the lawn

ONE Dig up the area with a shovel and loosen the soil with a rake.

TWO Remove all weed roots and smooth out the surface.

three Add a mixture of sand and compost to the area and sow the grass seed. Compress and water.

Organic pest control

A spray gun loaded with noxious chemicals may combat attacks by aphids, but at what price? Plants survived for years before pesticides appeared on the scene, and there are still natural ways to deal with pests.

You can fight aphids, flea beetles, wireworms, cabbage white caterpillars, slugs, snails, spider mites, voles and ants with plant-based sprays, or use other plants that will repel the unwanted invaders with their scent or the excretions from their roots.

Nasturtiums protect ornamental and vegetable plants from aphid.

ANTS

● Sprinkle ant trails that lead towards the house or patio with the leaves of fragrant herbs such as chervil, lavender, mint, thyme or juniper, with spices such as chilli pepper, or with salt. The pungency of the herbs not only repels the ants, it disrupts the scent trail that the scouts leave behind for other ants to follow.
● Plant ferns, lamb's lettuce or tansy and ants will also make a detour rather than enter your garden.

Fighting aphids with tansy

200g fresh or 30g dried tansy leaves
1 litre water

Boil the leaves in the water and let steep for 1 hour. Dilute the tea with water in a 1:1 ratio and spray on affected plants.
Or pour or spray a tansy slurry (made of 300g of fresh leaves and 10 litres of water) twice a week. Tansy is poisonous, so take special care around children and animals.

APHIDS

● Plant fennel or coriander between plants to protect shrubs. For roses, you can plant garlic, lavender or French marigolds.
● If plants are already infested, spray them in the morning with a strong jet of water or a mild detergent solution. A spray made from tansy or stinging nettle tea will also help.
● Encourage blue tits, beetles and ladybirds into the garden, all of which consume insect pests.

FLEA beetles

● Flea beetles attack brassicas of all kinds, including rocket and pak choi. They like to take up residence in planters, but if you stick a few matches (the wooden variety is best) headfirst into the soil, they disappear. The sulphur dissolved by watering drives them away without harming the plant.
● Tuck one or two cloves of garlic in potting soil to get the same effect.
● Sprinkle wood ashes or sawdust in the pot or flower box in dry weather.
● If you plant peppermint, lettuce or onions flea beetles should avoid your garden.
● Coat a piece of cardboard with Vaseline and run it through a row of plants. The flea beetles will stick to it.

WIREWORMS

Wireworms can eat their way into crops of carrots, turnips, beetroot and celery – but potatoes are most susceptible of all.
● Make carrot traps by cutting pieces of carrot and pushing them 5-10cm below the soil surface. Pull them up and dispose of them – and the wireworms they will have attracted – every two to three weeks.
● Harvest potatoes as soon as they mature rather than leaving them in the ground.

● Sow a crop of green manure, such as mustard, next to potatoes in late summer. It will attract wireworms away from the crop.

CARROT flies

These pests lay their eggs on young plants, then the developing larvae riddle the crop with tunnels. They locate plants by scent. As plants are vulnerable after thinning or weeding, do these jobs on dry days.
● Separate each row of carrots with two rows of onions or garlic. Mulch each row with grass cuttings.
● Mix carrot seed with a packet of mixed annuals, such as French marigolds, and sow. Or plant sage and rosemary close to carrots to help deter them.

CABBAGE white butterflies

The larvae of these butterflies are harmful to all types of cabbage and other plants in the mustard family, including horseradish, kale and broccoli.
● Collect the caterpillars and kill them if the infestation is not too severe. The worst time for these pests is from May to September.
● Plant mixed crops including mugwort, peppermint, sage, celery, thyme and tomatoes as a preventive measure, or protect vegetables with netting.

SLUGS and snails

These pests are a threat to all young and soft-leaved plants. Deal with them as soon as possible.

● Place scooped-out grapefruit halves cut side down in garden beds. Slugs will accumulate inside overnight.
● Sprinkle coffee grounds around garden beds – they find them distasteful.
● Collect them morning and evening and destroy them if there aren't too many.
● Use gravel or glass chippings on garden beds. The sharp edges cut slugs on their soft underbellies, killing them. Reapply it after every rainfall.

SPIDER mites

Fine gossamer on the underside of leaves indicates an infestation of red spider mites. They munch on the leaves of cucumbers and bean plants, but are most troublesome in greenhouses and on houseplants.
● Spray with stinging-nettle slurry to help rid plants of this pest.
● Use beneficial predator mites, available in well-stocked gardening centres. They will search plants for pest mites and kill them, then move on.

MICE and voles

These rodents nibble on bulbs and corms, devour newly sown peas and beans, and attack stored fruit and vegetables if not well protected.
● Cover newly planted seeds with twigs of gorse, holly, or some other thorny shrub to keep them away.
● Toss seeds in wood ash before you plant them. Give bulbs and corms the same treatment before storing.
● Plant mint near vulnerable plants – rodents do not like its aroma.
● Get a cat. Its mere presence may be more than enough to deter the pests.

HEDGEHOGS ARE great AT LIMITING slugs in the GARDEN.

Ornamental shrubs and trees

These undemanding plants add beauty and fragrance to a garden, providing a visual framework or drawing the eye to significant spots when grown singly on the lawn. Their foliage, flowers and berries can be used to provide colourful accents to any home.

Take your garden's light and soil conditions into account when you are planning where to put trees and shrubs. Otherwise, you will find you are continually having to transplant or replace them.

CHOICE and location

● Ornamental shrubs and trees that offer a visual treat all year round are perfect for smaller gardens that lack space for a larger selection of plants. A compact ceanothus produces a glorious display of blue flowers in late spring while a selection of hardy

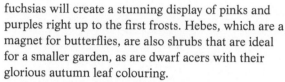

THE lush flowers of MAGNOLIA TREES delight in SPRING.

fuchsias will create a stunning display of pinks and purples right up to the first frosts. Hebes, which are a magnet for butterflies, are also shrubs that are ideal for a smaller garden, as are dwarf acers with their glorious autumn leaf colouring.

● Plants that thrive in the shade are red-flowering currant, beech, yew, dogwood, mahonia, skimmia, escallonia and privet.

● For damp locations, select alder buckthorn, white willow, hawthorn, eucalyptus, St John's wort or butcher's broom.

● Never plant birches, alders, willows or flowering cherries next to flowerbeds or vegetable plots, as these are shallow-rooted plants that make it difficult to cultivate the soil.

● Shrubs and trees draw water and nutrients from the soil, so make sure neighbouring plants can tolerate their proximity.

COLOUR for autumn and winter

● Find pretty red berries, even during the cold season, on the berberis, mountain ash, yew, spindle or cotoneaster.

● Rich blue fruits adorn the mahonia, purplish-pink-coloured fruits the beauty berry (Callicarpa).

● The bright russet red branches of dogwoods make an appealing sight in midwinter. Or, if space allows, grow a silver birch for its flaky bark in many shades of white and brown.

PLANTING

Spring and autumn are the best times to plant shrubs and trees, as they are either still in a state of rest or are preparing for winter after shedding their leaves.

● Loosen up matted rootballs with a fork before pulling the roots apart.

● Clip the roots and shoots to promote the growth of new and healthy plant parts. To prune a tree, lay it across a pair of sawhorses.

● Erect the supporting stake into the planting hole before inserting the tree to avoid damaging the roots. The stake will also be much easier to handle and insert without the tree or shrub in the way.

Deutzia is a flowering shrub that benefits from radical pruning.

● Thin out shrubs that have become scrubby and have stopped producing flowering branches.
● Prune back the bent branches of barberry and forsythia right behind the bend.
● Put layers of mulch around the base of shrubs and trees to keep the soil from drying out during the heat of summer and to provide protection against weeds.
● Protect the trunks of young trees from damage with a slit plastic bottle or chicken wire.
● Try adding wood ash to the soil if a shrub fails to flower. There may not be enough potash in the soil.

● Use old tights to tie the tree to the stake. They are cheaper than store-bought tree ropes and cords and, because of their softness and elasticity, they provide more protection, especially for younger trees.

PRUNING and cultivating

Ornamental trees and shrubs do need pruning to supply them with light and air.
● Prune in early spring, shortly before plants sprout. Don't prune during a severe frost. Prune the plants that flower in spring immediately after they have finished flowering, but wait until the following spring to prune those that flower after midsummer.
● Do not prune evergreen shrubs and conifers, laburnum, dogwood, magnolia or witch hazel regularly – it can do them harm.

Use figure-of-eight knots to tie a rope around a tree and its supporting stake.

Plant diseases

Using chemicals to combat plant diseases has become the norm. Now, however, many gardeners are returning to tried-and-trusted methods to counteract pests and keep plants healthy.

Certain chemicals used in fertilisers and pest controls have now been banned. Fortunately, nature can help combat many garden diseases.

CAUSES and prevention

Fungus in garden plants is often weather-related. Extremely damp or dry weather can encourage fungal growth. In addition, viruses and bacteria, nutrient deficiency or overabundance can cause health issues.
- Avoid overfertilising. Fortify plants with slurries.
- To prevent disease consider location and observe the rules of mixed cultivation and crop rotation.

APPLY HERBAL SPRAYS in the evening OR ON OVERCAST days.

CHLOROSIS

In this metabolic disorder the plant doesn't produce enough chlorophyll, so leaves or plant parts turn yellow and it withers. This occurs as a result of too much lime in the soil, dense soil or waterlogging. Raspberries, hydrangeas, roses, pelargoniums and rhododendrons are most likely to be affected.
- Water plants with rainwater or douse them occasionally with a stinging-nettle slurry.
- If chlorosis is already present, loosen up the soil and improve it by working in plenty of humus.

MILDEW

This fungal disease – subdivided into powdery mildew and downy mildew – is a common problem. It is characterised by a white or grey fungal growth.
- Powdery mildew is a threat to apples, apricots, peas, strawberries, cucumbers, peaches, delphiniums, roses, gooseberries and grapevines. It is prevalent in hot, dry conditions and appears as a whitish, floury layer.
- Downy mildew attacks strawberries, peas, lamb's lettuce, cabbages, lettuce, horseradish, radish, salsify, spinach and onions. It sets in during damp weather or in a moist greenhouse and spreads quickly. You will notice light spots on the top of the leaves and a greyish coating on the underside.
- Prevent both by purchasing mildew-resistant strains. Choose a sunny location to grow them and

The grape leaf (far left) shows chlorosis. The raspberry and blackberry leaves (middle and right) are afflicted with rust.

space the plants out well. Plant garlic between them or spray with a solution of 5g bicarbonate of soda in 1 litre of water.

● For a powdery mildew infestation, cut off sick leaves, shoots and branch tips; compost them.

● Spray every two to three weeks in spring and summer with garlic spray. Chop one or two garlic bulbs and cover with boiling water in a lidded jar. Soak overnight, strain and mix with 1 litre of diluted soap solution.

● To prevent or control downy mildew in seedlings, steep 7g of dried chamomile flowers in 1 litre boiling water. Cool the infusion and spray it on the seedlings.

● If plant parts are affected, remove the entire plant.

BOTRYTIS blight (grey mould)

A fungal disease, it attacks fruit and vegetables such as strawberries, sweet peppers, lettuce, tomatoes and grapevines. Ornamental plants can also be affected.

● Afflicted produce develops a greyish-white fungus and may rot and die.

● Space plants out well and prune to ensure good air circulation as prevention.

● Place straw underneath strawberry plants to keep the fruit from lying on the ground.

● If botrytis has gained a foothold, remove diseased plants and burn or bury them away from healthy ones.

BROWN rot

This fungal disease takes hold on trees and wood, as well as on stone fruit, especially peaches and plums. Infections on wood and flowers tend to occur when there is prolonged wet weather during flowering. Fruit rot occurs when the skin of the fruit is damaged.

● Blossom and twig blight occurs frequently with apricots, cherries, peaches and plums. The blossoms wither and the shoots die from the tip back. Affected leaves don't fall off.

● Fruit rot is characterised by small, squishy brown spots that rapidly grow in size until the whole fruit is rotten. It especially affects the fruits of apple, cherry and plum trees. The fruits don't always fall off.

● Prune endangered trees, open them up and ensure good ventilation. Fortify them with a tea made from horseradish leaves and roots; or try garlic spray (as for mildew) or a solution made by boiling 20g wild horsetail in 1 litre water for 30 minutes then cooling.

● If brown rot is present, discard all diseased fruit on both the tree and the ground. When blossom and twig blight are present, cut back diseased branches to the healthy wood. Both fruit and branches can go into the compost, where the heat will kill the fungus.

RUST fungus

A fungal infection, it appears on leaves and stems. Rust species tend to be host specific (for example, beans, leeks or asparagus), although some do change host plants. They spread by tiny spores and grow best in humid, moist environments.

● Orange or red pustules appear on the underside of leaves, as well as yellowish-red spots on the tops.

● For prevention, plant in a sunny, well-ventilated area and loosen the soil frequently. Wild horsetail solution may help fortify the plants.

● Remove diseased leaves immediately and add them to the compost pile.

● Garlic spray is also a good treatment for rust.

Moss extract

About 50g dried moss
1 litre water

Shred the moss and soak in the water for 24 hours. Strain it and put the solution in a spray bottle. Spray every couple of weeks to prevent powdery mildew. In case of infection, spray every couple of days. Sphagnum moss is the most effective moss but you can try using lawn moss (a good way of getting rid of it).

Planting for privacy

A garden should be a refuge in which you can unwind. But to achieve this sanctuary you will need visual cover, whether it's a wooden fence, a hedge or some artfully placed planting.

Apart from sheltering you from prying eyes, cover can be used to hide unattractive refuse bins or compost piles. Fences come in a variety of materials, styles and colours. Planting hedges is a more permanent cover option, and using potted plants as visual cover allows you to maintain maximum flexibility in your use of space. By replacing plants and moving pots around, you can come up with new combinations and surprising accents continually.

FENCES

● Natural materials such as wood, willow or bamboo work well in most gardens. Attach bamboo matting or visual screens made from woven willow to a newly erected fence for added privacy.
● Green up an existing or a new wood fence with climbing plants and hanging baskets.

Fast-growing bamboo, even if planted in tubs, helps to create visual cover on a patio.

TALL potted plants

Many plants grow well in planters, though they may not get as tall as they would in the garden.
● Opt for hydrangeas, cornelian cherry and weigela. They produce abundant flowers and thick foliage.
● Plant conifers in buckets for an evergreen visual cover.
● Elevate the plants by putting them onto a flat stone, a block of wood or an old, low bench. Or choose very tall containers.

PLANTED trellises

When planning visual cover in the form of a trellis, take into account the prevailing wind direction and angle of the sun, as they will play a role in determining whether the climbing plants will thrive. Excessively tall trellises sometimes cast unwelcome shadows.
● Wooden trellises or pergolas should be made from weatherproof, pressure-treated wood so that you don't have to use toxic wood preservatives to maintain them.
● Annual runner beans grow especially quickly on trellises. In just a few weeks they will produce a green wall with attractive red flowers. From late summer, you can eat the beans.
● Trellises provide perfect visual cover even in the winter, with evergreen climbing plants such as ivy and some varieties of honeysuckle.
● Make a trellis out of bamboo poles for annual climbing plants that are not too heavy. Place the poles on top of one another in a grid pattern and tie them together at each intersection with garden wire, then attach the trellis to the wall.
● A trellis thick with climbing roses produces abundant flowers and, depending on the variety of rose, a wonderfully fragrant privacy screen.
● Cut back trellis plants regularly so they retain their shape and don't become too heavy – which could result in the trellis breaking.
● Fragrant plants, including honeysuckle and jasmine, can be doubly useful. When trained on a trellis around the compost pile or in front of a refuse bin, they can cover up the view and simultaneously mask the smell.

HEDGES OFFER PRIVACY AND A sanctuary FOR wildlife.

HEDGES

Hedges act as a living fence to define a boundary and, depending on the plants chosen, can also provide visual cover.

● Plant a flowering hedge around a bench in the garden. Mix ornamental shrubs that flower at different times, such as lilac, forsythia, viburnum and pyracantha.

● Choose nearly impenetrable shrubs such as berberis, pyracantha, blackthorn, holly, hawthorn and wild roses for even more privacy and added security against intruders. They all have spines or thorns.

● Provide evergreen visual cover with box, yew, cherry laurel, privet, rhododendrons and spreading cotoneaster.

● Choose fast-growing hedge plants like hedge maple, cherry laurel and privet. Conifers such as western red cedar (*Thuja plicata*) and golden cypress (*Cupressus* 'Goldcrest') are also good choices.

● Use tall bushes or sunflowers for quick-growing visual cover. Ferns and tall ornamental grass varieties, such as giant feather grass (*Stipa gigantea*), also achieve vertical height quickly.

● Consider hardy evergreen bamboos, which make an ideal privacy hedge. Their dark-emerald – or even black – canes hold their colour even in the darkest winter. They grow quickly and reach a good height, and some varieties have especially thick foliage for providing visual cover. Use a combination of different-coloured leaves or stems to provide visual interest.

● Leaf hedges consisting of hornbeam or copper beech are an excellent choice. They don't lose all their leaves in the winter and they provide shelter for hedgehogs and other useful creatures. Birds are often drawn to privet and pyracantha.

Pyracantha keeps its leaves and provides shelter for birds.

Plant propagation

Propagation is a well-established method of producing plants for the garden, but different types of plant require different techniques. Also, propagation has to take place at the right time and the seeds or cuttings have to be planted in properly prepared soil.

DIVIDE SMALL perennials THEN plant in CLAY POTS.

You can sow seeds for lawn grass or sturdier plants outdoors in the spring. Plants that don't tolerate frost are best sown indoors or in a greenhouse.

Cuttings are shoots or parts of a shoot that have been cut off from the parent plant. You can place them in either water or soil to grow roots. A mixture of potting soil and coarse sand is ideal for cuttings.

PROPAGATION by seed

- Collect mature seeds during sunny weather and leave them to dry on paper towels where there is plenty of air, but try to avoid direct sunlight.
- Store seeds in a cool, dry, dark place. Use labelled paper bags, envelopes or plastic boxes for storage.
- Test the capacity of seeds to germinate. Place several in a glass of water. Those capable of germinating will sink to the bottom.
- Sow seeds that germinate in the dark, such as beans, about 2cm deep. Those that germinate in the light (such as basil and lobelia) should be sown on seed compost, then lightly pressed down in firm contact with the earth. You can lightly sprinkle soil on top of them using a sieve.

AIDS for sowing

- Make holes in the soil for sowing seeds with an old pencil or a screwdriver.
- Sow bigger seeds outdoors by putting them into a plastic bottle, then drill a hole in the cap, screw it back on and insert a drinking straw in the hole for air. The seeds will come out one at a time.
- When sowing small seeds outdoors, keep them from falling too close together by mixing them with sand or by sprinkling them onto the garden bed with a sieve.
- Use an empty plastic yoghurt pot or margarine tub as an economical alternative to commercial seed sprinklers. Wash them thoroughly and punch drainage holes in them.
- When sowing into pots or trays, mix compost with an equal quantity of vermiculite to improve drainage and prevent damping-off fungi from developing.

CARING for seedlings

- Outdoors, protect seeds by covering them with fleece. The fleece lets light pass through but softens the impact when the seeds are watered or pelted with rain. Fleece also retains moisture well.

Sowing in a pot

1 Place several seeds in a pot filled with potting compost. Insert a label to identify the plant.

2 Put a plastic bag over the pot. It will keep the soil moist and protect it from draughts.

Using sharp garden shears, clip softwood cuttings from fuchsias and place in water to grow roots.

- Transfer small plants outdoors in the evening or on overcast days. Once they are in the ground, water them carefully.
- Prevent seeds or young seedlings from being washed out of the soil by watering them with a mister or spray bottle rather than directly.
- Ensure seedlings get the light they need by placing the seeded pots on aluminium foil that reflects and intensifies the natural light.
- Put plastic bottles over young plants to protect them from snails, frost or dehydration. Cover them with horticultural fleece until they are well established.

CUTTINGS

- Prevent cuttings from drying out too quickly by placing them in a plastic bag.
- Ensure that softwood cuttings grow roots in a glass of water by holding them up with chicken wire or aluminium foil with holes poked in it and placed over the top of the glass. This will keep the cuttings from slipping too deep into the water.
- Promote the root growth of hardwood cuttings by adding sharp sand to the potting compost. Also add a little mycorrhizal powder – the fungi it contains will help promote fast and strong root growth.
- Make a small incision at the bottom of hardwood cuttings and stick a grain of rye or wheat in the notch. The nutrients produced by the grain will also encourage the root growth of the cuttings.

DIVIDING perennials

- Divide all perennials, as long as they have no taproots and more than one shoot.
- Each divided portion should have several roots and two or three vigorous shoots.
- Ornamental perennials, such as irises or black-eyed Susans, reach their peak after five to seven years, after which point they should be divided.

LAYERING plants

- Bend down the shoots to the soil to form roots. Once they have rooted, they can be clipped from the parent plant.
- Choose a one-year-old, flexible shoot that sits relatively close to the base of the parent plant.
- Layer many types of climbing plants, gooseberries and herbs such as rosemary.

DIVIDE THE ROOTBALL of a perennial WITH A SPADE to propagate.

Pot and tub plants

Plants in pots can beautify a balcony or patio with refreshing greenery and colourful flowers. But both the plants and their containers may need protection from hard frosts.

Plants will grow in anything that will hold soil, such as baskets, boxes, barrows or half a small wine barrel. You may need to modify the containers slightly to ensure plants remain healthy.

Plants will grow in anything that will hold soil.

THE selection
- Choose containers made from terracotta as they are permeable to water and air. Look for frost-resistant ones for plants that will be permanently outdoors. They are a bit more expensive but well worth it.

- Prevent the similar but much flimsier plastic pots from falling over in a strong wind by dropping a few heavy rocks into the bottom before adding soil.
- Use plastic pots to line metal and wood containers and prevent rust or rot. Alternatively, line the containers with plastic or aluminum foil.

PREPARING the containers
- Thoroughly clean all containers, including new ones, and rinse off any residue.
- Scrub off the unattractive white lime deposits on the outside of clay pots with vinegar and water, and feed plants with water that has aerated for 24 hours to prevent the deposits from returning.
- Use a hammer and a screwdriver to knock a drainage hole in the centre of the base of a clay pot that doesn't already have drainage holes, or use an electric drill to make several small holes in the bottom – but take great care or the pot will break.
- Prevent soil and nutrients from being washed away while watering by covering small drainage holes with broken crocks. Try placing a coffee filter over larger drainage holes – it allows the water to escape but keeps the soil in.
- In large tubs, prevent waterlogging of the roots by filling the pot to about a quarter full with coarse gravel before adding the potting compost.
- Be environmentally friendly: choose peat-free compost that avoids the depletion of natural peat resources.

GOOD TO KNOW ✓

The flea market: a place of discovery

You can find all kinds of things at flea markets that can be used as flowerpots: various colourful ceramics from other countries, tubs or pans, metal buckets or an old basin. Flowers can even feel at home in discarded cooking pots.

ADD 'SWELLGEL' AND FEED TO THE compost when planting.

● When planting, add a mixture of water-retaining gel such as SwellGel to aid water retention and granules of feed that will keep plants well nourished through the growing season.
● Ensure that the area where planters are kept has good water run-off. If necessary, place the containers on large stones or logs.
● Lift pots up to protect the roots on cold nights, when the chill from the ground could do some serious damage. If frost threatens outdoor pot plants, wrap them in fleece or bubble wrap.

MOST things thrive in a container

In addition to box and small conifers, contorted willow, winter jasmine and small, slow-growing ornamental shrubs thrive all year long in plant containers.
● Provide a splash of colour among winter-hardy evergreens with a frost-resistant Christmas rose, crocuses, or any plants that flower in early spring.
● Climbing plants such as passionflower, clematis or a potted rose can live comfortably on a balcony or patio, where they will disguise unattractive concrete walls.
● If your garden soil isn't hospitable for rhododendrons, azaleas or other acid-loving plants, put them into the appropriate soil in tubs and place them in the flowerbeds.
● If snails are a problem, grow dahlias, verbena, hydrangea or marigolds in tubs so that snails can't

get to them. Or secure copper bands around pots. Snails and slugs will be deterrred by the feel and taste of the metal.
● Choose box, impatiens, fuchsia, heather, hydrangeas, lobelia or ferns for shady or partially shady areas.

CARE and winter protection

● Stick your finger 1cm into the soil to see whether a plant needs water. If the soil feels dry, water the plant.
● Loosen up crusty topsoil regularly to allow water to penetrate better.
● As the soil contains fewer nutrients during the growing season, use a little more plant fertiliser on potted plants.
● Pour cooled water used for cooking vegetables and eggs, and any tea remnants, on potted plants. The liquid contains plenty of minerals that are good for flowering plants.
● Depending on the region in which you live, plants that are not winter-hardy should be indoors in their winter quarters before the first frost.
● Make sure that plants' winter quarters are bright and cool. The optimal temperature is between 5-8°C.
● Wrap the containers of even winter-hardy potted plants in fleece or bubblewrap if a frost is forecast. Tie up everything above the top edge of the pot.
● If you do water potted plants in the winter, make sure that it is on frost-free days because any cold winter winds will quickly dry out their roots.

If your garden soil isn't hospitable, plant azaleas in tubs.

Roses

The rose is perhaps the quintessential British garden flower. Used as a symbol of love, beauty, passion – even war – its countless colours and shapes, and delightful fragrance, still earn it a place in any garden, from the borders to a trellis or arbour.

Planting and caring for roses is not as difficult as it may seem. Armed with a sturdy pair of gardening gloves to protect you from their thorns, some garden shears and patience, your garden can soon include a wide variety of roses that will delight you and your visitors.

ROSE cultivation

● The original species and natural rose hybrids grow wild in most northern, temperate countries.
● Gardeners have long cultivated the rose, developing the double flower and, eventually, modern hybrids with a high, pointed centre in which ancestral floral characteristics are almost totally submerged.

PLANTING roses

Potted plants can be planted throughout the year, but roses may not always take root well.
● Be sure to plant bare-root roses after the last of the frosty days.
● Soak bare-root plants in a bucket of water for several hours before planting.
● Cut back rose plants before putting them in the ground. Shorten fairly small thread roots by half and larger roots by a third. In addition, cut back the branches to around 25cm.
● Make sure to plant roses so that the graft area is just below the surface of the ground. Rose growers will argue endlessly about the exact correct planting depth, but it really depends on where you live. If you live in a colder area, plant a bit deeper.

PRUNE CLIMBERS LIGHTLY – EVEN older shoots CAN FLOWER.

Maintenance involves regularly deadheading wilted flowers and dead plant parts.

- Use mycorrhizal rooting powder when planting or transplanting to give roses a great, natural boost.
- Pay attention to the distance between plants: it should usually be 40-45cm; for dwarf roses, 20cm is adequate; climbing roses need at least 1.5m.
- Provide roses with at least one botanical companion, as fragrant flowers or herbs will help to keep many pests away: lavender, rosemary and thyme repel aphids; French marigolds kill nematodes; chives helps prevent powdery mildew; and, the sulphur in garlic and onions wards off fungus growth.

ROSE care

- Remove any winter protection from roses on a cloudy, overcast day. Too much sun and warmth can give the plants a shock after a long winter's rest.
- Fertilise roses in early to mid spring.
- Use water-soluble fertilisers only when the weather is damp.
- Nurture roses with manure and water: add a large plant pot of dried horse manure to 8 litres of water and leave to soak for a couple of days. Dilute the liquid so it is the colour of weak tea and pour on the root area.
- Rake finely chopped banana skins into the soil to provide the plants with lime, magnesium, sulphur, nitrogen, potassium, phosphate and silicic acid.
- Water roses thoroughly but not daily during dry periods. Depending on the plant's size, during a drought a rose will need 10-20 litres of water per week to produce luxuriant flowers.

- Never water roses from above as this could result in fungal diseases.
- Prevent fungal diseases by spraying roses in the morning with a solution of 1 teaspoon of bicarbonate of soda in 4 litres of water. Adding 2-3 drops of soap blends the solution more effectively.
- Stop fertilising roses at least one month before the first frosts are likely to occur. Fertilising for too long into autumn encourages roses to produce tender new growth that will get nipped by the cold.
- Keep bush roses from lifting in the winter by covering the root crown with compost. Tie large bush roses or climbing roses with string to keep them from breaking in the wind or under a heavy load of snow.

CUTTING roses

Cut roses properly to encourage growth and the formation of buds.

- Wear gloves and use good secateurs. It is particularly easy to injure yourself when cutting climbing or bush roses.
- Twist off suckers that sprout from the ground right at the base. They will look different from the main plant.
- Remove prunings immediately from the garden as they can be a haven for insects and disease pathogens. Add this garden waste to the household rubbish rather than the compost pile, since many disease pathogens can withstand even the high temperatures inside the compost.
- Cut back bush roses by about a third in the autumn to produce bushier growth.

GOOD TO KNOW ☑

Colour into winter

Regional climate permitting, in the autumn a number of rose varieties, including *Rosa canina*, *Rosa rubiginosa* and *Rosa rugosa*, form large, shiny red rose hips – as long as the spring flowers aren't cut off. These fruits make excellent purées and fruit teas.

SEASON BY SEASON

There's a comforting rhythm to the cycle of the seasons and the corresponding chores that need to be undertaken in the garden. The reward for that year-round love and attention will be a horticultural haven.

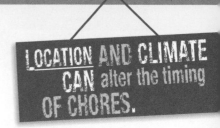

LOCATION AND CLIMATE CAN alter the timing OF CHORES.

Late spring

Plant shrubs in the ornamental garden, prune early-flowering bushes once the flowers have wilted, tie up climbing roses and move potted plants outdoors when the danger of frost is past. Plant biennials and those that flower early, as well as summer bulbs such as dahlias and gladiolus.

Add mulch or straw around strawberry plants. Thin out excess fruit on the trees and prop up branches overloaded with fruit using boards. Water fruit trees well if necessary.

Plant tomatoes, sweet peppers, chillies and other tender vegetables, as well as herbs like dill, oregano and basil.

Harvest early beets, radishes, rhubarb, lettuce, rocket and spinach, plus the first berries.

Spring

In ornamental gardens, plant or cut back roses, bushes and shrubs, divide and transplant existing perennials, and sow annual summer flowers such as pansies, snapdragons and sweet peas.

In the fruit garden, cut back autumn-fruiting raspberries and grape vines. Plant delicate fruit varieties such as apricots and peaches.

Plant cold-hardy vegetables such as peas, spinach, asparagus, broccoli, cabbage, kale, broad beans and onions.

When there is no further danger of frost, plant carrots, beetroot, Swiss chard, cauliflower, potatoes, celery and radishes.

Divide and move snowdrops.

Prune late-summer-flowering shrubs.

Late spring is the time to plant shrubs in the ornamental garden.

Colourful spikes of lupins embellish the early summer border, adding structure and cottage garden appeal.

Early summer

Trim deciduous hedges, remove wilted flowers, deadhead perennials, plant autumn-flowering bulbs and cut back to the ground the stalks of all early-summer-flowering plants. In addition, divide iris and lily of the valley bulbs once they have finished flowering. Wait to remove foliage of spring bulbs until it has died down.

Support heavily laden branches on fruit trees and, after the harvest, begin the first pruning and trimming to remove dead or diseased wood. Trim berry bushes after the harvest.

Plant pumpkin, squash, courgettes, leeks, onions, beetroot, sweetcorn and more in the vegetable garden so you will still have fresh vegetables in the autumn. Remove all the sideshoots from tomato plants (except for bush varieties) and stop growth once about five fruit clusters have formed.

Harvest the first potatoes and tomatoes, in addition to lettuce and many other vegetable varieties, berries, early apples and stone fruit.

Autumn and winter

Now you can plant the bulb plants for next spring and remove annuals entirely once they have faded.

Plant rose bushes in the flower garden and earth-up hybrid tea roses with soil, straw or peat moss to prevent them from freezing.

Remove dahlias, gladiolus, tuberous begonias and ranunculus from the ground after the first frost and store in a cool, dry place.

Prune ornamental shrubs and cover perennial gardens and borders with compost, bark mulch or leaf litter. Give evergreen bushes and shrubs a last good watering before the frost comes.

Plant fruit trees in the garden when there is no frost. Prune trees and remove dead or damaged wood on frost-free winter days.

Protect vegetable gardens from the first night frost. Cover any vegetables not yet harvested with fleece to help them ripen faster.

Plant garlic and overwintering varieties of broad beans and onions.

Late summer

Trim evergreen hedges in the ornamental garden, tie up late-flowering perennials and plant biennial summer flowers.

Plant berries and hazelnut bushes in the garden – new blackberry shoots must be tied up. Fruit trees shouldn't be cut any further.

Plant late crops of spinach, fennel, pak choi and radicchio. To prevent tomato blight, protect them with fleece.

Harvest nature's bounty: apples, pears, blackberries, plums, hazelnuts and walnuts. Enjoy tomatoes, leeks, late potatoes and late carrots.

Soil

A loose soil rich in humus and nutrients, with a slightly acidic to neutral pH value, is optimal for most garden plants. You will rarely encounter such ideal conditions but, fortunately, you have centuries of garden wisdom to call on to help improve the soil in your beds.

Soils can be divided into light sandy soil, heavy clay soil, medium loamy and chalky or limey soil on the basis of their sand, clay and loam content, plus lime and humus.

SOIL types

● Sandy soil is loose and easy to work and plant roots can spread out easily, but water and nutrients are poorly absorbed. Use compost to increase the amount of humus and mulch to prevent rapid drying.
● Clay soil makes it difficult for roots to spread. The soil is so tightly compacted that the roots of many plants can't penetrate it to reach water and nutrients so will wilt quickly. Loosen clay soil by adding sharp sand and compost. (Don't on any account use builders' sand.)
● Loamy soil offers the best gardening conditions. It stores water and nutrients effectively and the soil structure is loose enough for plants to root easily and reach the nutrients. By adding a little compost or organic fertiliser each year you can ensure that the soil doesn't become depleted over time.

SOIL analysis

Determine soil types before you get started on the gardening to ensure you choose the right one.
● Take a spade sample at several places in the garden, digging to a depth of about 50cm.
● Alternatively, determine soil conditions by studying the roots of plants currently growing in the garden. A small rootball and crooked, intertwined root strands point to impenetrable soil.
● Look for creepy-crawlies. The presence of many helpers such as woodlice, earthworms and millipedes in the soil is a sign of good soil quality.

TELL-TALE plants and weeds

So-called indicator plants prefer certain soils (although most are weeds). Their appearance makes it possible to draw conclusions about soil conditions.
● Compacted, heavy clay soil attracts creeping thistle, horsetail, lamb's foot, coltsfoot, polygonum, mullein and dandelion.
● Dry soil is preferred by yellow chamomile, wild poppies and speedwell.

HYDRANGEA FLOWERS ARE blue in acid soil, pink in alkaline.

Sandy soil is dry, grainy and crumbles easily; good loam soil first smears and then crumbles after some time.

- Acidic soils: summer-flowering heathers, azaleas, rhododendrons and blueberries.
- A high lime content is indicated by the presence of bellflowers, marigolds, delphiniums, cornflowers, lavender, alliums and spurge.
- Spray a little vinegar onto a clump of soil. The vinegar will bubble when it comes into contact with soil containing lime.

SOIL improvement

Adding the right substances improves most soils.
- Work garden gravel or coarse sharp sand into heavy clay soils to a depth of 30cm to increase permeability.
- Add compost or horse manure to sandy soil regularly. Plant St John's wort, ornamental grasses and tamarisks to help stabilise the soil.
- Dig garden lime, chalk or marl into acidic soil in autumn, and add phosphate and potash in the form of wood ash.
- Chalky, calcareous soil can often be modified only through repeated additions of large quantities of sulphur or peat. It is better, however, to plant beech, box, forsythia, viburnum, beans, cabbage or lettuce, which grow well in the soil. Regular use of organic fertilisers can help chalky soil improve in texture and quality and store nutrients more effectively.

- Soil with plenty of humus attracts nettle, dandelion and chickweed.
- High-nutrient soil lures nettles, thistles, goosefoot, wild radish, shepherd's purse, coltsfoot, nightshade, round-leaved dock (bitter dock) and chickweed.
- Clear indicators of soil low in nutrients include daisy, heather, common sorrel, dog daisy, common wood sorrel, pansy and white clover.

ACID and lime content

The health of garden plants may well depend on the acid content of the soil. Some plants grow well in acidic or alkaline soils, while others don't. Look into neighbours' gardens to see what plants thrive there.
- Alkaline soils are indicated by the presence of yarrow, Solomon's sea, bergenias and hellebores.

Determining pH value with a kit

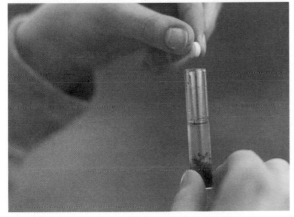

1 Put a soil sample with some water into the glass container from the kit and add a tablet. Close the container and shake well.

2 Determine the pH value by comparing the resulting colour change in the liquid to the colour chart on the side of the packet.

Tasks for winter

Indigenous plants are normally winter-hardy, but they still need to be ready for the colder weather. Prepare your garden well so that all the plants, including the more delicate ones, can survive the harshest winter.

With a bit of preparation, much can be done to help your garden survive the ravages of winter. This includes trimming trees, applying a coat of paint or even bringing part of the garden inside.

NATURAL protection

Opt for leaves, spruce and fir trimmings, straw, hessian and jute when covering or wrapping plants.
- If you can get them, consider using spruce trimmings for optimal winter protection. They gradually lose their needles and let more light reach the plants. Arrange cuttings like a small roof over the plants, but make sure that air can get to them or they may rot or become diseased.
- Check leaves for pests, fungus and other diseases before deciding on winter protection.
- Hessian is light and smooth, so it won't harm plants. For better protection, you can put straw between the plants and a hessian wrap over them.

PROTECTION from the cold

Plants from warmer regions and newly transplanted or young plants will make it through a tough winter only if they are well wrapped in sturdy hessian or spruce trimmings.
- Wrap fruit tree trunks in cardboard to keep them from cracking during major temperature changes.
- Make wigwams of bamboo canes around individual plants and pack them with straw.

- Protect delicate perennials and late-autumn plantings with dried leaves or straw, and use spruce cuttings to protect evergreen varieties.
- Tie tall ornamental grasses together in bunches to protect them from frost and snow and to provide winter shelter for helpful creatures.
- Press back down soil around perennials that may have buckled due to frost. This will protect the roots.
- Wrap lower branches of tender climbers and shrubs with straw held in place with hessian and string.

OTHER winter options

- If you have room, bring potted plants indoors for winter. They will bring colour and extra oxygen to the home. Just remember to give them adequate light and water, and be careful not to place them too close to the cold of the window or any heat sources.
- Create a windowsill herb garden. Use several small terracotta pots and some enriched earth to plant your favourites. You will have fresh herbs all winter long.

In a hard frost, place pots on wood so they don't freeze to the ground.

Transplanting

Redesigning your garden may involve transplanting perennials, shrubs or even trees. It is not a daunting task, but remember that timing is key because no plant should be kept out of the soil for too long.

The best times for transplanting are late summer or early autumn, which gives plants the chance to either put down roots and start growing before the onset of winter or, in spring, to start growing before the end of the season.

THE right timing

● Try to transplant during wet and cooler weather to reduce stress on the plants. Avoid hot spells.
● Transplant evergreen trees and shrubs in April or October, when the soil is moist and warm.
● Transplant shrubs that are sensitive to frost, such as mallow, hibiscus or hydrangea, in spring.

TRANSPLANTING large trees and shrubs

Trees and older shrubs should be prepared before being relocated or they won't tolerate the change. It takes two to three seasons to root-prune a tree fully but, in the end, a compact, well-branched root system will greatly increase a tree's chance of survival once it is moved.

1 In spring, dig a furrow 40cm deep and 25cm wide a third of the way around a tree, slightly closer to the trunk than you will eventually be digging when the tree is moved. By doing so, you can break long, unbranched roots, prompting the regrowth of new roots nearer the main trunk of the tree. Cut off the roots around the main shoots, at a slight angle toward the shrub or tree.

2 Fill the trench with fresh soil and compost and keep watering the plant well to promote the growth of new fibrous roots. Don't transplant until the following year.

3 Prune the shoots of the transplant by up to a third of their length to prevent harmful effects from damaged roots.

4 A handful of bone meal and some mature compost will give the tree or shrub a good start once it's positioned in its new hole.

5 Keep transplants, whether perennials, shrubs or trees, from dehydrating by spreading a thick layer of mulch on the ground around them.

TRANSPLANTING perennials

● Transplant perennials after they have finished flowering. If you transplant in the spring, don't expect the plants to flower that year.
● Transplant in late autumn so you can save yourself the labour of extensive watering.
● Add compost or humus to the hole for perennial plants to help them sprout new shoots.

A GRAFT SHOULD LIE INSIDE the soil WHEN a plant is moved.

Vegetable and salad beds

Plant greens in the sunniest and flattest part of the garden to ensure a good supply of vegetables throughout the summer and autumn. Plan vegetable and lettuce beds well so that they are easy to maintain, accessible for weeding and watering, and protected against garden invaders.

How you plan your beds will help you immensely in the long run. It will save time and effort, make it easier to harvest crops and could prevent you from developing a sore back.

MAKING a new bed

Keep vegetable beds no wider than 1.2m. To make sure they are accessible on all sides, don't place them right next to a wall or a hedge. The paths between individual beds should be about 30cm wide.
● Lay wooden boards on the paths to keep from sinking in the mud during rainy weather.

USE CARDBOARD egg cartons for STARTING SEEDLINGS.

● Remove all roots of perennial weeds before you plant. Dig the soil well.
● Divide the vegetable bed into two: one with things you harvest daily, such as carrots, radishes and lettuce; the other with permanent crops, such as herbs and horseradish, cabbages and with crops such as potatoes.
● Use berry and hazelnut bushes to provide attractive protection against cold wind.
● Plant potatoes and Jerusalem artichokes to loosen the soil and restrict weeds.
● Add basic fertiliser, preferably with compost, about three weeks before the first planting.
● Plant low-growing vegetables on the sunny side of taller ones so that they get adequate light.
● Allow for varying ripening times when you are laying out the bed so you can grow different plants close together.

SEED propagation indoors and sowing
● Use plastic starter trays. They are easy to clean, which makes it difficult for diseases to take hold.
● Don't forget to label individual starter trays with the name of the vegetable and the date of planting.

This small vegetable bed is located next to the compost heap, making it easy to fertilise.

Sow seeds in straight rows by stretching a string between two stakes at either end of the row.

- Don't plant too many seeds in one container. The fewer seeds you plant in a container, the less work you will have when it is time to thin out the plants.
- Sow larger seeds farther apart, singly or in pairs (just in case one of them fails to sprout).
- Scatter smaller seeds, such as spinach and chard, directly into shallow drills in the bed. Cover them lightly with soil.
- Mix very small seeds with a little sand so they don't fall so closely together when you are sowing them and are easier to spread.
- Soak seeds from legumes or vegetables (like squash or tomatoes, which are really fruit) in a milk marinade for 24 hours so they sprout faster.
- Warm the ground by covering it with a black tarpaulin or black plastic sheet for a week in advance and you can seed outdoors earlier.
- Plant courgette and cucumber seeds on their sides to speed germination.
- Make optimal use of space by planting slow-growing vegetables such as carrots between fast-growing lettuce.

BED maintenance

- Dig irrigation ditches between rows to let the water soak in slowly and reach the roots. This works well for any vegetable best watered from below, such as brussels sprouts.
- Make sure the water is warm to the touch or plants will get a real shock in hot weather.
- Dig green manure, such as clover, alfalfa and downy vetch, into the soil in the autumn. This provides a natural source of nutrients.
- In late autumn, enhance the soil in harvested beds by adding mulch or digging in manure.
- Plant aromatic herbs or flowers around the vegetable patch, such as dill, alyssum, cornflower and French marigold, to attract useful garden visitors such as ladybirds and hoverflies.

KEEPING out invasive pests

- Pick off harmful pests such as caterpillars and snails individually. Snails tend to lurk in grass, so keep the lawn around the vegetable patch cut short.
- If bushes or hedges near the garden are infested with aphids, protect vegetables with fleece.
- Install chicken wire or netting over a freshly sown bed to keep cats and birds at bay. Once the tender shoots have emerged, use it to make a fence around them.
- Make a scarecrow by hanging shiny objects such as CDs or tinfoil containers from branches or a forked pole stuck into the ground. They will reflect the sunlight as they move in the wind, helping to frighten away hungry birds.

A herbal extract for seedlings

50g stinging nettles
10g sage
10g rue
10g artemisia
20g ferns
20g onion peelings
10 litres water

Soak the herbs in water for 24 hours, then boil and strain. Dilute the completed slurry 10:1 and sprinkle onto the seedlings to strengthen them. Repeat this application weekly.

Watering

Regular watering – not too much, not too little – is important for all plants. Collecting rainwater in a barrel or butt for use during dry spells is a long-standing practice that is just as useful today, even if your garden is small.

Collecting rainwater is not only an environmentally friendly practice, it is better for plants than using tap water, which may contain small traces of chlorine.

HOW much water to apply

Look at a plant's leaves to see how much water to give.
● Small, leathery, thorny, shiny or fleshy leaves indicate a low need for water, as do leaves with a wax-like layer. Rock garden or Mediterranean plants do well without a lot of water, as do ivy, amaranth, nasturtium, sage, French marigolds and zinnia.
● Plants with soft, large or thin leaves tend to be thirsty, as do all flowering plants and those with solid rootballs and shallow root systems.

Water young plants carefully so that they don't lose their foothold.

A SOAKER OR OOZE HOSE RELEASE WATER slowly but STEADILY.

WHEN and how to water

● Water plants in both the flower and vegetable gardens before 9.30am. The leaves will dry off quickly, reducing the risk of a fungal infection. If you water plants at night in cool weather, the soil surrounding them remains wet, potentially causing the roots to rot – plus, the moisture could attract snails. But evening watering is highly recommended in hot weather as it reduces evaporation.
● Water infrequently but thoroughly. The pauses between watering leave time for a branching, deeper root system to form in the drier soil. Don't saturate plants in heavy soil as they will become waterlogged.
● Make an inexpensive soaker hose at home from an old garden hose. Drill some small holes into it, connect it to the main hose and you have got an effective, efficient watering system.

- Be particularly careful not to water the leaves of melons, sweet peppers or tomatoes. If you have time, it is even worth shaking rainwater from their leaves to ensure fungal diseases don't set in.
- Don't water directly onto flowers. In the sunlight, the little water droplets act like a magnifying glass, potentially burning delicate petals.
- During hot summer weather just spray the plants – but don't do it in bright sunshine.
- Young shoots need more water than plants that are several years old with deep root systems.
- Water thirsty plants at ground level so that maximum water gets to the roots.
- If you water intensively with a watering can every couple of days in the summer heat, repeat the process after about 30 minutes. The water will penetrate deeper into the soil.
- Water bushes and shrubs on the root area beneath the outer branches. During dry periods, large trees need plenty of water, especially fruit trees when they

Watering aids

one Dig a hole about 15cm deep next to a plant that needs lots of water.

two Place a flowerpot of the correct size into the hole, drain hole downwards, and cover it with a plate or piece of plastic. Disguise it with leaves or a little soil.

three Fill with water at regular intervals. It will trickle slowly into the earth and reach the roots directly.

are flowering or when fruit is ripening. Water for several hours.
- Always water plants well before you move them, and again after they are in their new positions.

COLLECTING water

Most plants tolerate rainwater better than water from a spring or tap. Collected rainwater is usually warmer and free of fluoride, chlorine and lime. Water that is high in lime leaves white spots on the leaves.
- Keep water clear by occasionally adding a little charcoal or activated carbon (for use in aquariums) to the barrel.
- Cover rain barrels so small animals can't get in.
- Collect the water in which vegetables have been cooked – which contains nutrients – for use in the garden.

OPTIMUM moisture

- Rake the soil regularly, including in the furrows. Loosening the soil gives water easier access to a plant's roots.
- After a heavy summer rain, loosen up the soil only on the surface or the moisture in it will evaporate too quickly.
- Lay a layer of mulch to hold the moisture in the soil longer and prevent plants from getting thirsty too quickly.

Connect a rain barrel directly to the downpipe of the roof gutter. If the barrel is raised up it is easier to fill the watering can.

WEATHER LORE

Humans have been forecasting the weather for almost as long as we've been on earth. Meteorologists acknowledge that many of the old weather rules hold up quite well scientifically – although they are not infallible.

A CHANGE IN WIND, A CHANGE in weather IS AN old adage.

Animals predict the weather

Animals are rooted so firmly in nature and its events that they sense changes in the weather – their lives often depend on it. Migratory birds, for instance, instinctively sense when winter is coming and prepare for their departure. Gardeners should pay particular attention to birds. Research has proven this lore to have more than a grain of truth in it.

- Swallows and bats fly close to the ground before a rain. Birds flying low, expect a blow.
- If the goose honks high, fair weather; if the goose honks low, foul weather. South or north, sally forth; west or east, travel least.
- Seagull, seagull, sit on the sand; it's a sign of rain when you are at hand.

By observing the behaviour of other animals, you can draw additional conclusions about the weather that are worth taking note of, even if they don't always match up to the creatures' forecasts.

- If garden spiders forsake their webs, it means rain is coming.
- When a cow thumps her ribs with an angry tail, look for thunder, lightning and hail.
- Cats scratch a post before wind; wash their face before rain; and sit with their back to the fire before snow.
- The louder the frog, the more the rain.

Wind, clouds and leaves

'A change in wind, a change in weather,' goes the old farmers' adage. And it's true that a change in the weather usually does follow a rapid change in wind direction. The traditional sayings below are also often accurate.

- A wind from the south has rain in its mouth. A wind from the east is not good for man or beast.
- When smoke descends, good weather ends.
- A sunny shower won't last an hour.
- When leaves show their backs, it will rain.

If a rooster crows on going to bed, you may rise with rain on your head.

Rings around the moon are a good predictor of weather change.

Flowers and the seasons

Certain plants in the garden and countryside indicate the seasons.

- Early spring: snowdrops begin to flower.
- Springtime: apple blossoms.
- Early summer: elder flowers.
- High summer: full flowering of the small-leaved lime.
- Late summer: start of the wheat harvest.
- Early autumn: full flowering of autumn crocuses.
- Autumn: deciduous leaves change colour; start of general leaf fall.

The sun and moon as weather predictors

Meteorologists call rings around the moon and sun halos. They have long been used as predictors of changes in the weather. The following sayings about the sun and moon have been handed down for generations, and every gardener should be familiar with them. To avoid taking chances with the weather, however, it is just as well to keep an eye on the thermometer.

- A halo around sun or moon means rain or snow very soon.
- Clear moon, frost soon.
- Pale moon rains, red moon blows, white moon neither rains nor snows.
- Plant your beans when the moon is light; you will find that this is right.
- Plant potatoes when the moon is dark, and to this line you'll always hark.

Colours in the sky

Colours in the sky – red in the morning and evening, or rainbows – tell us different things about the weather. They are the result of water vapour in the lower layers of the atmosphere. The more water vapour there is, the stronger the colours. A colourful sunrise indicates rain, but in the evening it merely produces dew on the ground. Gardeners may be helped by these sayings.

- Red sky at night, shepherd's delight; red sky in the morning, shepherd's warning.
- Evening red and morning grey set the traveller on his way. But evening grey and morning red will bring rain down upon his head.
- Rainbow in the morning gives you fair warning. A rainbow to windward, rain ahead. A rainbow to leeward, rains end. A rainbow at noon, more rain soon.
- If there is enough blue sky to make a sailor a pair of trousers, the weather will soon clear.
- When clouds look like black smoke, a wise man will put on his cloak.

Weed control

'The best shade in the garden is the gardener's own shadow,' counsels an ancient Chinese proverb, which is why our grandparents were out in the garden every day pulling weeds. But if you're pressed for time, here are a few preventative measures you can try to keep botanical invaders at bay.

DIG OUT THE TAP ROOTS OF dandelions WITH a trowel.

Weeds are plants that grow and reproduce quickly and easily but aren't welcome in the garden because they edge out the more delicate flora and make the beds look untidy. However, noting which weeds thrive allows you to evaluate the composition of the soil, and their roots may ventilate it and enrich it with nutrients. Certain weeds serve as the basic element in liquid fertilisers or compounds used to prevent and combat garden pests and diseases, while herbs such as common horsetail – a menace in the garden – and coltsfoot are renowned for their medicinal properties.

COMMON weeds

● Creeping thistle is not only tough, it attracts pests. This root-spreading weed grows up to 1.2m. Remove its flowers first, then dig up the plant, roots and all if possible. Don't dispose of its remnants on the compost heap as the roots could multiply.

● Common horsetail, also called bottlebrush or cat's tail, has a far-spreading root system that is virtually impossible to pull out. The plant breeds by means of spores and roots. Since it likes moisture, you can effectively combat it only by draining the soil or through sporadic liming of the ground. Again, don't throw it on the compost heap as it will continue to multiply there.

● Stinging nettles like nitrogen-rich soil. If left unchecked, they become rampant and form clusters. They are shallow-rooted plants, so must either be sheared back once a week (the roots will wither away within two years) or pulled out singly from the ground along with the roots. Wear gloves.

● Speedwell thrives in loose, nitrogen-rich soil. This weed is spread by seeds and usually flowers between March and July. It has to be removed by hand or when young with a hoe.

● Coltsfoot grows in even small cracks between paving stones or on house walls. Dig out each plant singly before it flowers. The weed, the flowers of which resemble the dandelion, grows in clusters. It propagates by seeds and can be composted.

Be especially careful when hoeing weeds among single plants in a vegetable bed.

- Chickweed flowers from March until October, establishing itself not only in vegetable patches or shrub beds but also in the lawn. Before the seeds form, pull the plant out of the ground by hand.
- Convolvulus or bindweed is a perennial weed that, left unchecked, will wind its way around flowers and fruit bushes. Its underground roots are hard to remove, but regular hoeing and weeding will gradually weaken plants. It should never be composted.
- Ground elder also has a persistent root system that is hard to eradicate, so don't compost this weed, either. Only regular hand weeding will get rid of it.

WEED removal

Make sure that weeds are pulled out regularly as they spread quickly and rob other plants of nutrients.
- Weed on dry days after a rain when the soil is loose, allowing plants to be pulled up easily.
- In dry weather, let the weeds decompose at the edge of the bed.
- Weed again a few days after breaking up the soil of a bed. When you turn the soil, the seeds of weeds lying on the bottom may come to the top and start to shoot.
- Cut off flowerheads or seedpods before weeding or hoeing to prevent seeds from getting into the soil.
- Make short work of weeds growing in cracks and fissures between paving stones or on house walls by pouring boiling water on them.
- Pour salt on lawn weeds or sprinkle them with a solution of one part vinegar and one part water.

PREVENTING weeds

Chemical weed and pest killers are a bad idea. The substances they contain harm the environment and run the risk of damaging the soil so much that the plants you have nurtured die, too. With so many natural ways to control weeds available, there is no need to use them.
- Plants that rob weeds of light and nutrients through their own growth are an environmentally friendly way to control weeds. Ground-covering plants are especially useful.
- Use dense-growing ground cover-plants such as wild strawberries, periwinkles and violets for shady areas.
- Prevent weeds in sunny beds by planting sedums, euphorbias and ground-cover roses.
- Mulch between plants to prevent unwanted weeds from coming to the surface. For mulch, use freshly

Lady's mantle thrives even in shady spots and its dense growth helps banish weeds.

cut grass or wood chips; in rock gardens you can also use gravel.
- Sow plants that nourish the soil. This also suppresses weeds and will later contribute to healthy growth in the flowerbed.
- Lay black plastic mulch over larger or inaccessible areas and cut crosses into the mulch through which plants can emerge (or into which they can be planted).

GOOD TO KNOW ☑

The culinary side of weeds

Weeds are generally despised as invaders, to be pulled swiftly from the garden. But many of our ancestors would have seen them as food. Daisies, fat hen (goosefoot) and chickweed are tasty in salads. Ground elder and sorrel make a delicious soup or, just like the leaves of nettles, you can serve them as a vegetable. The deep-fried, fresh leaves of stinging nettles are a surprisingly tasty treat.

INDEX

PICTURE CREDITS

CONSULTANTS

Ruth Binney MA (Cantab)

Vince Forte BA (Cantab) MB BS(Lond)
MRCGP MSc DA

Ann Godsell BSc FTOPRA

Amanda Cutbill BSc MNIMH

FOR VIVAT DIRECT

Project editor Rachel Warren Chadd

Editor Diane Cross

Art editor Simon Webb

Designer Nicola Liddiard

Editorial director Julian Browne

Art director Anne-Marie Bulat

Managing editor Nina Hathway

Trade books editor Penny Craig

Picture resource manager Eleanor Ashfield

Pre-press technical manager Dean Russell

Product production manager Claudette Bramble

Senior production controller Jan Bucil

Colour origination by FMG
Printed and bound in China

ISBN 978-1-78020-139-9
Book code 400-611-UP0000-1

Traditional Household Hints

This paperback edition published in 2012 in the United Kingdom by Vivat Direct Limited (t/a Reader's Digest), 157 Edgware Road, London W2 2HR.

First published in 2011 under the title *Traditional Wisdom Rediscovered*

Traditional Household Hints is owned and under licence from The Reader's Digest Association, Inc. All rights reserved.

We are committed both to the quality of our products and the service we provide to our customers. We value your comments, so please do contact us on **0871 351 1000** or visit our website at **www.readersdigest.co.uk**

If you have any comments or suggestions about the content of our books, email us at **gbeditorial@readersdigest.co.uk**